Mechanical Properties of Porous and Cellular Materials

MATERIALS RESEARCH SOCIETY SYMPOSIUM PROCEEDINGS VOLUME 207

Mechanical Properties of Porous and Cellular Materials

Symposium held November 26-27, 1990, Boston, Massachusetts, U.S.A.

EDITORS:

K. Sieradzki
The Johns Hopkins University, Baltimore, Maryland, U.S.A.

D.J. Green
Pennsylvania State University, University Park, Pennsylvania, U.S.A.

L.J. Gibson
Massachusetts Institute of Technology, Cambridge, Massachusetts, U.S.A.

MRS MATERIALS RESEARCH SOCIETY
Pittsburgh, Pennsylvania

Single article reprints from this publication are available through University Microfilms Inc., 300 North Zeeb Road, Ann Arbor, Michigan 48106

CODEN: MRSPDH

Published by:

Materials Research Society
9800 McKnight Road
Pittsburgh, Pennsylvania 15237
Telephone (412) 367-3003
Fax (412) 367-4373

Library of Congress Cataloging in Publication Data

Mechanical properties of porous and cellular materials : symposium held November 26-27, 1990, Boston, Massachusetts, U.S.A. / editors, K. Sieradzki, D.J. Green, L.J. Gibson.

p. cm. — (Materials Research Society symposium proceedings, ISSN 0272-9172 ; v. 207)

Includes bibliographical references and index.

ISBN 1-55899-099-2

1. Porous materials—Mechanical properties—Congresses. I. Sieradzki, Karl. II. Green, D.J. (David J.) III. Gibson, Lorna J. IV. Series: Materials Research Society symposium proceedings ; v. 207.

TA418.9.P6M43 1991 91-21285
620.1'1692—dc20 CIP

Manufactured in the United States of America

Contents

PREFACE ix

MATERIALS RESEARCH SOCIETY SYMPOSIUM PROCEEDINGS x

PART I: MECHANICS AND
FAILURE OF CELLULAR MATERIALS

FRACTURE BEHAVIOR OF OPEN CELL CERAMICS 3
Rasto Brezny and David J. Green

MULTIAXIAL FAILURE CRITERIA FOR CELLULAR MATERIALS 9
T.C. Triantafillou and L.J. Gibson

THE MECHANICAL BEHAVIOR OF MICROCELLULAR FOAMS 15
M.H. Ozkul, J.E. Mark, and J.H. Aubert

MECHANICAL STRUCTURE-PROPERTY RELATIONSHIPS OF
MICROCELLULAR, LOW DENSITY FOAMS 21
James D. LeMay

THERMAL SHOCK BEHAVIOR OF OPEN CELL CERAMIC FOAMS 27
Robert M. Orenstein, David J. Green, and
Albert E. Segall

MECHANICAL BEHAVIOR OF A CELLULAR-CORE CERAMIC SANDWICH
SYSTEM 35
Eric J. Van Voorhees and David J. Green

THE INDENTATION AND NAILING OF CELLULAR MATERIALS 41
M. Fátima Vaz and M.A. Fortes

ANISOTROPY AND SIMPLE MECHANICS OF THE FLESH OF APPLES 61
Julian F.V. Vincent

FRACTURE PROPERTIES OF AN ANISOTROPIC BIOLOGICAL CELLULAR
MATERIAL - APPLE FLESH 65
Ali A. Khan and Julian F.V. Vincent

PART II: MICROSTRUCTURAL INFLUENCES
ON ELASTICITY AND FRACTURE

HIGH STRENGTH, POROUS, BRITTLE MATERIALS 71
J.S. Haggerty, A. Lightfoot, J.E. Ritter, and
S.V. Nair

FRACTURE AND YIELD OF A POROUS COMPOSITE OF
TRINITROTOLUENE AND CYCLOTRIMETHYLENE TRINITRAMINE 77
Donald A. Wiegand and J. Pinto

*REPRESENTATION OF THE MICROSTRUCTURAL DEPENDENCE OF THE ANISOTROPIC ELASTIC CONSTANTS OF TEXTURED MATERIALS 83
Stephen C. Cowin

DIGITAL-IMAGE-BASED STUDY OF CIRCULAR HOLES IN AN ELASTIC MATRIX 95
A.R. Day, M.F. Thorpe, K.A. Snyder, and E.J. Garboczi

RIGIDITY OF LAYERED RANDOM ALLOYS 103
Wei Jin, S.D. Mahanti, and M.F. Thorpe

DENSIFICATION MODELS FOR PLASMA SPRAYED METAL MATRIX COMPOSITE FOILS 109
D.M. Elzey, J.M. Kunze, J.M. Duva, and H.N.G. Wadley

PART III: PROCESSING AND CHARACTERIZATION OF CELLULAR MATERIALS

*MICROCELLULAR FOAMS PREPARED FROM DEMIXED POLYMER SOLUTIONS 117
J.H. Aubert

SYNTHESIS AND CHARACTERIZATION OF CELLULAR SiO_2 MATERIALS BY FOAMING SOL GELS 129
Josephine Covino and Allen P. Gehris, Jr.

IMPROVEMENT OF A POROUS MATERIAL MECHANICAL PROPERTY BY HOT ISOSTATIC PROCESS 135
Atsushi Takata, Kozo Ishizaki, and Shojiro Okada

CHARACTERIZATION OF POROUS CELLULAR MATERIALS FABRICATED BY CHEMICAL VAPOR DEPOSITION 141
Andrew J. Sherman, Brian E. Williams, Mark J. Delarosa, and Raffaele Laferla

STRESS-DENSITY VARIATIONS IN ALUMINA SEDIMENTS: EFFECTS OF POLYMER CHEMISTRY 151
C.H. Schilling, J.J. Lannutti, W.-H. Shih, and I.A. Aksay

ENERGY ABSORPTION BY EXPANDED BEAD POLYSTYRENE FOAM: DEPENDENCE ON FRACTURE TOUGHNESS AND BEAD FUSION 157
P.R. Stupak and J.A. Donovan

TESTING OF IMPACT LIMITERS FOR TRANSPORTATION CASK DESIGN 163
A.K. Maji, S. Donald, and K. Cone

MECHANICAL ANALYSES OF WIPP DISPOSAL ROOMS BACKFILLED WITH EITHER CRUSHED SALT OR CRUSHED SALT-BENTONITE 169
Ralph A. Wagner, G.D. Callahan, and B.M. Butcher

*Invited Paper

PART IV: SCALING APPROACHES OF MECHANICS IN DISORDERED SOLIDS

*SCALING THEORY OF ELASTICITY AND FRACTURE IN DISORDERED NETWORKS 179
P.M. Duxbury and S.G. Kim

A COMPARISON OF MECHANICAL PROPERTIES AND SCALING LAW RELATIONSHIPS FOR SILICA AEROGELS AND THEIR ORGANIC COUNTERPARTS 197
R.W. Pekala, L.W. Hrubesh, T.M. Tillotson, C.T. Alviso, J.F. Poco, and J.D. LeMay

SCALING LAWS FOR TRANSPORT, MECHANICAL AND FRACTURE PROPERTIES OF DISORDERED MATERIALS 201
Muhammad Sahimi and Sepehr Arbabi

THE POROSITY DEPENDENCE OF MECHANICAL AND OTHER PROPERTIES OF MATERIALS 229
Roy W. Rice

OPTIMAL SELECTION OF FOAMS AND HONEYCOMBS IN PACKAGING DESIGN 235
J. Zhang and M.F. Ashby

AUTHOR INDEX 247

SUBJECT INDEX 249

MATERIALS RESEARCH SOCIETY SYMPOSIUM PROCEEDINGS 251

*Invited Paper

Preface

This symposium successfully brought scientists together from a wide variety of disciplines to focus on the mechanical behavior of porous and cellular solids composed of metals, ceramics, polymers, or biological materials.

For cellular materials, papers ranged from processing techniques through microstructure-mechanical property relationships to design. In an overview talk, Mike Ashby (Cambridge Univ.) showed how porous cellular materials can be more efficient than dense materials in designs that require minimum weight. He indicated that many biological materials have been able to accomplish such efficiency but there exists an opportunity to design even more efficient, manmade materials controlling microstructures at different scale levels.

In the area of processing, James Aubert (Sandia National Laboratories) discussed techniques for manipulating polymer-solvent phase equilibria to control the microstructure of microcellular foams. Other papers on processing discussed the production of cellular ceramics by CVD, HIPing and sol-gel techniques.

Papers on the mechanical behavior of cellular materials considered various ceramics microcellular polymers, conventional polymer foams and apples. There were also contributions that considered optimum design procedures for cellular materials. Steven Cowin (City Univ. of New York) discussed procedures to match the discrete microstructural aspects of cellular materials with the continuum mechanics approach to their elastic behavior.

Other papers considered the role of pores in the ductile failure of metals and the mechanical response of porous composites during HIPing. David Srolovitz (Univ. of Michigan) discussed techniques for simulating microstructural features, such as porosity, in the computer modelling of elastic-brittle spring networks. Another interesting group of papers considered the deformation processes in power beds and other particle networks.

The final session of the symposium was held in conjunction with Symposium W and was devoted to the mechanics of disordered solids. Phil Duxbury (Michigan State) gave an overview of recent advances in scaling theories of elastic and fracture behavior in disordered solids. Several other papers considered this topic in detail.

Support

ICI Polyurethanes
Alcan International Ltd.
Mobay Corporation
Pennsylvania State University

K. Sieradzki
D.J. Green
L.J. Gibson

May 1991

Volume 180—Better Ceramics Through Chemistry IV, C.J. Brinker, D.E. Clark, D.R. Ulrich, B.J.J. Zelinsky, 1990, ISBN: 1-55899-069-0

Volume 181—Advanced Metallizations in Microelectronics, A. Katz, S.P. Murarka, A. Appelbaum, 1990, ISBN: 1-55899-070-4

Volume 182—Polysilicon Thin Films and Interfaces, B. Raicu, T.Kamins, C.V. Thompson, 1990, ISBN: 1-55899-071-2

Volume 183—High-Resolution Electron Microscopy of Defects in Materials, R. Sinclair, D.J. Smith, U. Dahmen, 1990, ISBN: 1-55899-072-0

Volume 184—Degradation Mechanisms in III-V Compound Semiconductor Devices and Structures, V. Swaminathan, S.J. Pearton, O. Manasreh, 1990, ISBN: 1-55899-073-9

Volume 185—Materials Issues in Art and Archaeology II, J.R. Druzik, P.B. Vandiver, G. Wheeler, 1990, ISBN: 1-55899-074-7

Volume 186—Alloy Phase Stability and Design, G.M. Stocks, D.P. Pope, A.F. Giamei, 1990, ISBN: 1-55899-075-5

Volume 187—Thin Film Structures and Phase Stability, B.M. Clemens, W.L. Johnson, 1990, ISBN: 1-55899-076-3

Volume 188—Thin Films: Stresses and Mechanical Properties II, W.C. Oliver, M. Doerner, G.M. Pharr, F.R. Brotzen, 1990, ISBN: 1-55899-077-1

Volume 189—Microwave Processing of Materials II, W.B. Snyder, W.H. Sutton, D.L. Johnson, M.F. Iskander, 1990, ISBN: 1-55899-078-X

Volume 190—Plasma Processing and Synthesis of Materials III, D. Apelian, J. Szekely, 1990, ISBN: 1-55899-079-8

Volume 191—Laser Ablation for Materials Synthesis, D.C. Paine, J.C. Bravman, 1990, ISBN: 1-55899-080-1

Volume 192—Amorphous Silicon Technology, P.C. Taylor, M.J. Thompson, P.G. LeComber, Y. Hamakawa, A. Madan, 1990, ISBN: 1-55899-081-X

Volume 193—Atomic Scale Calculations of Structure in Materials, M.A. Schluter, M.S. Daw, 1990, ISBN: 1-55899-082-8

Volume 194—Intermetallic Matrix Composites, D.L. Anton, R. McMeeking, D. Miracle, P. Martin, 1990, ISBN: 1-55899-083-6

Volume 195—Physical Phenomena in Granular Materials, T.H. Geballe, P. Sheng, G.D. Cody, 1990, ISBN: 1-55899-084-4

Volume 196—Superplasticity in Metals, Ceramics, and Intermetallics, M.J. Mayo, J. Wadsworth, M. Kobayashi, A.K. Mukherjee, 1990, ISBN: 1-55899-085-2

Volume 197—Materials Interactions Relevant to the Pulp, Paper, and Wood Industries, J.D. Passaretti, D. Caulfield, R. Roy, V. Setterholm, 1990, ISBN: 1-55899-086-0

Volume 198—Epitaxial Heterostructures, D.W. Shaw, J.C. Bean, V.G. Keramidas, P.S. Peercy, 1990, ISBN: 1-55899-087-9

Volume 199—Workshop on Specimen Preparation for Transmission Electron Microscopy of Materials II, R. Anderson, 1990, ISBN: 1-55899-088-7

Volume 200—Ferroelectric Thin Films, A.I. Kingon, E.R. Myers, 1990, ISBN: 1-55899-089-5

Volume 201—Surface Chemistry and Beam-Solid Interactions, H. Atwater, F.A. Houle, D. Lowndes, 1991, ISBN: 1-55899-093-3

Volume 202—Evolution of Thin Film and Surface Microstructure, C.V. Thompson, J.Y. Tsao, D.J. Srolovitz, 1991, ISBN: 1-55899-094-1

Volume 203—Electronic Packaging Materials Science V, E.D. Lillie, R.J. Jaccodine, P. Ho, K. Jackson, 1991, ISBN: 1-55899-095-X

Volume 204—Chemical Perspectives of Microelectronic Materials II, L.H. Dubois, L.V. Interrante, M.E. Gross, K.F. Jensen, 1991 ISBN: 1-55899-096-8

Volume 205—Kinetics of Phase Transformations, M.O. Thompson, M. Aziz, G.B. Stephenson, D. Cherns, 1991, ISBN: 1-55899-097-6

Volume 206—Clusters amd Cluster-Assembled Materials, R.S. Averback, D.L. Nelson, J. Bernholc, 1991, ISBN: 1-55899-098-4

Volume 207—Mechanical Properties of Porous and Cellular Materials, L.J. Gibson, D. Green, K. Sieradzki, 1991, ISBN-1-55899-099-2

Volume 208—Advances in Surface and Thin Film Diffraction, P.I. Cohen, D.J. Eaglesham, T.C. Huang, 1991, ISBN: 1-55899-100-X

Volume 209—Defects in Materials, P.D. Bristowe, J.E. Epperson, J.E. Griffith, Z. Liliental-Weber, 1991, ISBN: 1-55899-101-8

Volume 210—Solid State Ionics II, G.-A. Nazri, R.A. Huggins, D.F. Shriver, M. Balkanski, 1991, ISBN: 1-55899-102-6

Volume 211—Fiber-Reinforced Cementitious Materials, S. Mindess, J.P. Skalny, 1991, ISBN: 1-55899-103-4

Volume 212—Scientific Basis for Nuclear Waste Management XIV, T. Abrajano, Jr., L.H. Johnson, 1991, ISBN: 1-55899-104-2

Volume 213—High Temperature Ordered Intermetallic Alloys IV, L. Johnson, D.P. Pope, J.O. Stiegler, 1991, ISBN: 1-55899-105-0

Volume 214—Optical and Electrical Properties of Polymers, J.A. Emerson, J.M. Torkelson, 1991, ISBN: 1-55899-106-9

Volume 215—Structure, Relaxation and Physical Aging of Glassy Polymers, R.J. Roe, J.M. O'Reilly, J. Torkelson, 1991, ISBN: 1-55899-107-7

Volume 216—Long-Wavelength Semiconductor Devices, Materials and Processes, A. Katz, R.M. Biefeld, R.J. Malik, R.L. Gunshor, 1991, ISBN 1-55899-108-5

Volume 217—Advanced Tomographic Imaging Methods for the Analysis of Materials, J.L. Ackerman, W.A. Ellingson, 1991, ISBN: 1-55899-109-3

Volume 218—Materials Synthesis Based on Biological Processes, M. Alper, P.C. Rieke, R. Frankel, P.D. Calvert, D.A. Tirrell, 1991, ISBN: 1-55899-110-7

Earlier Materials Research Society Symposium Proceedings listed in the back.

PART I

Mechanics and Failure of Cellular Materials

FRACTURE BEHAVIOR OF OPEN CELL CERAMICS

RASTO BREZNY * AND DAVID J. GREEN
The Pennsylvania State University, Steidle Bldg., University Park, PA 16802

ABSTRACT

Cellular ceramics constitute a class of materials where the properties may be optimized against the weight. These materials may be potentially useful in both structural and non-structural applications. In order to broaden the utilization of the currently- available materials, the mechanical properties must be improved. The objective of this paper is to review several recent studies on the strength and toughness of open cell brittle materials. The work was aimed at understanding the mechanisms involved in the failure of these materials and the relationship between the properties and the microstructure.

INTRODUCTION

Cellular ceramics have been fabricated in two-dimensional (honeycombs) and three-dimensional (foams) macrostructures. Three-dimensional structures can be further sub-classified as open cell or closed cell depending on whether or not the individual cells possess solid faces. It is useful to consider the structure of cellular materials at two levels. The macrostructure is the large scale arrangement of polygons making up the body whereas microstructure refers to the fine scale structure of pores and single crystal grains within the solid portion. This overall project has focused on the mechanical behavior of open cell ceramic foams where the void space is connected through the cell faces making the material permeable. The ceramic beams which compose the open cell foam will be referred to as struts. The materials used in this study had relative densities in the range of 0.05 to 0.3. The relative density is the bulk density of the foam (ρ) divided by the density of the struts (ρ_s). In some cases, relative density plays an important role in producing a structure having an optimum stiffness or strength with respect to the weight of a body [1]. In addition to the relative density, the cell size may also play a pivotal role in cellular ceramics where flaws can dramatically limit the strength. Brezny and Green [2] have considered a number of parameters for cellular ceramics and have shown that these materials can offer advantages in structural applications over dense ceramics.

Currently, ceramic foams are not nearly as common as their polymeric counterparts, but this review will show that they could play an important role in structural applications. The application of cellular ceramics may require the optimization of properties other than strength and stiffness, such as chemical and thermal stability, permeability, thermal conductivity, thermal shock resistance, dielectric constant, etc; however, this review will focus on the fracture behavior, in particular, the strength and toughness.

In order to understand the mechanical behavior of cellular materials, and be able to tailor the properties to specific applications, it is critical to understand

---

* Now with W.R. Grace & Co. - Conn., 7379 Route 32, Columbia, MD 21044

the relationships between the properties and the foam parameters, such as density and cell size. The scientific approach to predicting the behavior has been to identify a suitable unit cell geometry, analyze its deformation behavior and scale this to provide bulk mechanical properties. The resulting relationships describe the properties of the bulk foam in terms of the solid material properties and foam parameters. For this study, the theoretical approach by Gibson and Ashby [3] will be used to analyze experimental data on open cell ceramics. As a first approximation the mechanical properties of these materials will be assumed to be isotropic but one should be aware that cellular materials can show differing degrees of anisotropy [4].

FACTURE TOUGHNESS BEHAVIOR

The fracture toughness of brittle foams was theoretically analyzed by Maiti et al., [5] using two types of analyses. Both approaches were based on the unit cell considered by Gibson and Ashby [6]. The first approach was derived from a linear elastic fracture mechanics argument by considering the stresses around a crack in the material and the resulting moments applied to struts in front of the crack tip. The other approach was a simple fracture energy balance between the energy to fracture a unit cell and that to fracture a single strut of material. Recent work on the fracture toughness behavior of an open cell, glassy carbon has shown that only the fracture mechanics approach is consistent with experimental data [7]. The fracture mechanics approach predicts the toughness of a brittle, open cell foam by

$$K_{IC} = C_1 \sigma_{fs} \sqrt{(\pi L)} \left[\frac{\rho}{\rho_s}\right]^{3/2} \quad (1)$$

where σ_{fs} is the strength of the struts, L is the cell size and C_1 is a geometric constant [5]. This constant was determined empirically as 0.65 for a number of materials; however, this value could be very sensitive to the assumed values for strut strength. In flaw sensitive materials, like ceramics, the strength may vary with the volume of material under stress. This phenomenon stems from the increased probability of finding a flaw of a critical size in a larger volume of material [8]. For this reason it is feasible that the strut strength in brittle foams could be dependent on strut size.

In order to critically evaluate the fracture toughness behavior, it has been necessary to devise techniques to measure the strut strength. Techniques based on mechanical testing of individual struts [9] and quantitative fractography of strut fracture surfaces [10] have been developed to measure the strut strength. These studies demonstrated that although the strut strengths exhibit wide variability, values in excess of 1 GPa are feasible [9,10]. The strength is limited by large flaws within the struts and thus the implications of these observations is that substantial improvements in the fracture toughness of cellular ceramics are possible. These could be realized through improved processing techniques aimed at reducing the variability and magnitude of flaws within the struts. The fracture toughness of a foam can thus depend on cell size, as well as density.

The fracture toughness behavior of open cell, alumina-based ceramics has been measured as a function of relative density [11]. It was critical to quantitatively assess the macrostructure in the materials, especially the presence of hollow struts. The struts remain hollow following the burnout of the polymer foam used as a sacrificial substrate during processing [9]. Once these factors and the strut

Wed Aug 01 2001

8/13

Please retrieve the following items to be placed on the hold shelf.

Circulation Department
Library Media Services
Kent State University
Kent OH 44242-0001

MAIN see bldg guide
CALL NO: TA418.9.P6 M43 1991
AUTHOR:
Mechanical properties of porous and
BARCODE: 31850012467974
REC NO: i25245466
Kelvin Smith Library Service Desk

AI LIU
CWRU- MTLS SCI & ENG
WHITE BLDG RM 438
CLEVELAND, OH 44106
INSTITUTION: Case Western
LOCATION: Patro

???-???-????

-iu, A

Wed Aug 01 2001

strength were considered the density exponent agreed with that given in Eq. 1. In some materials, the presence of closed faces and/or filled cells led to deviation from Eq. 1 but overall normalization of K_{Ic} by $\sigma_{fs}\sqrt{L}$ resulted in a similar density variation for all the materials tested.

The value of the geometric constant ($C_1 \sim 0.13$) estimated for these materials is much lower than suggested by Maiti et al. [5] and may be attributed to an incorrect assumption for the value of strut strength in the original work.

For the alumina-based, open cell ceramics the normalized toughness was found to depend on cell size in a different way than predicted by Eq. 1. This was related to the presence of split struts. In some cases, failure could occur by the coalescence of strut cracks rather than the successive bending failure of individual struts as assumed in the model [10]. Furthermore, the macrostructure varied from entirely open cell in large cell size specimens to partially closed in the finest cell samples. The fracture toughness of an open cell, vitreous carbon was measured as a function of cell size because the macrostructure remained open for all cell sizes. The data were in agreement with Eq. 1 and the value of C_1 calculated for the carbon foam was 0.14 [10].

TENSILE STRENGTH BEHAVIOR

The relationship between the tensile strength (σ_{ft}) and the critical foam parameters of a brittle, open cell material was derived by Maiti et al., [5] by using the linear elastic fracture mechanics relationship between fracture stress and fracture toughness

$$\sigma_{ft} = K_{Ic} / [Y \sqrt{c}] \qquad (2)$$

where ,c, is the critical flaw size and Y is a geometric factor. Applying this to Eq. 1 results in the expression for the tensile strength of a brittle foam for $c \geq L$.

$$\sigma_{ft} = C_2 \sigma_{fs} \left(\frac{L}{c}\right)^{1/2} \left[\frac{\rho}{\rho_s}\right]^{3/2} \qquad (3)$$

The cell size represents the lower limit of the flaw size; however, the presence of broken struts or other macroscopic defects can increase this to several cells. These macroscopic flaws act to concentrate stresses in the vicinity of the flaw and lower the tensile strength.

The results of strength measurements made under flexural and axial loading gave good agreement with Eq. 3 in both the alumina-based and glassy carbon materials [4,10,11]. The strength of the alumina-based samples exhibited the same trend with cell size as observed in the fracture toughness experiments; however, the cell size effect on the glassy carbon resulted in an increasing strength with decreasing cell size. This behavior is due to both an increasing strut strength and decreasing macroscopic flaw size in the finer cell materials.

COMPRESSIVE STRENGTH BEHAVIOR

The compressive failure of a foam, as put forth by Gibson and Ashby [3], is governed by the bending failure of the struts due to the applied external load. This criteria was used to derive the expression for the compressive strength (σ_{fc}) of a brittle, open cell material.

$$\sigma_{fc} = C_3 \sigma_{fs} \left[\frac{\rho}{\rho_s}\right]^{3/2} \quad (4)$$

This property depends on the strut strength and relative density in the same way as described for the toughness and tensile strength. This is due to the fact that all of the expressions were derived by assuming catastrophic flexural failure of the struts. Comparing Eqs. 3 and 4, it is interesting to note that in the theoretical lower limit of the flaw size ($c = L$), the predicted tensile and compressive strengths should be equal in cellular ceramics.

The compressive strength was measured as a function of cell size in the glassy carbon material [10]. A cell size effect is not implicit from Eq. 4 if the strut strength is invariant with cell size. An increasing compressive strength with decreasing cell size was measured and attributed to the increasing strut strength with decreasing strut size measured in this material. This effect was accurately predicted by Eq. 4.

The strength values measured in compression were very sensitive to the method of load application, that is the uniformity of load distribution over the entire contact surface. The compressive strength was measured at four relative densities in the alumina-based foam using a variety of loading surfaces including steel, neoprene and an epoxy bond layer [4]. The empirically determined value of the exponent on the relative density (Eq. 4) varied dramatically with the method of loading and was attributed to variations in load uniformity. The best method for compressive testing of these materials involved bonding steel platens to both loading surfaces using a high strength epoxy. The epoxy infiltrated the first layer of cells and acted to distribute the load over all of the struts within the contact area.

The theoretical model predicts the same density dependence in tension and compression because the failure mode assumed in both analyses is the same (Eq. 3 and 4). The alumina-based samples, however, exhibited significantly different behavior as a function of density and this was attributed to a difference in the failure mode occurring in the two loading directions. Furthermore, the magnitude of the compressive strength was several times greater than the tensile value. This is a common behavior in brittle materials which tend to be much more sensitive to flaws under tensile stresses than compressive.

FAILURE MODE ANALYSIS

A combination of acoustic emission and slow motion video photography was implemented to study the mode of failure occurring in cellular materials subjected to tensile and compressive loading [4]. The signals from individual failure events were characterized to enable the interpretation of data collected on bulk specimens. The signals obtained from tensile and compressive specimens were quite different, as well as those obtained on the two types of materials tested (alumina and vitreous carbon foams).

Regardless of the loading direction, individual strut failure throughout the entire sample volume was detected prior to catastrophic fracture of the specimen. This was attributed to failure of the weakest struts within the body, which was expected as the struts were known to exhibit a wide distribution of strengths [10]. As the load was increased the highest number of acoustic emission events were detected prior to maximum load in compression, whereas the peak number of events occurred at the fracture load in tension. A change in compliance of the specimen was detected at a load corresponding to the peak in acoustic emission during compressive loading. Thus it was concluded that compressive failure in brittle cellular materials was associated with a damage accumulation process whereas tensile failure occurs by the propagation of a single critical flaw. These results were verified by the visual observation of failure processes using the video recording system.

Although the compressive failure process was similar in both materials, the ultimate failure event was different. The alumina-based material contained a large number of preexisting cracks and defects within the struts resulting from the fabrication. Compressive loading of this material occurred by the linking of these preexisting defects into damaged regions of material followed by breaking away of these regions prior to collapse of the specimen. Failure often occurred along acute angles with respect to the direction of loading. This type of failure process has been described for the compressive failure of dense ceramics as well [12].

Due to the relatively defect free nature of the struts within the glassy carbon material, the accumulation of damage was much more localized. The ultimate failure was associated with a single catastrophic event as several layers of cells collapsed under the applied load. The damage was contained within a narrow band of material perpendicular to the loading direction. The individual acoustic events, detected in the glassy carbon foam, were characterized as catastrophic strut failures rather than the stable growth of preexisting microcracks [4].

SUMMARY AND CONCLUSIONS

The fracture toughness, bend strength, tensile strength and compressive strength have been analyzed as a function of density and cell size in two types of brittle, open cell materials. The model of Gibson and Ashby [3] has been identified as most accurately describing the mechanical behavior of these materials as it is the only approach which has considered the bending failure of the struts in deriving relationships for the strength and toughness. The compressive strength as a function of density was observed to deviate from the predicted behavior and this disagreement was attributed to the dramatically different failure mode observed in the alumina material from that assumed in the model. Under compressive loading of the alumina material, failure was characterized by a damage accumulation process. The tensile failure was found to agree with a fracture mechanics type of approach as described by the model.

In order to apply the theoretical analyses to cellular ceramics, it is critical to have a detailed understanding of the macro- and microstructure of the foams. Furthermore, measurements of the strut properties with density and cell size must be incorporated into the analysis. Techniques have been introduced to measure this property. Improvements in the strut strength should focus on reducing the size and distribution of flaws, such as cracks and pores, through improvements in the processing. Such measures should lead to improved properties in the bulk cellular materials.

A comparison of the experimental results obtained from the vitreous carbon and alumina open cell foams makes an important point when working with brittle cellular solids. The models developed to describe the behavior of these materials are very general and show only the basic relationships between the parameters. When working with materials whose properties are extremely sensitive to the microstructure, like ceramics, one must be careful when using these simple models to predict the mechanical behavior. The parameters in the model can vary substantially, depending on the details of the actual microstructure. To make accurate predictions it is important to thoroughly understand the type of microstructural and macrostructural variability which can develop in these cellular materials.

REFERENCES

1. M.F. Ashby, Acta Metall., 37, 1273-1293 (1989)

2. R. Brezny and D.J. Green, in Materials Science and Technology - a comprehensive treatment, to be published, (VCH Publishers, Germany).

3. L.J. Gibson and M.F. Ashby, Cellular Solids: Structure and Properties, (Pergamon Press, New York, 1988).

4. R. Brezny, PhD thesis, Penn State University, 1990.

5. S.K. Maiti M.F. Ashby and L.J. Gibson, Scripta Metall., 18, 213-217 (1984).

6. L.J. Gibson and M.F. Ashby, Proc. R. Soc. London, Ser. A382, 43-59 (1982).

7. R. Brezny and D.J. Green, J. Am. Ceram. Soc. To be published.

8. W. Weibull, J. Appl. Mech., 18, 243 (1951).

9. R. Brezny, D.J. Green and C.Q. Dam, J. Am. Ceram. Soc., 72, 885-889 (1989).

10. R. Brezny and D.J. Green, Acta Metall., to be published.

11. R. Brezny and D.J. Green, J. Am. Ceram. Soc., 72, 1145-1152 (1989).

12. M.F. Ashby and C.G. Sammis, Cambridge University Engineering Department Report #CUED/C-MATS/TR 144, July (1988).

MULTIAXIAL FAILURE CRITERIA FOR CELLULAR MATERIALS

T. C. TRIANTAFILLOU* and L. J. GIBSON**
*Assistant Professor, **Associate Professor
Department of Civil Engineering, Massachusetts Institute of Technology, Cambridge, MA 02139

ABSTRACT

Cellular materials are increasingly used in engineering. Proper design requires an understanding of the response of the materials to stress; and, in real engineering design, the stress state is often a complex one. In this paper we model the elastic buckling, plastic yield, and brittle fracture of cellular solids under multiaxial stresses to develop equations describing their failure surfaces. Comparison of the analysis with data shows that the models describe the main features of the multiaxial behavior of foams well.

INTRODUCTION

Materials with a cellular structure are increasingly used in modern engineering. Aluminum, paper and polymeric honeycombs are used as cores for high-performance sandwich panels; plastic and metal foams absorb energy in packaging and safety padding; and natural cellular materials-particularly woods-are widely used in structures.

The behavior of cellular solids in simple uniaxial loading is tolerably well understood (see, for example, [1]). But loads in real engineering structures are often multiaxial. Then it is not the uniaxial stress but the combination of stresses causing failure which is important to the designer. "Failure", of course, can mean different things, depending on the design. It could mean "excessive elastic deformation"; more usually it means "elastic collapse", "the onset of plasticity" or "fracture". Each such mechanism can be characterized by a failure surface: a surface in stress-space describing the combination of stresses which cause failure. The mechanism of failure itself may depend on stress state: a foam may be plastic in compression but brittle in tension, for instance. Then the failure surface is the inner envelope of the intersecting surfaces for the individual mechanisms.

In this paper we present equations describing each mode of failure for foams under multiaxial stresses. Typical failure envelopes for biaxial and axisymmetric stress states, showing the intersection of the surfaces for different failure mechanisms, are plotted. The results are then compared with experiments. It is shown that the models describe the main features of material behavior well.

PLASTIC COLLAPSE AND THE YIELD SURFACE

We now model the failure of elastic-plastic foams under multiaxial loads. Consider the isotropic, open cell of Fig. 1. Each cell wall has length l and cross-section t^2. The relative density of the foam (its density ρ^*, divided by that of the cell wall material, ρ_s) is proportional to $(t/l)^2$. The uniaxial plastic collapse stress is described as

$$\sigma^*_{pl} = C_1(\rho^*/\rho_s)^{3/2}\sigma_{ys} \quad (1)$$

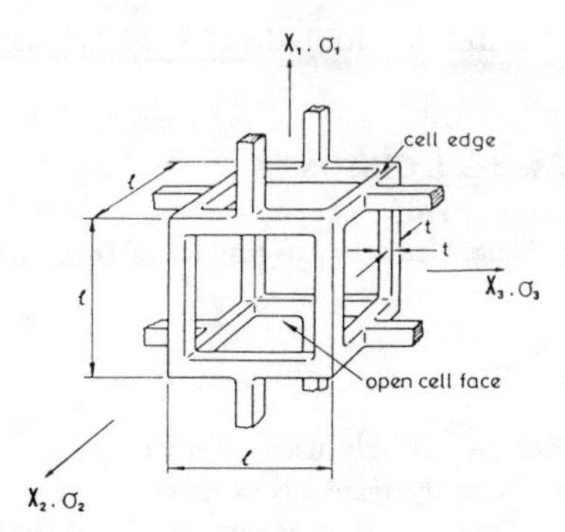

Fig.1 Three-dimensional, open cubic cell under triaxial loading (after [1]).

where the constant C_l is 0.3 and σ_{ys} is the yield stress of the cell wall material [2]. Equation (1) is obtained based on the assumption that plastic collapse occurs when the maximum moment in the cell walls reaches the plastic moment, M_p. Now consider triaxial loading. If the three principal stresses acting on the cell are all equal, the cell walls are subjected to purely axial tension or compression; then plastic collapse occurs when the axial stress in the cell wall exceeds σ_{ys}, or:

$$\sigma_a = \sigma_{ys} = \frac{I_1}{(\rho^*/\rho_s)} \qquad (2)$$

where $I_1=\sigma_1+\sigma_2+\sigma_3$ is the *first invariant* of the *stress tensor.* Under any other combination of loads the cell walls suffer bending as well as axial deformation. The average bending moment is proportional to the deviatoric stress (square root of the *second invariant*, J_2, of the *deviatoric stress tensor*) times the cube of the cell size:

$$M \propto \sqrt{J_2}\, l^3 = \sqrt{\frac{1}{2}\left[(\sigma_1 - \sigma_2)^2 + (\sigma_2 - \sigma_3)^2 + (\sigma_3 - \sigma_1)^2\right]}\; l^3 \qquad (3)$$

But the axial load on the cell wall modifies the moment required to bend it plastically, according to the following equation:

$$M_p \propto \sigma_{ys} t^3 \left[1 - \left(\frac{\sigma_a}{\sigma_{ys}}\right)^2\right] \qquad (4)$$

Equating this to eqn. (3), using the fact that $(\rho^*/\rho_s) \propto (t/l)^2$, considering eqn. (1), and calibrating the result for the case of simple compression gives:

$$\pm \frac{\sqrt{3}\sqrt{J_2}}{\sigma^*_{pl}} + 0.09\left(\frac{\rho^*}{\rho_s}\right)\left(\frac{I_1}{\sigma^*_{pl}}\right)^2 = 1 \qquad (5)$$

The yield surface described by eqn. (5) is illustrated in Fig.2.

BRITTLE COLLAPSE AND THE FRACTURE SURFACE

The cell struts of a brittle foam fail when the maximum bending tensile stress at the extreme fiber equals the modulus of rupture of the solid cell wall material, σ_{fs}. From this observation,

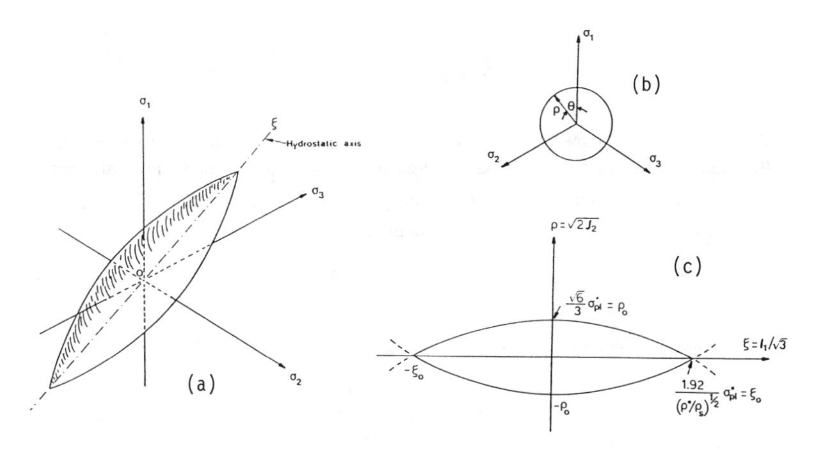

Fig. 2 (a) The plastic collapse surface of an isotropic foam in principal stress space. (b) Projections on the deviatoric plane of the coordinate axes. (c) The meridians.

the uniaxial brittle crushing strength, σ^*_{cr}, has been found to be [2]:

$$\sigma^*_{cr} = - C_2(\rho^*/\rho_s)^{3/2}\sigma_{fs} \tag{6}$$

where the constant C_2 is expected to be approximately 0.2 [3] and compression is taken to be negative. The fracture strength in tension is governed by the largest macroscopic crack in the material. Tensile failure can be treated by using the methods of linear elastic fracture mechanics; the mode I fracture toughness of the foam, K^*_{IC}, in terms of the fracture strength of the cell struts, σ_{fs}, and the relative density ratio, ρ^*/ρ_s, is [2]:

$$K^*_{IC} = C_3\sigma_{fs}\sqrt{\pi l}\,(\rho^*/\rho_s)^{3/2} \tag{7}$$

where l is the cell size. The tensile fracture stress is:

$$\sigma^*_{fr} = \frac{K^*_{IC}}{\sqrt{\pi a}} \tag{8}$$

where 2a is the length of the largest crack in the sample. [3]and [4] suggest that C_3=0.2.

Under a general state of stress, axial as well as bending stresses in the cell walls are important. *Tensile rupture* of the cell struts occurs when the extreme fiber tensile stress in a strut equals its modulus of rupture:

$$\sigma_{fs} - \sigma_a = \frac{Mt}{2I} \tag{9}$$

where σ_a is the axial stress in the cell strut (given by eqn. (2) with σ_{ys} replaced by σ_{ts}, the tensile strength of the cell strut material), t is its thickness, I is its moment of inertia ($\propto t^4$), and the moment M (given by eqn. (3)) can be either positive or negative. Combining eqn. (9) with these observations, considering eqn. (6), noting that the relative density is proportional to $(t/l)^2$, and calibrating the result for the case of simple compression we obtain:

$$\pm\frac{\sqrt{3}\sqrt{J_2}}{\sigma^*_{cr}} + 0.2\left(\frac{\rho^*}{\rho_s}\right)^{1/2}\left(\frac{\sigma_{fs}}{\sigma_{ts}}\right)\frac{I_1}{|\sigma^*_{cr}|} = 1 \tag{10}$$

Under some stress states, the cell struts can fail by *compressive crushing.* This is the case when the maximum compressive stress at the extreme fiber reaches the compressive crushing strength of the cell strut material, σ_{cs}. The analysis is identical to that given above for tensile rupture with two differences: first, the failure condition now becomes:

$$\sigma_{cs} + \sigma_a = \frac{Mt}{2I} \tag{11}$$

and second, when the material is subjected to uniform compression, we have

$$\sigma_a = \sigma_{cs} = \frac{I_1}{(\rho^*/\rho_s)} \tag{12}$$

The failure criterion for this case is:

$$\pm\frac{\sqrt{3}\sqrt{J_2}}{\sigma^*_{cr}} - 0.2\left(\frac{\rho^*}{\rho_s}\right)^{1/2}\frac{I_1}{|\sigma^*_{cr}|} + \frac{\sigma_{cs}}{\sigma_{fs}} = 0 \tag{13}$$

Note that according to our sign convention both σ^*_{cr} and σ_{cs} are negative because they are compressive. The brittle collapse surface described by eqns (10) and (13) is illustrated in Fig. 3.

If the foam contains cracks or flaws (which is often the case) the failure surface is truncated by brittle fracture in the tensile octant. Fracture occurs when the maximum principal tensile stress reaches σ^*_{fr}: the failure surface is a box bounded by planes of constant principal stress, corresponding to the uniaxial tensile strength of the material [1, 3, 5].

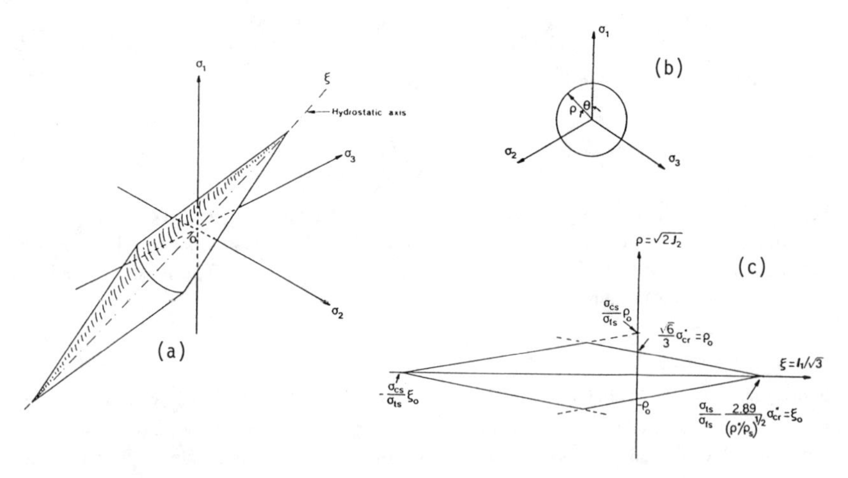

Fig. 3 (a) The brittle collapse surface of an isotropic foam in principal stress space. (b) Projections on the deviatoric plane of the coordinate axes. (c) The meridians.

ELASTIC BUCKLING SURFACE AND FAILURE IN ANISOTROPIC FOAMS

In the compressive octant of the stress space both the yield surface and the brittle collapse surface are truncated by an almost box-like cutoff corresponding to the elastic buckling of the cell walls. The buckling cutoff is described in [3, 6, 7]. Furthermore, the failure criteria described herein can be generalized to orthotropic foams by distorting the collapse surfaces so that they intersect the six axes of the generized stress space at the points of uniaxial strengths [3, 4, 7].

EXPERIMENTAL RESULTS

Uniaxial, biaxial, and triaxial (axisymmetric) mechanical tests were carried out on flexible, rigid, and brittle foams and the failure load was recorded. Details of the experimental program can be found in [4, 7, 8]. A comparison between typical experimental results and the analytical failure envelopes is illustrated in Figs 4 and 5.

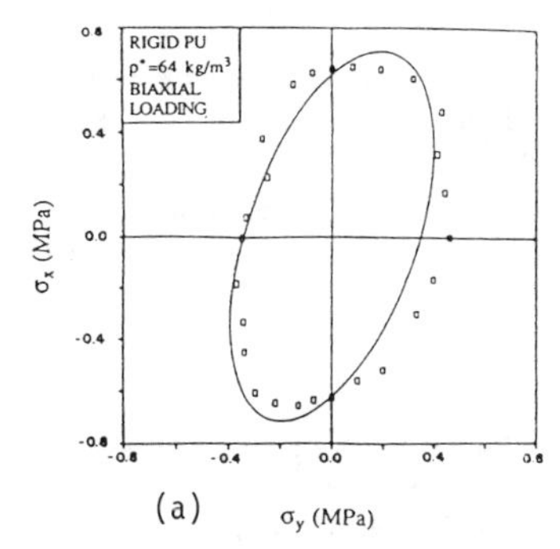

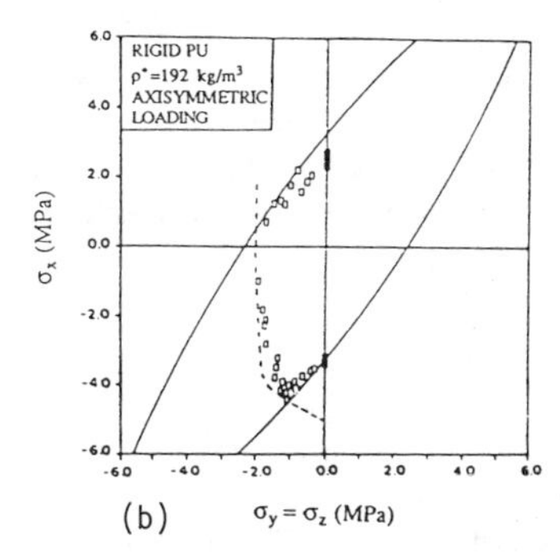

Fig. 4 Data for the failure of two densities of rigid polyurethane foam: (a) biaxial loading in the anisotropic X-Y plane; (b) axisymmetric loading. The analytical yield and elastic buckling surfaces are indicated by the solid and dashed lines, respectively.

CONCLUSIONS

The failure surface for cellular materials is more complicated than that for more conventional materials. It is the inner envelope of the surfaces corresponding to three distinct mechanisms: elastic collapse by cell wall buckling, plastic collapse by cell wall bending, and brittle crushing or fracture by cell wall breakage.

The elastic buckling surface exists only in the quadrants in which at least one principal stress is compressive. There is no buckling mode when all stresses are tensile; here other surfaces appear, which can (depending on relative density and cell wall properties) lie inside the buckling surface in other quadrants as well. The simplest is that for plastic yielding which has the shape of a distorted ellipsoid elongated along the hydrostatic axis. The brittle failure surface is a little more complicated; it depends on whether or not a crack larger than the cell size pre-exists in the material. If no large crack is present, the failure surface is described by a dual failure criterion associated with either tensile rupture or compressive crushing of the cell struts. A pre-existing crack reduces the strength of the foam in tension; failure then is described by a maximum principal tensile stress criterion. The results of an extensive experimental program confirm that the models describe the main features of the multiaxial behavior of cellular materials well.

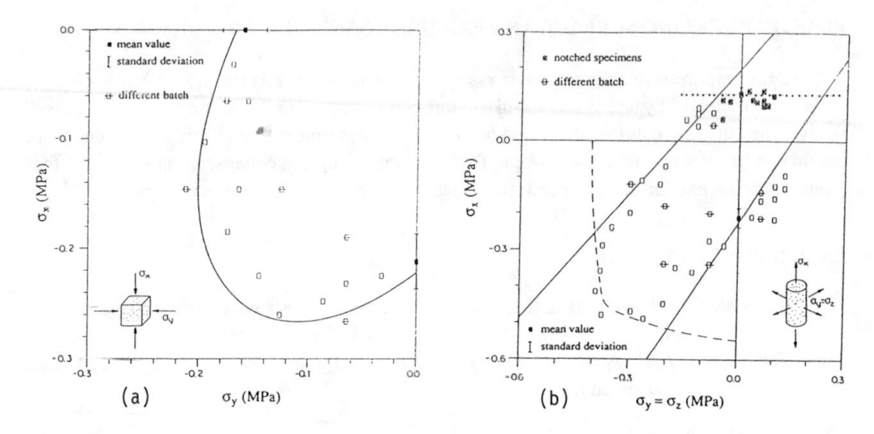

Fig. 5 Data for the failure of reticulated vitreous carbon foam (ρ^*=48 kg/m^3): (a) biaxial compression in the anisotropic X-Y plane; (b) axisymmetric loading. The crushing envelope given by the model is indicated by the solid line; the tensile cutoff is indicated by the dotted line; and the elastic buckling cutoff is indicated by the dashed line.

REFERENCES

1. L. J. Gibson and M. F. Ashby, Cellular Solids: Structure and Properties, Pergamon Press, Oxford (1988).
2. S. K. Maiti, L. J. Gibson and M. F. Ashby, Acta Met. 32, 1963 (1984).
3. L. J. Gibson, M. F. Ashby, J. Zhang and T. C. Triantafillou, Int. J. Mech. Sci., 31, 635 (1989).
4. T. C. Triantafillou and L. J. Gibson, Int. J. Mech. Sci., 32, 479 (1990).
5. F. A. McClintock and A. S. Argon, Mechanical Behavior of Materials, Addison-Wesley, Mass., U.S.A. (1966).
6. J. Zhang, Ph.D. Thesis, Cambridge University Engineering Dept., Cambridge, U.K. (1989).
7. T. C. Triantafillou, Ph.D. Thesis, Dept. of Civil Engineering, Massachusetts Institute of Technology, Cambridge, MA (1989).
8. T. C. Triantafillou, J. Zhang, T. L. Shercliff, L. J. Gibson and M. F. Ashby, Int. J. Mech. Sci., 31, 665 (1989).

THE MECHANICAL BEHAVIOR OF MICROCELLULAR FOAMS

M. H. Ozkul*, J. E. Mark*, and J. H. Aubert**
*Department of Chemistry and the Polymer Research Center
University of Cincinnati, Cincinnati, OH 45221-0172
**Sandia National Laboratories, Albuquerque, NM 87185

ABSTRACT

The mechanical behavior of microcellular open-cell foams prepared by a thermally induced phase separation process are investigated. The foams studied were prepared from isotactic polystyrene, polyacrylonitrile, and poly(4-methyl-1-pentene) (rigid foams), and polyurethane and Lycra (elastomeric foams). Their densities were in the range 0.04-0.27 g/cm3. Conventional polystyrene foams were used for comparison. The moduli and collapse stresses of these foams were measured in compression and compared with the current constitutive laws which relate mechanical properties to densities. A reinforcement technique based on the in-situ precipitation of silica was used to improve the mechanical properties.

INTRODUCTION

Microcellular foams have primarily been developed for their use in inertial confinement fusion in high energy physics laboratories [1]. Recently, however, the possibility of using these foams in biomedical and drug release applications has also been discussed [2]. The preparation technique and pore structure of microcellular foams are different than those of conventional foams. These differences are a result of using a thermally-induced phase separation technique. First, the polymer and solvent are heated above their critical point to form a homogenous solution, then phase separation is induced by lowering the temperature, and finally the solvent is removed by either extraction or by vacuum sublimation, to produce a foam. The cells thus formed are open and have dimensions of 0.1 to 20 μm which are 10 to 100 times smaller than that of conventional foams having the same density.

The constitutive laws for conventional foams have been previously formulated [3,4] and are well understood. Under small strains, the wall elements perpendicular to the applied force bend and the relative modulus depends on the relative density according to the following equation for open-cell foams:

$$\frac{E_f}{E_s} = C_1 \left(\frac{\rho_f}{\rho_s}\right)^2 \qquad (1)$$

where E_f and E_S are the Young's moduli and ρ_f and ρ_s are the densities of the foam and cell wall polymer, respectively. Here C_1 is a constant which was found experimentally to be nearly equal to 1 [4]. A theoretical model [5] predicted C_1=0.91. When the applied force increases, the walls parallel to the force begin to buckle reversibly in the case of the elastomeric foams. The collapse mechanism and the resulting collapse stress are different for the rigid polymers; plastic hinges are formed at the adjoining points of vertical and horizontal wall elements and the collapse stress σ_{pl} is given by;

$$\frac{\sigma_{pl}}{\sigma_{ys}} = C_2(\frac{\rho_f}{\rho_s})^{3/2} \tag{2}$$

where σ_{ys} is the yielding strength of the polymer and C_2 is a constant which was found experimentally to be 0.30 [4]. For the closed-cell foams the membrane forces and the pressure of the gas in the cells should also be considered.

The elastic mechanical properties of microcellular foams have been previously studied [6,7] by the models given above, and the constant C_1 in eq.(1) was found to be smaller than the predicted value. In this study, an attempt was made to understand the discrepancy between the theoretical and experimental results by introducing defects into the structure of the foams by compressing the rigid conventional ones in the plastic region. Large decreases in the modulus were observed during the reloading experiment.

The polymers used to make the foams are isotactic polystyrene (IPS), polyacrylonitrile (PAN), and poly(4-methyl-1-pentene) (TPX) (rigid foams), and polyurethane (PU) and Lycra (elastomeric foams). The densities of foams are in the range 0.04-0.27 g/cm^3.

A technique based on the in-situ precipitation of silica, which has been widely used for the reinforcement of elastomers [8], was applied and the effect of the reinforcement on the mechanical properties was also investigated.

EXPERIMENTAL

The foams were prepared by using the solvents listed in Table I. The test samples were prepared by putting them between two parallel plates and cutting the part of the sample protruding with a blade perpendicular to the plates. The irregularities on the surface were removed by rubbing the sample between similar plates where each surface of the plates were covered with sand paper. The dimensions of samples were approximetly 10X10X8mm.

Compression testing of foams was carried out using an Instron Testing Machine (1122 model) with a 1000 lb load cell. The crosshead speed was 0.02 inch/min, and the strain was calculated from the displacement of cross-head. The tests were made at a temperature of 20°C ±1°C.

The effect of sample height/lateral dimension ratio on the results was tested and none was observed. The samples were also tested in different directions and no obvious difference was obtained, indicating that the foams are isotropic. Although the samples showed long term relaxation, the effect of strain rate on the modulus over a range of an order of magnitude was not seen.

Extruded closed-cell polystyrene foams (provided by the E and C Company, Cardena CA) were compressed to produce defects in the structure. For each compression ratio a different sample was used, and after preloading the sample was left for recovery, and then reloaded. The moduli and collapse stresses were determined from the loading and reloading experiments.

Silica Reinforcement

The TPX, PU, and Lycra samples were held immersed in tetraethoxysilane (TEOS) for 24 hours. The other samples were placed in TEOS vapor for the same period of time. The hydrolysis of TEOS (with diethyl-amine as catalyst) was carried out according to the reaction

$$Si(OC_2H_5)_4 + 2H_2O \rightarrow SiO_2 + 4C_2H_5OH \quad (3)$$

which results in the precipitation of silica particles into and onto the polymer. The amount of silica precipitated was calculated from the difference between the densities before and after the treatment.

Table.I Foam and polymer properties

Foam	Foam			Polymer		
	ρ_f(g/cm^3)	Solvent	Method	ρ_s(g/cm^3)	E_s (MPa)	σ_{ys} (MPa)
IPS 350	0.162	1-chlorodecane	Gelation/with extraction	1.11 [9]	5600 [10]	148 [11]
IPS 351	0.155					
IPS 352	0.168					
IPS 353	0.094					
IPS 354	0.099					
IPS 444	0.137	1-chlorodecane	Gelation/extrac.			
PAN 221A	0.040	Maleic anhyride	Sublimation	1.18 [7]	3400 [7]	83 [12]
PAN 247	0.092					
PAN 389	0.081	Dimethylformamide and ethylene slyrol	Gelation/ extraction			
PAN 391	0.058					
PAN 444	0.052					
TPX 101	0.046	Decalin and 1-dodecanol	Gelation/ extraction	0.83 [7]	1250 [7]	
Lycra-23	0.225	Dimethylacetamide /H_2O	Gelation/ extraction	1.20 [13,4]	45 [13,4]	
PU	0.234			1.20 [4]	45 [4]	
PU	0.274					

TEST RESULTS and DISCUSSION

A double logarithmic plot of relative modulus versus relative density for the microcellular foams is given in Fig.1, together with the prediction of the theory (eq.1) for conventional foams. A least squares linear regression analysis gives a slope of 2.29 and the constant C_1=0.38, which are predicted by the theory to be 2 and 1, respectively.The similar plot of the relative collapse stress versus relative density for rigid foams is given in Fig.2. Linear regression gives a slope of 1.85 and C_2=0.15. The theoretical curve of eq.(2), which has slope of 1.5 and C_2=0.3, is also given in this figure.

The experimental data lie well below the theoretical curves in both Figures 1 and 2. The E_s and σ_s values given in Table II were taken from the literature. The importance of the value of E_s chosen has been previously mentioned [7], where the higher the value of E_s, the lower the experimental data fall relative to the theoretical prediction, this is

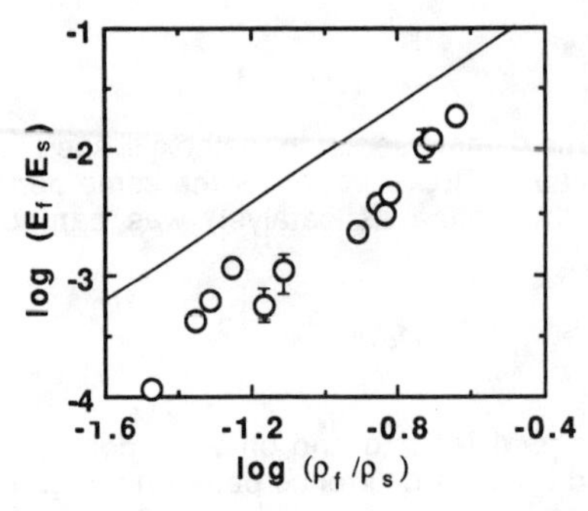

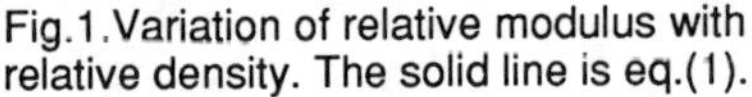
Fig.1.Variation of relative modulus with relative density. The solid line is eq.(1).

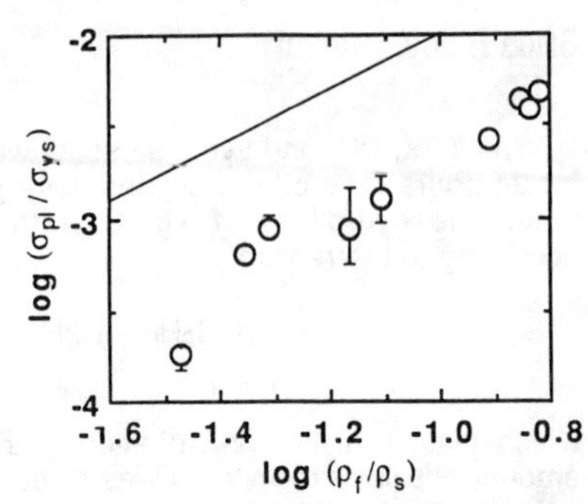

Fig.2.Variation of relative collapse stress with relative density.The solid line is eq.(2).

also true for the σ_S values, but it appears that choosing even the lowest values for E_S and σ_S is not enough to approach the predicted curves. Williams [6] measured the moduli of microcellular TPX foams by using the penetration method, and he explained the lower values of the modulus by a fraction of non-contributing material, and the ineffiency of the contributing mass. He also observed that the foams having a loose and randomly distributed structure have lower values of the modulus than those which have an orderly distributed polymer structure. Jackson et al. [7], using the dynamic tension-compression measurements, found the constant C_1 to be 0.16 in eq.(1). They concluded that the imperfect cell geometry and the inefficient use of polymer in making up the cell microstructure were responsible for the discrepancy from the theory. They also added that the friable foams might have been damaged during the cutting and mounting process.

The moduli and collapse stresses of preloaded extruded polystyrene foams are given in Table II, together with the prestrain values and permanent strains obtained after the preloading. Permanent strains appear to be small enough not to cause a significant densification. When the amount of prestrain increases, so does the number of defects produced in the structure, which results in a decrease in the modulus . For a prestrain of 65%, the modulus drops to 13% of its original value. The initial linear portion of the stress-strain curve shows a break before reaching the collapse stress, indicating a kind of buckling, and under large prestrains, the collapse stress disappears. Under moderate prestrains, the collapse stresses were not influenced very much by this process.

In Fig. 3 the stress-strain curves for the microcellular foams are given. For both rigid and elastomeric foams, the plateau regions of the curves show no horizontal or flat portion. Instead, they increase continuously with the strain. Similar behavior has been observed for the highly preloaded extruded polystyrene foams described in Fig. 4. The length of the horizontal portions of the curves decrease when the applied prestrain increases, and finally for higher values of prestrain it disappears entirely.

The dramatic decrease in the modulus of preloaded conventional foams and the similarity between the stress-strain curves of these foams and the microcellular ones

Table II. The Change of Modulus and Collapse Stress with Prestrain

Prestrain(%) :	11.0	15.2	20.2	30.1	39.9	51.4	65.0
Permanent strain(%) :	5.0	7.4	9.1	9.8	15.3	16.1	16.8
$E_{final}/E_{initial}$:	0.86	0.60	0.49	0.22	0.19	0.16	0.13
$\sigma_{final}/\sigma_{initial}$:	0.99	1.00	0.98	1.03	1.03	1.13	-

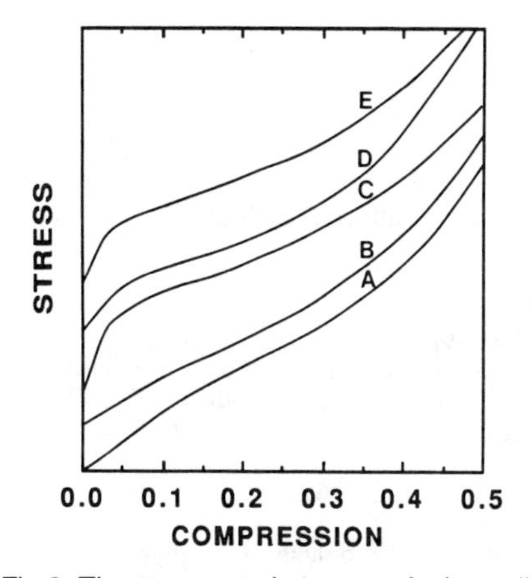

Fig.3. The stress- strain curves of microcellular foams. A:PU; B:Lycra; C:TPX 101; D:PAN 391; E:IPS 351.The scale of the vertical axis is arbitrary

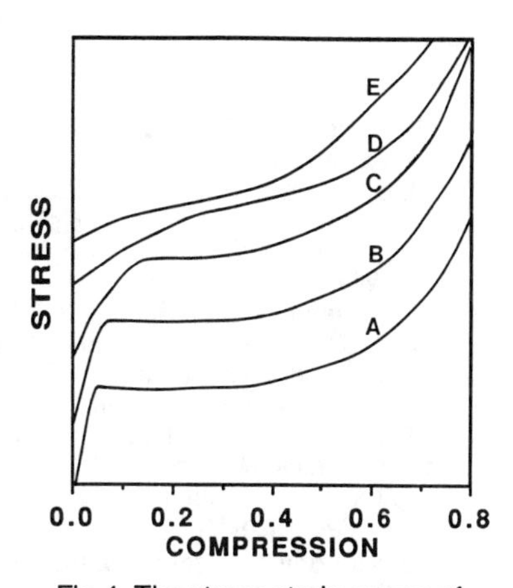

Fig.4. The stress-strain curves of preloaded extruded polystyrene foam Prestrain(%): A;0; B:11; C:20.2; D:30.1; E:65.0

indicate that the lower values obtained in the mechanical response of microcellular foams might be due to the defects produced in the structure during production. These defects might appear as nonuniform wall elements, cracked and broken elements, or plasticaly deformed elements which are formed because of the non-uniform temperature distribution present in the foam during the cooling process. Also any shrinkage that occurred during the removal of the solvent might be another reason. The explanation based on the damage occurring during the shaping of samples might be true for only the lower density and friable foams, but it is not the case for the elastomeric or high density rigid ones which are tough enough to be shaped without causing any damage.

The modulus and collapse stresses of the microcellular foams which are reinforced by the in-situ silica precipitation method are given in Table III. It appears that both moduli and collapse stresses increase by the treatment, indicating that the modulus and yield strength of rigid polymers, and the modulus of elastomeric ones are increased by this method. However, the ratio of the increase in the mechanical properties with the increase in the densities is greater for the elastomers than for the rigid ones. The samples showed shrinkage in different extents during the treatment process, and the increase in the densities after the treatment is partially due to this shrinkage.

CONCLUSIONS

The moduli and collapse stresses of microcellular foams were compared with the current constitutive laws of conventional foams. Although the square dependence of the modulus and the 1.5 power dependence of the collapse stress on relative densities predicted by the theory appear satisfactory, the experimental data lie well below the predicted curves. It is concluded that the defects formed during the production are responsible for the discrepancies from theory. In-situ silica reinforcement is a convenient method to increase both the moduli and collapse stresses of microcellular foams.

REFERENCES

1. J.H. Aubert and R.L. Clough, Polymer, 26 2047 (1985)
2. J.H. Aubert, A.P. Sylwester, and P. Rand, Polymer Preprints, 30, 447 (1989)
3. L.J. Gibson and M.F. Ashby, R.Soc.Lond. A, A382, 43 (1982)
4. L.J. Gibson and M.F. Ashby, Cellular Solids. Structure and Properties, (Pergammon Press, Oxford, England, 1988)
5. W.E. Warren and A.M. Kraynik, J. Appl. Mech. 55, 341 (1988)
6. J.M. Williams, J. Mat. Sci. 23, 900 (1988)
7. C.L. Jackson, M.T. Shaw,and J.H. Aubert (accepted by Polymer)
8. J.E. Mark and D.W. Schaefer in Polymer Based Molecular Composites, edited by D. W. Schaefer and J.E. Mark (M ater.Res.Soc. Proc. 171, Pi t t sburg, PA 1990)
9. J. Brandrup and E.H. Immergut, Polymer Handbook (3rd Ed., JohnWiley and Sons Inc , NewYork, NY 1989) p.V/82.
10. H.A. Lanceley, J. Mann,and G. Pagany in Composite Materials, edited by L. Holliday (Elsevier Pub. Comp.1966) p.224
11. The ratio of yield stress/modulus is calculated for the polystyrenes given in Modern Plastics Encyclopedia, edited by J.Agranoff, (McGraw-Hill Inc., 61, NewYork, NY 1984-1985) p.453 , and an average ratio of 0.0265 is obtained; by using this ratio, and the modulus given in Ref.10, the yield stress of IPS is calculated to be 148 MPa.
12. Ref.11, p.353.
13. The Lycra has been mentioned as a polyurethane in W.M. Phillips, W.S. Pierce, G. Rosenberg and J.H.Donachy in Synthetic Biomedical Polymers. Concepts and Application, edited by M. Szychen and W.J.Robinson (Technomic Publishing Co. Company, Inc.,Westport, CT 1980) p.41, therefore, the values given in Ref.4 for the polyurethane are also used for the Lycra.

Table III. Properties of Silica Reinforced Foams

Foam	Density (g/cm3)	Shrinkage (vol.%)	SiO2 (wt.%)	E_t^*/E_f	σ_{ct}^*/σ_c
IPS350	0.296	43.0	5.2	4.2	4.9
	0.271	36.4	5.2	3.7	4.6
IPS351	0.244	34.3	4.9	1.9	2.5
	0.175	7.7	5.1	1.5	1.5
IPS352	0.284	31.0	13.0	3.3	-
IPS353	0.120	20.1	2.3	1.5	1.5
	0.119	17.4	2.9	1.8	2.5
IPS354	0.145	24.3	15.2	4.4	3.6
PAN 389	0.257	60.0	26.7	9.9	17.0
	0.142	31.0	23.3	3.1	3.1
	0.113	13.8	11.0	1.9	1.7
	0.107	8.0	7.5	2.0	1.9
PAN391	0.122	45.0	19.1	2.4	2.2
TPX101	0.267	57.0	174.0	12.6	13.3
PU	0.474	33.6	13.9	5.4	6.4
	0.345	17.7	16.3	3.6	4.0
	0.332	23.3	4.3	2.6	2.6
	0.270	10.0	2.2	1.9	1.5
LYCRA	0.344	21.0	29.2	6.1	5.3
	0.330	8.0	43.5	5.7	
	0.303	10.4	25.0	2.8	3.2
	0.269	11.1	7.0	2.6	2.7

* Substcript "t"shows the treated foam properties and σ_c shows the collapse stress.

MECHANICAL STRUCTURE-PROPERTY RELATIONSHIPS OF MICROCELLULAR, LOW DENSITY FOAMS.

JAMES D. LEMAY
Lawrence Livermore National Laboratory, Chemistry and Materials Science Department, Livermore, CA 94550, U.S.A.

ABSTRACT

High energy physics applications at the Department of Energy National Laboratories require unique low-density foams of demanding homogeneity specifications (cell sizes on the order of 10 μm or smaller). These delicate and fragile foams are machined and shaped into specimens to exacting tolerances. In this work, the mechanical properties of a variety of these low density microcellular foams are reported as functions of foam density and morphology.

INTRODUCTION

Materials characterized by both low density and high homogeneity (i.e., derived from a microcellular morphology) are used at the National Laboratories to study hot and expanded states of matter in high energy density laser experiments. Of particular interest are materials made from carbon and silica polymers having densities in the range 0.030-0.300 g/cc and cell sizes in the range 0.010-10 μm. Because these materials are not readily available from commercial sources, they have been researched and developed at the National Laboratories over the last decade.[1-17]

Low density microcellular materials (LDMMs) were not developed to be structural materials; their use in high energy physics experiments requires only that they be reasonably resistant to damage during handling and machining. These attributes are largely qualitative, and not easily defined by simple mechanical property specifications. A number of other reasons exist for evaluating the mechanical properties of LDMMs: (1) mechanical testing provides guidelines for process optimization, (2) knowledge of the mechanical properties of different LDMMs facilitates their selection for physical applications, and (3) LDMMs are unique materials of scientific interest. The LDMMs evaluated in this study are listed in Table I. Other mechanical property evaluations of LDMMs are cited in references 18-21.

MECHANICAL PROPERTY TESTING

LDMMs at moderate densities (say >0.150 g/cc) are relatively robust, but at lower densities they are inherently weak and soft. Characterization of their mechanical properties can be a challenge. That is, specimens cannot be readily gripped or bonded to fixtures without being damaged (e.g., for tensile and shear testing) and they are difficult to machine into complex shapes (e.g., tensile dogbones). A desirable test method is one that utilizes simple specimen shapes and does not require that specimens be gripped.

In this work, mechanical tests were performed in uniaxial compression on small rectangular prisms (usually cubes), or right circular cylinders. The specimens were machined from as-produced LDMM "bricks" using a high speed circular saw and specially designed vacuum chucks. Great care was exercised to ensure that the specimens were machined with flat and plane-parallel opposing faces. In all cases, the surfaces of the as-produced LDMM "brick" were removed to eliminate possible contributions by high density surface skins.

Table I. Low Density Microcellular Materials Evaluated

LDMM	Source	Process	Morphology Description (Open Cell)	Cell Size	Ref(s)
Poly(4-methyl-1-pentene)	LLNL/LANL	A	Flakelike, weakly connected	10-50 μm	2,3
Polystyrene, isotactic	SNLL	A	Weblike struts, well connected	0.1-2 μm	5
Polystyrene	LLNL	B	Reticulated, well-defined spherical cells, well-connected	1-10 μm	13,14
Carbon, NaCl replica	LLNL	C	Ribbonlike sheets with holes	10-40 μm	6
Carbon, PS bead replica	SNLL	C	Reticulated, spherical cells	10-40 μm	
Resorcinol-formaldehyde (RF)	LLNL	D	Chains of colloidal size clusters	0.01 μm	8
Carbonized RF	LLNL	D	Chains of colloidal size clusters	0.01 μm	9
Melamine-formaldehyde (MF)	LLNL	D	Chains of colloidal size clusters	0.01 μm	10
Silica aerogel Base catalyzed	LLNL	D	Chains of colloidal size clusters	0.01 μm	28
Silica aerogel Acid catalyzed	LLNL	D	Entangled polymerlike chains	0.01 μm	28
Silica aerogel, condensed and ultra-low density	LLNL	D	Mixed elements of both acid and base catalyzed aerogels	0.01 μm	15,16

Sources:

LLNL = Lawrence Livermore National Laboratory

LANL = Los Alamos National Laboratory

SNLL = Sandia National Laboratory, Livermore

Processes:

A = Thermally induced phase separation from solution

B = Inverse emulsion

C = Replication of removable pore-former

D = Sol-gel process (chemically induced phase separation)

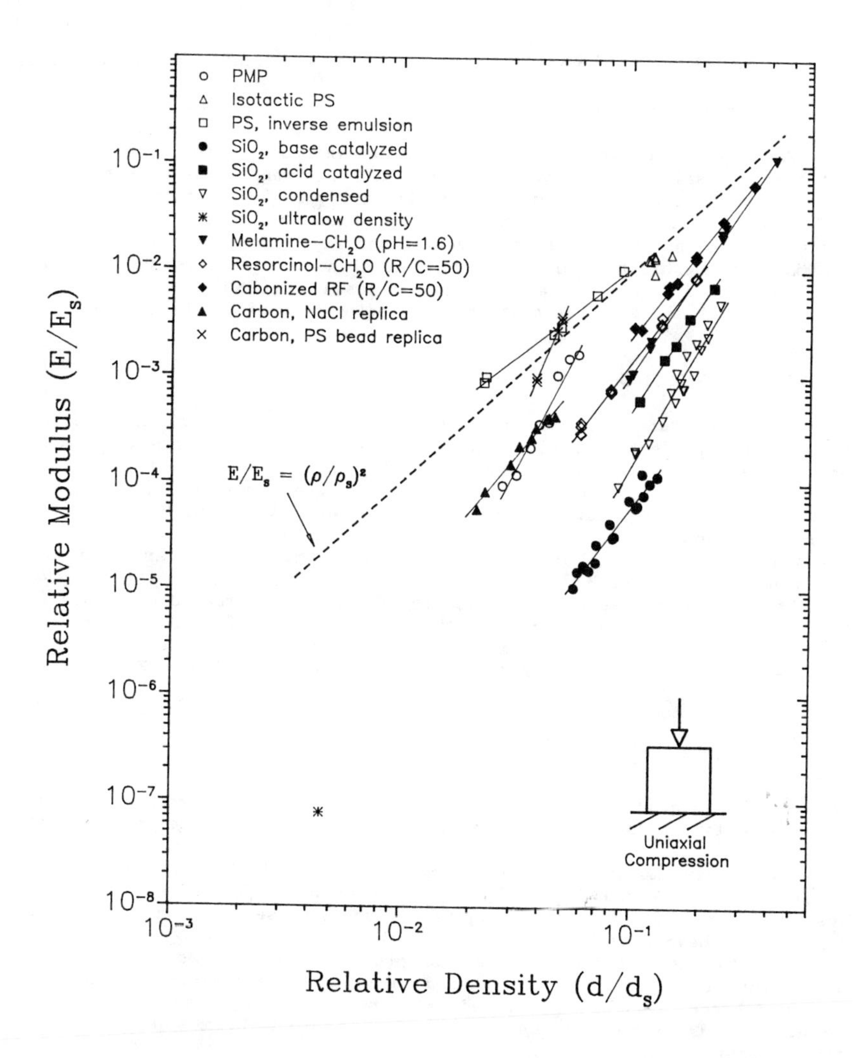

Figure 1. Relative modulus vs relative density (E/E_s vs d/d_s) for various LDMMs tested in uniaxial compression. The dashed line shows Equation 1 with n=2. Normalization constants: poly(4-methyl-1-pentene) (PMP), d_s=0.84 g/cc, E_s=1.2 MPa; isotactic polystyrene d_s=1.04 g/cc, E_s=1.4 MPa; cross-linked polystyrene (inverse emulsion LDMMs), d_s=1.10 g/cc, E_s=3.0 MPa; resorcinol-formaldehyde (RF) and melamine formaldehyde (MF), d_s=1.31 g/cc, E_s=3.4 MPa; all silica aerogels, d_s=2.20 g/cc, E_s=72 MPa; all carbon LDMMs (amorphous), d_s=1.50 g/cc, E_s=24 MPa.

The modulus (stiffness) and strength were measured at an initial strain rate of 0.1%/s. The tests were performed under ambient conditions. The relative humidity generally was 50-70% during testing. No special precautions were taken to prevent moisture adsorption by the specimens.

The apparent compressive modulus was determined from the linear region of the load-displacement curve. The compressive strength was determined at the point where the curve deviates from linearity by 0.2% strain. There is no fundamental reason for choosing this offset value as a measure of strength; we simply find it to be a convenient way to compare a wide variety of LDMMs which exhibit diverse responses at higher compressive strains. We have observed in cyclic loading experiments that LDMMs usually do not exhibit significant hysteresis at stresses below the offset value.

RESULTS AND DISCUSSION

The literature contains numerous models and theoretical analyses of the structure-property relationships of foams and cellular materials (see, for example, reviews by Gibson and Ashby [22], Hilyard [23], and Meinecke & Clark [24]). In the limit of low-density, however, most of these models predict a simple power-law dependence of stiffness and strength on density, and no dependence on cell size. In general, these analyses are based on specific geometric cell models that have little resemblance to the morphologies of real LDMMs.

The primary parameter in all models of cellular or foamed materials is the volume fraction of the matrix, d/d_s, where d is the foam density and d_s the density of the solid matrix material. For isotropic, low-density foams, 3-dimensional models [22-24] predict the scaling relationship in Equation 1

$$P/P_s = (d/d_s)^n \tag{1}$$

where P is a property of the foam (e.g., modulus, strength, conductivity), P_s the associated property of the matrix material, and n a scaling exponent (often "2").

Compressive modulus data for a wide variety of LDMMs are plotted log-log in Figure 1 in the form of Equation 1. The dashed line represents Equation 1 with n=2. With the exception of the inverse emulsion polystyrene LDMMs, which have a well connected reticulate-like morphology, the dashed line significantly overestimates the properties of LDMMs. On closer examination, it also appears that the modulus data are described by scaling constants larger than the predicted value n=2 (i.e., they exhibit steeper slopes). Although not shown, LDMM strength data show similar characteristics.

We offer the following discussion as an explanation of why models fail to accurately predict the modulus of LDMMs. The stiffness of a porous material is determined by the degree of interconnectivity of the solid material comprising its structure (i.e., the relative amount of the solid material that actually supports the load). Simple geometric cell models typically assume that all the solid material is connected and load supporting. Scanning and transmission electron micrographs of LDMMs, however, reveal their morphologies are generally quite poorly connected. It also seems reasonable that the fraction of solid matrix that is "connected" may diminish with decreasing density, especially as the morphology is pushed towards the low density limit. This could explain why the experimental data tend to describe apparent scaling exponents larger than "2".

A useful feature of normalizing LDMM properties per Equation 1, is that different LDMMs can be compared in terms of their "structural efficiency." That is, LDMMs that make better use of the available solid matrix in a given volume, exhibit larger normalized stiffness at equivalent d/d_s. From Figure 1 we observe that the reticulate, spherical cell morphology of inverse emulsion derived LDMMs is superior to structures obtained via phase separation and sol-gel routes. Aerogels, in particular, appear the least structurally efficient (although we note that their ranking depends strongly on processing conditions).

Recently, Warren & Kraynik modeled open cell low-density materials as three-dimensional structures of randomly oriented tetrahedral-joints, in which relative displacements of the strut mid-points are affine and dominated by bending.[25-27] They derived expressions for the foam's elastic constants in terms of the relative density, and analyzed different testing modes (e.g., pure shear, and uniaxial extension). One of their results is a prediction that the modulus is proportional to the square of the relative density. Interestingly, when Warren & Kraynik replace their model with a cubic structural joint, they determine a linear dependence (Equation 1 with n=1). Warren & Kraynik argue that the tetrahedral joint is more realistic for materials in which surface tension forces were prevalent during cell formation (like most LDMMs). They also conclude that low-density materials having more realistic, energetically probable morphologies are "softer" than more idealized joint morphologies (such as cubic joints). The data for LDMMs in Figure 1 support this conclusion.

ACKNOWLEDGMENTS

This work performed under the auspices of the U.S. Department of Energy by Lawrence Livermore National Laboratory under Contract #W-7405-ENG-48.

REFERENCES

1. J.D. LeMay, R.W. Hopper, L.W. Hrubesh, and R.W. Pekala, MRS Bulletin, December, 1990, in press.
2. A.T. Young, D.K. Mareno, and R.G Marsters, J. Vac. Sci Technol., 20 1094 (1982).
3. A.T. Young, J. Cellular Plastics, 23, 55 (1987).
4. J.H. Aubert and R.L. Clough, Polymer, 26, 2047 (1985).
5. J.H. Aubert, Polym. Preprints., 28, 147 (1987).
6. R.W. Pekala, R.W. Hopper, J. Mat. Sci., 22, 1840 (1987).
7. R.W. Pekala and R.E. Stone, Polym. Preprints, 29(1), 204 (1988).
8. R.W. Pekala, J. Mater. Sci.,24, 3221 (1989).
9. R.W. Pekala and F.M. Kong, Polym. Preprints, 30(1), 221 (1989).
10. R.W. Pekala and C.T. Alviso in Better Ceramics Through Chemistry IV, edited by C.J. Brinker, D.E. Clark, D.R. Ulrich and B.J. Zelinski (Mat. Res. Soc. Proc., 180, Pittsburgh, PA, 1990) p. 791.
11. J.D. LeMay, Polym. Mat. Sci. and Eng. 60, 695 (1990).
12. L.M. Hair, S.A. Letts, and T.M. Tillotson, Polym. Mat. Sci. Eng., 59, 749 (1988).
13. F.M. Kong, R. Cook, B. Haendler, L. Hair, and S. Letts, J. Vac. Sci.Technol., A6(3) 1894 (1988).
14. J.M. Williams and D.A. Wrobleski, Langmuir, 4, 44 (1988); ibid., p. 656.
15. T.M. Tillotson, L.W. Hrubesh, and I.M. Thomas in Better Ceramics Through Chemistry III, edited by C.J. Brinker, D.E. Clark and D.R. Ulrich (Mater. Res. Soc. Proc. 121, Pittsburgh, PA, 1988) p. 685.

16. T.M. Tillotson and L.W. Hrubesh in, Better Ceramics Through Chemistry IV, edited by C.J. Brinker, D.E. Clark, D.R. Ulrich and B.J. Zelinski (Mat. Res. Soc. Proc., 180, Pittsburgh, PA, 1990) in press; L.W. Hrubesh, T.M. Tillotson, and J.F Poco, ibid.
17. A.P. Sylvester, J.H. Aubert, P.B Rand, C. Arnold, Jr., and R.L. Clough, Polym. Mat. Sci. Eng., 57 (1987) p. 113.
18. R.W. Pekala, C.T. Alviso and J.D. LeMay, J. Non-Cryst. Solids., in press.
19. T. Woignier, J. Phalippou and R. Vacher, J. Mat. Res., 4(3), 688 (1989).
20. J.D. LeMay, R.W. Pekala and L.W. Hrubesh, Pacific Polym. Preprints, 1, 295 (1989).
21. J.D. LeMay, T.M. Tillotson, L.W. Hrubesh, R.W. Pekala in Better Ceramics Through Chemistry IV, edited by C.J. Brinker, D.E. Clark, D.R. Ulrich and B.J Zelinski (Mat. Res. Soc. Proc, 180, Pittsburgh, PA, 1990) p. 321.
22. L.J. Gibson and M.F. Ashby, Cellular Solids - Structure and Properties (Pergamon Press, Oxford, 1988).
23. N.C. Hilyard, editor Mechanics of Cellular Plastics (Macmillan, New York 1982).
24. E.A. Meinecke and R.C. Clark, Mechanical Properties of Polymeric Foams (Technomic, Westport, CT, 1973).
25. W.E. Warren, A.M. Kraynik, Mechanics of Materials, 1(1), 27 (1987).
26. W.E. Warren, A.M. Kraynik, Proc. 11th Canadian Cong. Appl. Mech., 1, (Edmonton, Alberta, 31 May 1987) p. a46.
27. W.E. Warren, A.M. Kraynik, J. Appl. Mech., 110, 341 (1988).
28. C.J. Brinker and G.W. Scherer, J. Non-Cryst. Solids, 70, 301 (1985).

THERMAL SHOCK BEHAVIOR OF OPEN CELL CERAMIC FOAMS

ROBERT M. ORENSTEIN*, DAVID J. GREEN** AND ALBERT E. SEGALL***
*General Electric Company, Power Generation Engineering, 1 River Rd., Schenectady, NY 12345
**The Pennsylvania State University, Department of Materials Science and Engineering, University Park, PA 16802
***Center for Advanced Materials, 410 Walker Bldg., University Park, PA 16802

ABSTRACT

Specimens of heated alumina foams were thermally shocked by immersion in water or oil. Two distinct temperature profiles were found to exist during liquid quenching: a macroscopic gradient due to heating of the fluid during infiltration into the foam and a microscopic gradient across each individual strut. Thermal stresses were calculated using a semi-empirical, finite element model. The peak thermal stresses were coupled with the two-parameter Weibull distribution of the strut strengths to calculate the probability of crack extension on a strut for eight cellular geometries. Experimentally, the thermal shock resistance increased with increasing relative density and increasing cell size. The predicted critical temperature differences were in fair agreement with measured values.

INTRODUCTION

Solids having a foam-like structure are used in numerous applications, including filters, catalyst supports, thermal insulation, energy storage batteries, lightweight structural laminates and energy damping structures. The properties of a solid foam depend on both the cellular geometry, e.g., the density, cell size and shape, etc., and the properties of the material comprising the solid phase. Solid foams can have either open or closed cells. Open cell foams consist of a reticulated network of interconnected beams, or "struts." Closed cell foams consist of a similar network, but the struts are bridged by thin faces which isolate the individual cells. In both types of structure, the struts act as the primary load bearing members [1].

When an open cell ceramic foam is subjected to a thermal transient, two distinct temperature gradients must be considered; a temperature profile across the solid phase, and a temperature profile across the bulk foam (solid plus void). These temperature gradients give rise to stress profiles across the bulk and through the individual struts. In this paper, the response of alumina foams to severe thermal shock is described in terms of the probability of crack extension on a strut and a critical temperature difference for the bulk foam. The damage to the foam was quantified using the reduction in the elastic modulus that resulted from the thermal shock. Particular emphasis in the study was placed on the dependence of the stresses on cellular geometry and type of quenching medium.

BACKGROUND

The most common description of the resistance of a material to damage initiation resulting from a thermal stress is the parameter *R*, which is defined as

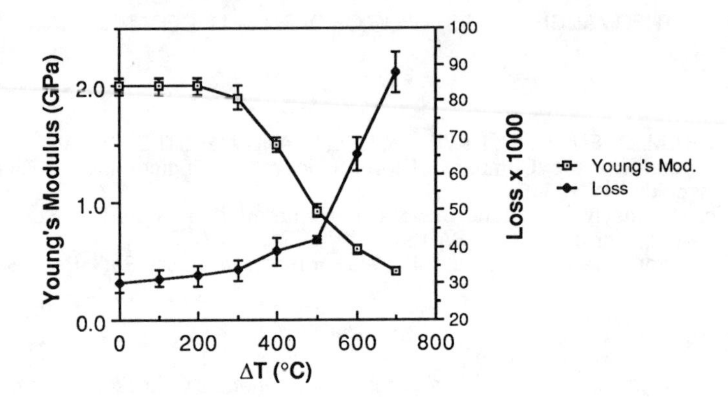

Figure 1. Young's modulus and internal friction (loss) as a function of quench temperature difference for water-quenched 2011 material. Error bars represent the 0.95 confidence levels for the means of 35 values.

$$R = \sigma_f (1-\upsilon)/E\alpha \tag{1}$$

where σ_f is the failure stress of the material, υ is the Poisson's ratio, E is the Young's modulus, and α is the linear coefficient of thermal expansion. The parameter R is the most fundamental description of the smallest temperature difference that will result in stresses of sufficient magnitude to extend cracks [2] and is commonly referred to as the critical temperature difference, ΔT_C. The reduction in Young's modulus as a function of thermal shock severity was chosen to characterize the onset and degree of damage in the materials studied. Fig. 1 shows the response of Young's modulus and internal friction as functions of thermal shock severity (ΔT). Both properties are sensitive to the presence of damage in the material; however, a discrete value of ΔT_C is not discernible. Therefore, the temperature difference for which the material experienced a 10% reduction in Young's modulus, ΔT_{10}, was used as the thermal shock resistance parameter. This reduction in Young's modulus correlated well with reductions in bending and compressive strengths for a series of thermally shocked materials of a single cellular geometry [3].

As thermal shock leads to damage in these open cell ceramics, it is useful to consider this damage in terms of the probability of strut failure. The two-parameter Weibull strength distribution was used to predict the probability of strut failure for a given thermal stress. A useful form of the Weibull probability function is [4]

$$P = 1\text{-}\exp\{-((1/m)!)^m (\sigma_t/\sigma_f)^m\} \tag{2}$$

where P is the failure probability, m is the Weibull modulus and σ_t Is the tensile thermal stress. The Weibull modulus, m, was approximately 2.0 for the struts and 7.0 for the bulk cellular materials [3]. Thermal stresses were predicted using a finite element model by considering the microscopic thermal gradient across a strut and a macroscopic thermal gradient due to the fluid heating [3]. This latter gradient was approximately modeled using experimentally measured values of the quench medium temperature at the surface and the center of the specimen [3].

Previous work indicated that when thermal stresses act on an open cell ceramic foam, cracks extend from microscopic flaws on the strut surfaces rather than macroscopic crack propagation within the bulk material [3,5], i.e., the damage is primarily localized to individual struts. Using the Weibull strength distribution of the struts and the calculated thermal stresses, the failure probability of the struts exposed to a particular thermal shock can be predicted using Eq. (2).

EXPERIMENTAL PROCEDURE

A single composition of a commercially available 92% alumina foam[1] in 8 different cellular geometries was used. The relative density ranged from 8 to 22%, and the cell sizes, expressed as the number of pores per linear inch (ppi), ranged from 7 to 65 ppi. Hereafter, the materials are referred to by 4 digit designations (e.g. 0711); the first two digits indicate the pores per inch (07 ≡ 7 ppi) and the second two digits indicate the nominal relative density (11 ≡ 11% dense material). Figure 2 shows a typical microstructural geometry for these materials.

Thermal shock experiments were performed by rapidly transferring the heated specimen from a resistance furnace to a quenching bath containing either water or oil. The temperature difference recorded was the furnace temperature minus the temperature of the bath, which remained at room temperature. Transfer of the specimen, using metal tongs, required approximately one second.

The probability of crack extension following thermal shock was determined by microscopic observation of at least 100 struts per thermal shock condition. The number of struts with visible cracks was counted and crack lengths were measured. Failure probability was calculated by dividing the number of cracked struts per unit (viewing) area by the number of struts in the area. This value was normalized by the percentage of cracked struts present in the as-received material. Young's modulus of the bulk specimens prior to and following thermal shock was measured using a dynamic resonance method. Further details of the experimental procedure are given in ref. 3.

RESULTS AND DISCUSSION

The damage morphology observed following thermal shock supports the hypothesis that extension of preexisting flaws in the struts was responsible for the observed reductions in strength and stiffness. Fig. 3 shows typical damage; i.e., the cracks ran predominantly in directions parallel to the length of the struts.

Failure probability predictions were made for the cases of microscopic (due to strut ΔT) and macroscopic (due to bulk ΔT) stresses individually. In both cases, the predicted failure probabilities were less than the experimental failure probabilities. However, both thermal stress mechanisms can act simultaneously on the struts during quenching, such that the total thermal stress as a function of time is determined by superposition. The most conservative approach to the analysis is to assume the maximum stresses completely overlap in time. This is equivalent to summing the peak values of each stress, the result of which produced the upper failure probability curve shown in Fig. 4. The least conservative approach to the analysis is to assume the stresses are completely separated in time. This is equivalent to using the highest individual stress. For the 2011 material, the stress due to the bulk temperature gradient was calculated to be higher of the two and

---

[1] Reticel™ Reticulated Ceramics, Hi Tech Ceramics Inc., Alfred NY

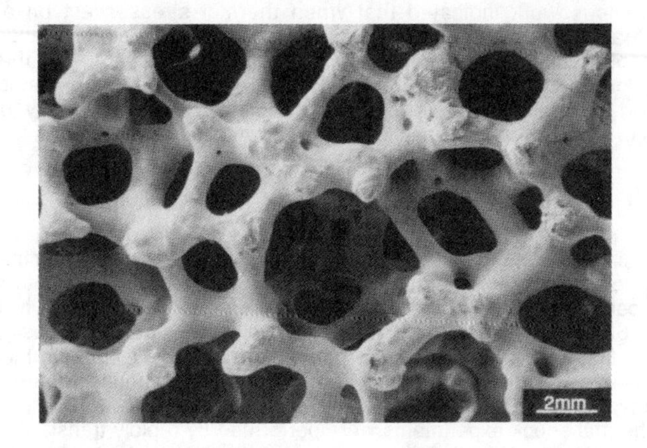

Figure 2. Scanning electron micrograph of 0716 material, showing the open cell structure.

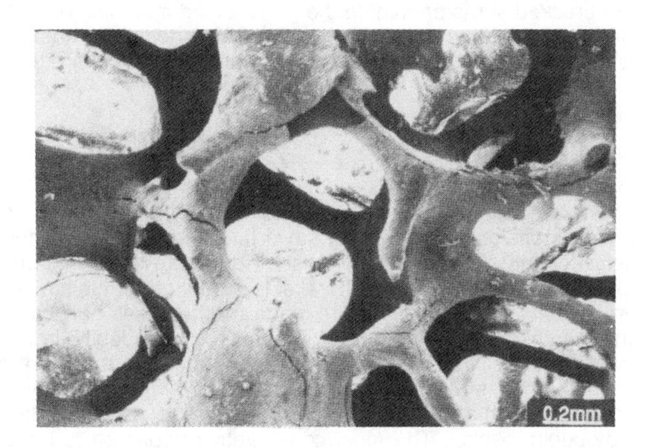

Figure 3. Scanning electron micrograph of 6511 material quenched from 1020°C into 20°C water.

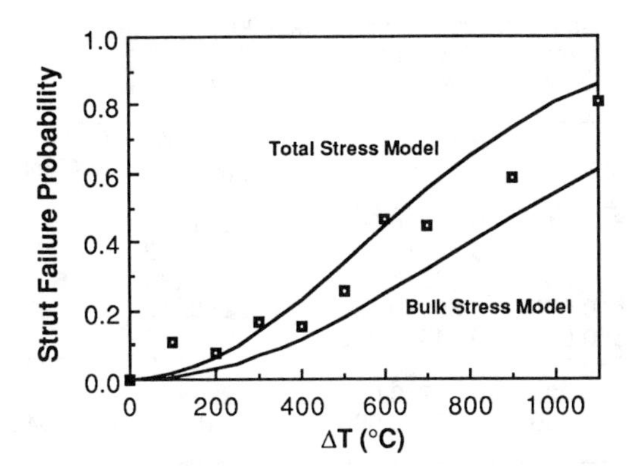

Figure 4. Failure probability analysis on water-quenched 2011 material using the combined peak stresses in the thermal stress model (upper bound) and the temperature profile across the bulk foam in the thermal stress model (lower bound). Squares are normalized measured probabilities of crack extension obtained by microscopic observation.

resulted in the lower failure probability curve shown in Fig. 4. [Note: The finite element model predicts varying degrees of stress overlap in time; the amount ofoverlap increases as strut diameter increases.] It can be seen that most of the normalized experimental strut failure probabilities are effectively bounded by the two curves. This was not the case when the Weibull modulus of the bulk material was used in the calculations (m = 7.0), whereby the change in predicted failure probability with increasing ΔT was considerably steeper. This lends further support to the hypothesis that the flaw population on the struts, rather than the macroscopic flaw population, controls the thermal shock resistance in brittle open cell materials.

Comparison of the retained stiffness data (for 300°C water quenches and 600°C oil quenches) and the ΔT_{10} data is given in Table I. The retained stiffness was obtained by dividing the Young's modulus of the shocked specimen by its as-received value. Both the retained stiffness and ΔT_{10} data show the thermal shock resistance increased with increasing relative density and cell size for the quenching conditions in this study.

The magnitudes of the predicted stresses due to the strut temperature profiles are much higher for the water quenches than the oil. This is a result of the large difference in convective heat transfer coefficient between water and oil quenching. By contrast, the magnitudes of the predicted stresses due to the bulk temperature profiles are similar. This is because the bulk temperature difference, which results from fluid heating during infiltration, is a function of the heat transfer coefficient, fluid viscosity, fluid heat capacity, and cellular geometry. The differences in fluid properties tend to cancel for the two quench conditions, with the result that the cellular geometry rather than the fluid properties most strongly influenced the bulk stress magnitude. The best correlations between measured and predicted values of ΔT_{10} were obtained when the stresses due to the bulk temperature profiles were used to predict ΔT_{10} [3]. The plot for water quenching is shown in Fig. 5.

It is concluded that the thermal stresses due to the bulk temperature difference in an open cell ceramic foam rapidly quenched into a liquid are relatively more

Table I. Retained stiffness for the 300°C water and 600°C oil quenches, and the temperature difference, ΔT_{10}, for water and oil quenching. Values are presented as the mean ± the 0.95 confidence level for 4 values.

MATERIAL	$E/E_{initial}$		ΔT_{10} (°C)	
	[water]	[oil]	[water]	[oil]
0711	0.938±0.018	0.959±0.007	321±14	1096±157
0716	0.970±0.011	0.971±0.029	379±35	1120±174
2008	0.881±0.057	0.865±0.032	287±36	525± 57
2011	0.884±0.075	0.890±0.054	288±43	560± 68
2016	0.905±0.019	0.948±0.042	299±10	739±122
2022	0.909±0.042	0.970±0.021	304±35	830± 68
6511	0.758±0.029	0.717±0.039	231±11	333± 53
6516	0.813±0.050	0.772±0.026	240±14	354± 43

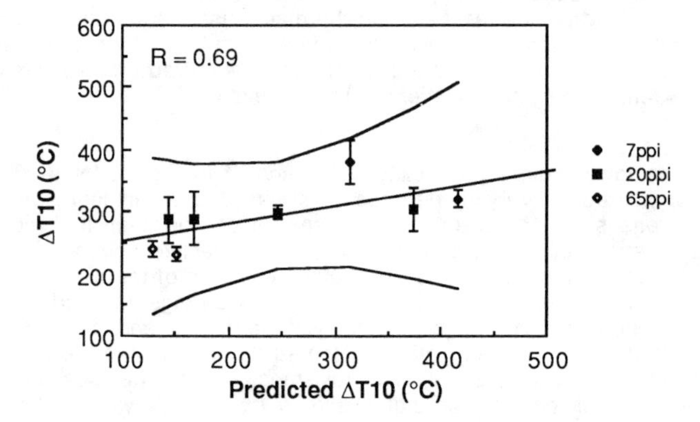

Figure 5. Measured versus predicted temperature difference for a 10% reduction in Young's modulus for the water-quenched materials, bounded by the 0.95 confidence of the regression. Error bars represent the 0.95 confidence levels of the means of 4 values.

important than the stresses due to the temperature profiles across the individual struts. These latter stresses can, however, be substantial under conditions of rapid heat transfer. This conclusion is supported by Fig. 6, which shows a strong correlation between thermal shock resistance and strut diameter. A similar correlation is obtained for oil quenching. The trends are the reverse of those expected if the microscopic stresses controlled the thermal shock resistance, because Biot modulus, and therefore, the microscopic thermal stresses, increases with increasing strut diameter. It is apparent from Fig. 6 that cellular geometry does strongly influence the thermal stress state. This is attributed to the increased heating of the quenching liquid during infiltration into the foam with decreasing strut size (e.g., increasing number of struts causes increasing flow tortuousity or offers a larger surface area for heat transfer). The inability to predict more exactly these trends from the stress model may be explained by the omission of the variations in fluid heat and mass transfer with cellular geometry from the analysis.

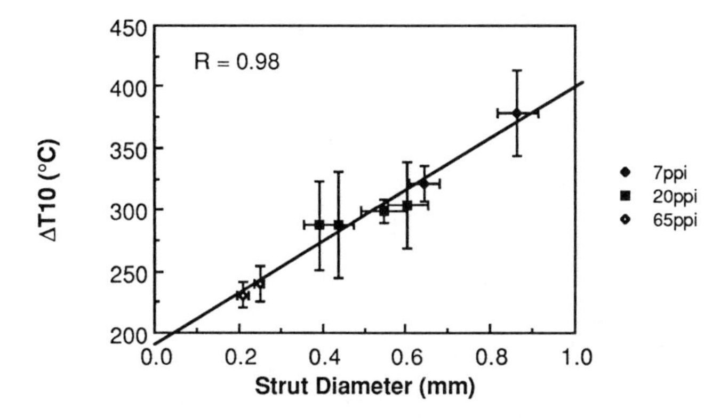

Figure 6. Temperature difference for a 10% reduction in Young's modulus as a function of strut diameter for the water-quenched materials. Error bars represent the 0.95 confidence levels of the means.

CONCLUSIONS

- Thermal shock resistance in brittle solid foams can be quantified by the temperature difference required for a given reduction in strength or stiffness (at a particular heat transfer condition). This approach demonstrated there was a gradual reduction in the mechanical properties of open cell ceramics during thermal shock.

- The probability of flaw extension on a strut can be predicted by a thermoelastic stress analysis, using the Weibull strength distribution of the struts. This approach gave reasonable agreement with the number of observed strut failures.

- Accurate models of the time-dependent bulk thermal profiles are required to increase the accuracy of the stress analyses and to allow more accurate predictions of thermal shock resistance as a function of cellular geometry.

ACKNOWLEDGEMENTS

This work was supported by the Gas Research Institute under contract number 5084-238-1302.

REFERENCES

1. M.F. Ashby, Metall. Trans. A, 14A, 1755-69 (1983).
2. G. Ziegler, *Z. Werkstofftech.*, 16, 12-18 (1985).
3. R.M. Orenstein, Thermal Shock Behavior of Open Cell Ceramic Foams, MS Thesis, The Pennsylvania State University, May 1990.
4. P. Stanley, H. Fessler and D. Sivill, Proc. Br. Ceram. Soc., 22, 453-87 (1973).
5. R.M. Orenstein and D.J. Green in Proc. of the 3rd Int. Symposium on Ceramic Materials and Components for Engines, Edited by V.J. Tennery (Am. Ceram. Soc. Proceedings, Westerville, OH 1989) pp. 641- 650.

MECHANICAL BEHAVIOR OF A CELLULAR-CORE CERAMIC SANDWICH SYSTEM

ERIC J. VAN VOORHEES* AND DAVID J. GREEN**
*Lanxide Corporation, 1300 Marrows Road, P.O. Box 6077, Newark, DE 19714-6077
**Pennsylvania State University, 118 Steidle Building, University Park, PA 16802.

ABSTRACT

The failure behavior and stress distribution of a novel ceramic sandwich system was investigated. A highly porous, open-cell Al_2O_3 was used as the core material with relative densities ranging from 9% to 20%. The faceplates were dense Al_2O_3 of varying thickness. The composites were tested in flexure and failure loads of the cellular core were shown to compare favorably to those predicted by theory when Weibull statistics were incorporated. Stresses in the tensile faceplates were also measured. For composites with relatively thick faces and low modulus cores, the stress distribution was significantly altered by a local bending effect. These stresses were predicted using a refined model.

BACKROUND

In producing materials that minimize stiffness or strength with minimum weight, it is often useful to combine porous and dense materials into composites. In these cases, the dense material with its higher stiffness can carry more of the load and thus protect the weaker, porous component. An excellent example of this philosophy is to form a composite in which the porous material is "sandwiched" between denser faces.

In bending, the sandwich composite is roughly analogous to an I-beam. The two thin, but relatively stiff faces are separated from a common center of bending with a thick, but very lightweight core. This structure greatly increases the moment of inertia of the beam with respect to its overall weight and is therefore efficient at resisting bending moments.

The stiffness of a sandwich beam is quantified through the equivalent flexural rigidity, $(EI)_{eq}$. This term represents a sum of the products of the elastic (Young's) modulus and moment of inertia of each component. Using the parallel axis theorem and the conventional assumptions of bending theory, the equivalent flexural rigidity is [1,2]:

$$(EI)_{eq} = \frac{E_f bt^3}{6} + \frac{E_c bc^3}{12} + \frac{E_f btd^2}{2} \qquad (1)$$

where b is the beam width, c is the core thickness, t is the face thickness and d is the distance between the midplanes of the faces. Also, E_f and E_c are the elastic moduli of the face and core, respectively. It should be noted that the third term of eqn. 1, representing the bending stiffness of the faceplates about the centroid of the sandwich, constitutes the majority of sandwich stiffness.

In flexure, the normal stresses of a sandwich composite are largely carried by the faces while the core supports a significant amount of shear stress. The equivalent flexural rigidity term is used in the calculation of the normal and shear stresses in each component of the sandwich composite. Again using the conventional assumptions of bending theory, the strain at some distance y from the neutral axis is simply $My/(EI)_{eq}$ where M is the applied moment. Using Hooke's law, the component stress can therefore be computed by simply multiplying by the appropriate elastic modulus. The normal stresses in the face (σ_f) and core (σ_c) are therefore [1]:

$$\sigma_f = \frac{MyE_f}{(EI)_{eq}} \qquad (2)$$

for $c/2 < y < (c+2t)/2$ and $-(c+2t)/2 < y < -c/2$;

$$\sigma_c = \frac{MyE_c}{(EI)_{eq}} \quad (3)$$

for $-c/2 < y < c/2$. The shear stress in a sandwich core can be written [2]:

$$\tau_c = \frac{V}{(EI)_{eq}}\left[\frac{E_f dt}{2} + \frac{E_c}{2}\left(\frac{c^2}{4} - y^2\right)\right] \quad (4)$$

where V is the shear force. As both shear and normal stresses can be significant in a sandwich core, it is necessary to consider the principal stress (σ_{1c}) when determining failure loads. The principal stress in a sandwich core is given by [2]:

$$\sigma_{1c} = \frac{\sigma_c}{2} + \left[\left(\frac{\sigma_c}{2}\right)^2 + \tau_c^2\right]^{1/2} \quad (5)$$

The principal stress in the faceplates, conversely, is nearly equal to the normal stress (eqn. 2) in practical composites as the shear stress in the faces is usually negligible.

The most important difference between a sandwich beam and a typical monolithic beam is that shear stresses in the core, and therefore shear deflections, become significant in bending. Indeed, the deflection of a sandwich under transverse loading is comprised of a bending component and a shear component. Associated with the shear component is a "local bending effect," in which the faces must bend about their own axes on either side of the load point in order to compensate for a local discontinuity [1]. This mechanism reduces the amount of shear deflection at the expense of creating an extra bending moment in the faces. The magnitude of this extra bending moment increases as the ratio of shear stress to normal stress in the core increases.

EXPERIMENTAL PROCEDURE

A 92 wt% alumina-8 wt% mullite open cell ceramic was used as the sandwich core material in the form of 114.3 x 25.4 x 12.7 mm bars. Five nominal relative densities were tested: 0.10, 0.13, 0.16, 0.19 and 0.22, at a cell size of 30 pores per inch. The flexural strength of the cellular bars was determined by the three-point method and at least 10 samples per density were tested. The flexural strength data were fitted to a two-parameter Weibull distribution. Elastic moduli were determined from vibrational spectra and fundamental resonance frequencies.

The sandwich faceplates were comprised of 96 wt% Al_2O_3 substrate material with a length of 114.3 mm and width of 25.4 mm. Four plate thicknesses were tested: 0.381, 0.508, 0.762 and 1.02 mm. The faceplates were bonded to the cellular core using conventional cement. The failure loads of the sandwich composites were also measured in three-point flexure and at least 10 samples were tested per combination of core density and face thickness.

The magnitude of strain in the tensile faceplate was monitored through the use of strain gages. A 3.175 mm gage length strain gage was placed along the tensile axis of the bar, centered opposite the load point. Axial strains in the tensile face were recorded at 44.4 N intervals until failure for at least one specimen per combination of face thickness and core density.

RESULTS AND DISCUSSION

The initial sandwich failure mechanism observed in the composites tested was almost exclusively a single crack propagating through the core thickness. The orientation and position of these cracks depended on the relative ratio of shear stress to normal stress (τ_c/σ_c) in the core. For low values of τ_c/σ_c, the crack can be characterized as vertical or slightly angled within 5 mm of the central load point. When shear stresses become predominant, conversely, the cracks become angled towards 45° and were observed throughout the distance between the supports. The varied locations of these cracks is not unexpected because the shear stress, unlike the tensile stress, is approximately constant throughout the core volume in three-point bending.

The average failure loads for each combination of core density and face thickness are listed in Table I. Also included are the measured relative densities (ratio of bulk density, ρ, to theoretical density, ρ_t) of the cellular core material. It is evident from this table that increasing the face thickness for a given core density does not substantially increase the load carrying capability. This indicates that the composites become more and more overdesigned with increasing face thickness because composite weight increases with relatively little increase in failure load.

The predicted core failure loads listed in Table I were calculated by assuming that failure occurs when the maximum principal stress in the core (σ_{1c}) equals the measured flexural strength of the foam. When calculating σ_{1c} in eqn. 5, the maximum normal and shear stresses in the core were determined with eqns. 3 and 4, respectively. When the failure loads were first calculated by this method, it was observed that the experimental failure loads were significantly lower than predicted. It was postulated that this was due to the difference in stress distribution between the sandwich core and the cellular material when tested as a separate component. That is, the effective volume of cellular material being acted on by tensile stresses increases when it is used as a core material. Weibull statistics were therefore used to scale up the flexural strengths measured in three-point bending (σ_{cell}) to those that can be expected in a sandwich core (σ_{sand}). This scaling is done through the relation [3]:

Table I. Summary of observed and predicted core failure loads.

Face Thickness (mm)	Observed Failure Load (N)	$(\sigma_{sand}/\sigma_{cell})$	Predicted Failure Load (N)
ρ/ρ_t=0.091			
0.381	232.5 ± 34	1.90	252.1
0.508	233.8 ± 31	2.00	265.0
0.762	264.2 ± 51	2.11	280.1
1.016	250.7 ± 23	2.17	290.4
ρ/ρ_t=0.118			
0.381	373.4 ± 40	1.43	352.3
0.508	364.3 ± 40	1.51	383.9
0.762	400.0 ± 42	1.61	419.8
1.016	437.9 ± 42	1.67	441.9
ρ/ρ_t=0.153			
0.381	540.7 ± 45	1.72	455.7
0.508	451.6 ± 67	1.84	512.3
0.762	528.6 ± 62	2.02	580.9
1.016	599.7 ± 78	2.15	623.9
ρ/ρ_t=0.167			
0.381	529.9 ± 56	1.34	565.8
0.508	505.8 ± 97	1.39	601.0
0.762	644.9 ± 74	1.49	779.6
1.016	716.2 ± 117	1.57	853.6
ρ/ρ_t=0.202			
0.381	625.3 ± 46	1.39	603.9
0.508	607.7 ± 77	1.45	717.4
0.762	720.4 ± 109	1.56	877.2
1.016	732.5 ± 75	1.65	981.3

$$\frac{\sigma_{sand}}{\sigma_{cell}} = \left[\frac{(KV)_{cell}}{(KV)_{sand}}\right]^{1/m} \qquad (6)$$

where m is the Weibull modulus, K is a dimensionless load factor describing the type of loading and V is the total volume under load. The value of $(KV)_{cell}$ is simply that for a monolithic beam in three-point bending; which is well known. As no predetermined loading factor exists for the core material in sandwich composites, the value of $(KV)_{sand}$ for each combination of core density and face thickness was determined numerically. This analysis was based on the following relationship for the risk of rupture (R) [3]:

$$R = \int_0^V \left(\frac{\sigma_1}{\sigma_0}\right)^m dV = (KV)_{sand}\left(\frac{\sigma_{1,max}}{\sigma_0}\right)^m \qquad (7)$$

where σ_0 is the characteristic strength and $\sigma_{1,max}$ is the maximum principal stress. Rearranging:

$$(KV)_{sand} = \int_0^V \frac{\sigma_1^m dV}{(\sigma_{1,max})^m} \qquad (8)$$

Once the value of $(KV)_{sand}$ is known, the value of $\sigma_{sand}/\sigma_{cell}$ can be calculated with eqn. 6. This ratio is listed in Table I for each core density-face thickness combination. The corrected failure loads are calculated by simply dividing the core tensile strengths by $\sigma_{sand}/\sigma_{cell}$. These loads correlate well with the average failure loads measured experimentally, considering the variability in the cellular strengths. The average error is 10.6% and all but four of the predicted loads were within 15% of experimental. The most significant source of error in these calculations are the measured Weibull moduli, which were determined with only 10 samples per density.

The remainder of the paper will deal with the stress distribution in the faces of the composites. The modulus of the faces (E_f) was measured to be 323 GPa. The stress in the faces (σ_f) during flexure was thereby calculated from the measured strain using Hooke's Law. Table II presents a summary of the experimental and theoretical face stress data. Because face stress for all the combinations of core densities and face thicknesses was approximately linear elastic, Table II presents the data as face stress per unit load. The theoretical values of σ_f were determined by substituting $y = (c+2t)/2$ into eqn. 2 and computing $(EI)_{eq}$ with eqn. 1.

For the three highest core densities, the measured and predicted face stresses agree well: all are within 9%. However, significant deviations (as high as 38%) between the measured and predicted values are evident in the two lowest core densities. These data reveal a definite trend: as relative core density decreases and face thickness increases, measured face stresses become increasingly greater than those predicted by eqn. 2. This trend exactly follows that expected by the local bending effect associated with shear deflections. As previously discussed, this effect would be expected in composites constructed with thick faces and low modulus cores. For these cases, the faces must bend about their own centroidal axes due to the large difference in stiffness between the two components. This extra bending moment would lead to higher than expected face stresses beneath the load point. Allen [1] has approached this problem mathematically and has developed the following relationship describing the maximum stress in a sandwich face subjected to three-point loading:

$$\sigma_{f,max} = \frac{PL}{4}\left[\frac{(c+2t)\varphi}{2I} + \frac{t(1-\varphi)}{2I_f}\right] \qquad (9)$$

Table II. Summary of measured and predicted stresses in tensile face plate during flexure. Corrected face stresses are given only when simple model predictions are significantly in error.

Face Thickness (mm)	Measured Face Stress per Unit Load (MPa/N)	Predicted Face Stress per Unit Load (MPa/N)	Difference Between Measured and Predicted (%)	Corrected Face Stress per Unit Load (MPa/N)	Difference Between Measured and Corrected Prediction (%)
ρ/ρ_t =0.091					
0.381	0.208	0.196	-6.0		
0.508	0.185	0.147	-20.4	0.164	+11.3
0.762	0.147	0.098	-32.9	0.152	-3.2
1.016	0.119	0.074	-37.7	0.124	-4.0
ρ/ρ_t=0.118					
0.381	0.195	0.193	-1.2		
0.508	0.167	0.146	-12.7	0.149	+10.8
0.762	0.127	0.098	-20.0	0.120	+5.5
1.016	0.104	0.074	-28.9	0.112	-7.1
ρ/ρ_t=0.153					
0.381	0.174	0.182	4.2		
0.508	0.133	0.140	5.3		
0.762	0.106	0.100	-5.8		
1.016	0.085	0.077	-8.9		
ρ/ρ_t =0.167					
0.381	0.171	0.184	7.4		
0.508	0.129	0.141	8.6		
0.762	0.090	0.096	6.5		
1.016	0.073	0.073	0		
ρ/ρ_t =0.202					
0.381	0.165	0.173	4.6		
0.508	0.128	0.135	2.5		
0.762	0.094	0.094	0		
1.016	0.201	0.190	-5.7		

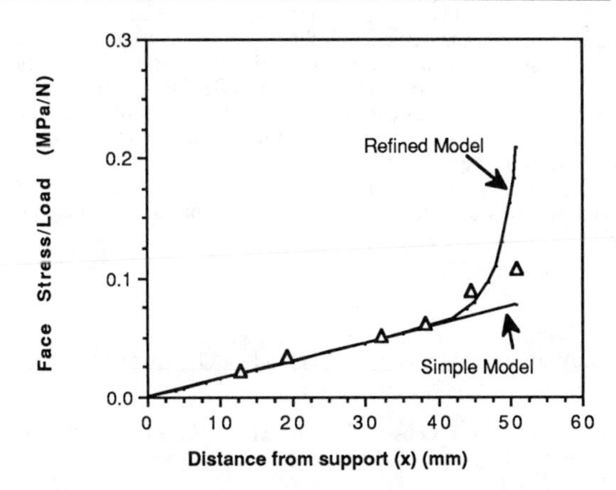

Figure 1. Comparison of experimental face stress/unit load and stresses predicted by the simple and refined models.

where I is the moment of inertia of the faces about the sandwich centroid and I_f is the moment of inertia of the faces about their own centroids. The parameter φ accounts for the ratio of the core shear stiffness to the bending stiffness of the faces. When local bending is not important, this term goes to unity and eqn. 9 reduces to eqn. 2.

In order to prove the occurrence of local bending, strain gages were placed along the distance between the central load point and outer support. Fig. 1 plots the measured face stress per unit load as a function of distance (inward) from the outer support, for the sandwich with the lowest core density (0.091) and the thickest faceplate (1.02 mm). Also plotted are the stresses predicted by eqn. 2 (referred to as the "simple model") and those predicted by eqn. 9 (referred to as the "refined model"). There are several important observations to be made from this plot. The simple model predicts the tensile face stresses to increase linearly from the support to the mid-span. Conversely, the refined model predicts a severe stress gradient near the load point. The maximum stress predicted by the refined model is much higher than that predicted by the simple model.

It is also evident from the plot that the measured stresses are higher than predicted by the simple model only near the load point. This is consistent with local bending. However, the maximum stress predicted by the refined model is significantly higher than the measured maximum stress. This difference is most likely because the measured data represents an average strain across the strain gage length. As evident from the plot, there is, theoretically, a large strain gradient across the gage length near the beam center. Individual points in this gradient could not be measured accurately with the strain gages used in this study. In order to compare the maximum face stress predicted by the refined model to that recorded by the strain gage, it is necessary to average the stresses predicted by the refined model across the gage length. It is these averages that are reported in the "corrected face stress per unit load" column in Table II. It is clear that agreement between the refined model and the measured stresses is good for the sandwiches where localized bending is significant.

CONCLUSIONS

The mechanical behavior of a ceramic sandwich system incorporating a porous, open cell alumina core was investigated. Initial failure mechanisms observed in the sandwich constructions were almost exclusively core fracture. The orientation and location of the crack varied considerably, depending on the angle of maximum principal stress in the core. It was determined that composite failure loads can be predicted by equating the maximum principal stress in the core with the core flexural strength. It is necessary to incorporate Weibull statistics in order to scale up independent flexural strength measurements for the core material to those that can be expected in the sandwich. The face stresses of an all-ceramic sandwich composite can be predicted with existing sandwich theory. For composites with a large difference in stiffness between the face and core, however, maximum face stresses can be substantially higher than those predicted by simple models due to a localized bending of the faces about their own centroidal axes. A refined analysis is required to successfully account for this behavior.

This work was supported by the National Science Foundation under grant No. DMR-8603878. We acknowledge the students and staff of the Materials Science and Engineering Department who assisted in this study.

REFERENCES

1. H.G. Allen, Analysis and Design of Structural Sandwich Panels, (Pergammon Press, Oxford, England, 1969).

2. T.C. Triantafillou and L.J. Gibson, "Failure Mode Maps for Foam Core Sandwich Beams," Mat. Sci. Eng., 95, 37-53 (1987) .

3. G.J. DeSalvo, "Theory and Structural Design Applications of Weibull Statistics," Westinghouse Electric Corporation, WANL-TME-2688, May 1970.

THE INDENTATION AND NAILING OF CELLULAR MATERIALS

M. Fátima Vaz and M. A. Fortes
Departamento de Engenharia de Materiais. Instituto Superior Técnico, Av. Rovisco Pais, 1000 Lisboa, PORTUGAL

ABSTRACT

An experimental study was carried out on the indentation of cellular materials under quasi-static conditions. The study concentrated on a closed cell rigid polyurethane foam, but other cellular materials were tested. Load (F) – penetration depth (x) curves were obtained for various types of indenters which fall into two main types: sharp indenters (e.g. normal nails) and blunt indenters (e.g. flat ended nails). The F(x) curves are similar for the two types with an initial high slope region followed by a lower slope region in which F increases linearly with x. Nevertheless, the mechanisms of indentation are completely different, with a blunt indenter pushing the material under it and a sharp indenter pushing it radially as it penetrates. The effect of dimensions and shape of the indenters on the F(x) curves was investigated and scaling relations could be derived. Penetration-removal cycles clarify the differences between the two types of indenters, and give additional information on the mechanics of indentation. The experiments were complemented with scanning microscope observations of sections of the indented material.

Two approaches are advanced for modelling indentation: a discrete approach in which the cellular structure is taken into account and a continuum approach.

INTRODUCTION

The penetration of indenters in cellular materials is a quite common operation, examplified by the nailing of wood, the introduction of a cork screw in a bottle stopper or of a fork in a potato, the cutting of bread or of a plastic foam and the penetration of arrows, darts or simple pins in a board.

There is a resisting force, F, to indentation which in general increases with penetration depth, x, and which depends on the geometries of the indenter and of the indented specimen and on the mechanical properties of the material that is indented (the indenter behaves in most cases as a completely rigid material, but otherwise its mechanical properties must also be considered). The resisting force also depends on penetration rate.

The main purpose of this paper is to interpret experimental force-penetration depth, or F(x), curves, relating their particularities to simple mechanical properties of the indented material. We shall consider only indentation under quasi-static conditions but will discuss the effect of penetration rate. We also restrict the study to indenters with uniform section (perpendicular to the direction of penetration), except, eventually, at their extremities, as in normal nails. Blades may be included in this category, but all experiments to be reported were done with indenters similar to nails, with fairly equiaxed cross sections. The term nail will be used in this paper in a broad sense to include normal nails with sharp tips (e.g. a conical tip in a round nail) and flat ended indenters. The terminology "sharp" and "blunt" indenters will also be used. The discussion will concentrate on indentations that lead to fracture of the material, i.e. no special attention will be given to the initial stage of indentation, in which the material is deformed but does not fracture. A study of light indentation in cellular materials was undertaken by Wilsea et al [1].

The resistance to quasi-static penetration of sharp nails in wood was previously studied by Salem et al [2]. They found a linear variation of the force, F, with penetration depth, x, following an initial transient associated with penetration of the tip of the nail:

$$F = F_0 + \lambda x \quad (1)$$

where F_0 is the load at the end of the transient (when x=0) and λ is the slope of F(x). The increasing force was attributed to the friction between the nail shank and the material. Salem et al [2] found in experiments with wood that F_0 is proportional to the area of the cross section of both round and oval nails. They also predicted that λ should be proportional to the radius of round nails but this was not experimentally confirmed.

The wood penetration curves published by Salem et al [2] show profuse load serrations. Similar serrations in the indentation of a rigid open cell foam with a flat indenter were reported by Gibson and Ashby [3] and attributed to the successive bending and fracture of the cell edges. In this case the resisting force, measured at the force peaks, was found to be proportional to $A^{3/2}$ and not to A, where A is the area of the indenter cross section.

The force of extraction of a nail is an important engineering property. It has been experimentally measured for nails of various geometries [4 - 6]. In first approximation this force should be equal to λx in eq. 1, but stress relaxation of the material around the nail may lead to a lower value while rusting of steel nails may increase the extraction force. Other topics that have been discussed in relation to nailing are the dynamics of penetration under impact (Salem et al [2]) and the scaling relation in nails which is related to the resistance to buckling [2,7].

In the present study most experiments were carried out with a rigid closed cell polyurethane foam rather than with wood. The main reason for this choice was to avoid the scatter in experimental results that occurs with natural materials. The penetration experiments were on the line of those reported by Salem et al [2] and covered sharp nails with various tip angles and shank diameters and round flat ended nails of various diameters. Rate effects were investigated. Load-depth curves were also obtained in extraction of the indenter and in a subsequent penetration-removal cycle to the same depth. The experiments were complemented with electron scanning microscope (SEM) observations of the indented material.

In addition to the rigid foam, experiments were carried out, but only with normal round nails, with other cellular materials, including pine wood, cork, a flexible foam and a potato.

Experimental

The rigid polyurethane foam had closed cells and a density of 34 kg/m^3. The average cell diameters, determined by the linear intercept method in SEM images, were as follows: parallel to the rise direction: 420 μm; perpendicular to the rise direction: 310 μm. A stress-strain curve for this foam at a strain rate $\dot{\epsilon} \simeq 8.8x10^{-4}s^{-1}$ is shown in Fig. 8a. The buckling stress is around 0.2 MPa at a strain $\epsilon \simeq 0.07$ giving an average slope of the initial part of the stress-strain curve $E \simeq$ 3MPa. The strain at the end of the plateau region is around $\epsilon_b \simeq 0.7$. The slope of the final part of the curve is $E' \simeq 7E$.

The specimens for indentation were in the form of parallelipipeds, in most cases with the three edge lengths equal to 3.5 cm. The specimens had one edge parallel to the rise direction and all indenters were introduced with their axes parallel to that direction to a depth of 50 to 75% of the specimen height. The tests were conducted in a tensile machine with the specimens supported on the lower plate. An effect of specimen dimensions on the penetration curves was not detected for the dimensions investigated, between 2 and 12 cm parallel to the indenter and between 3.5 and 8 cm across. The penetration rate was in most cases of $3.3x10^{-3}cm\ s^{-1}$, but experiments at higher rates up to $3.3x10^{-1}cm\ s^{-1}$ were also carried out. In other experiments the machine crosshead speed was suddenly changed and the effect on force recorded. Finally, a number of experiments were carried out to obtain the extraction F(x) curve and subsequent penetration-extraction curves.

Depending on the materials and indenters, two load cells, respectively of 5000N and 100N, were used. The 100N load cell (sensitivity 0.1N) is adequate to study the load serrations in the polyurethane foam.

The SEM observations were done in non-indented and indented specimens. Sections containing the indenter axis and transverse sections were observed, with the indenter placed in the specimen. A special wood indenter was used, which consisted of two halves that could be separated in the preparation of the axial sections with a razor blade; the indenter was simply sectioned in the preparation of transverse sections.

The other materials tested included pine wood, cork, a flexible closed cell polyurethane foam and a fresh potato. Properties of these materials are indicated in Table II.

Steel nails were used in all cases. The round nails had diameters in the range 0.9 - 5.4 mm, both with flat ends and with conical tips. The tip angle was varied between 4° and 33°. Most commercial nails have a tip angle around 20°. Experiments were also made with flat and cone ended round nails which had their shank diameters reduced by machining. The main types of round nails are those indicated schematically in Fig.1. Nails of square section were also tested and in other experiments two round nails, placed side by side, were simultaneously driven in a specimen.

RESULTS

We indicate first the results for the rigid polyurethane foam, under two headings: sharp nails and blunt nails. At the end of the section, results obtained with sharp nails in other cellular materials will be reported.

Sharp nails

Load-penetration depth curves

The load-penetration depth curves for round nails with a sharp conical tip are as shown in Fig. 1a. Serrations are shown at higher magnification in Fig. 1e. The curves are of the type reported by Salem et al [2]. The initial region corresponds to tip penetration and is not linear. We found that it can be approximated by a parabolic relation

$$F = \xi x^2 \qquad (2)$$

where x=0 corresponds to the first contact with the specimen surface and ξ is a constant for each nail. The force F_0 at the end of tip penetration corresponds to $x=h=R \text{ ctg } \alpha$, where R is the shank radius and α is the apperture of the conical tip. The continuum model to be discussed in the Section 3 gives the following relation for F_0:

$$\frac{F_0}{\pi R^2} = \sigma_n (1 + \mu \text{ ctg } \alpha) \qquad (3)$$

where σ_n and μ are constants for a given indenter and specimen materials. σ_n is the pressure on the nail surface and μ is the coefficient of sliding friction between the nail and the indented material. The experimental results compiled in Fig. 2a show that $F_0/\pi R^2$ is indeed independent of R for a given α. The dependence of F_0 on α predicted by eq. 3 was also experimentally confirmed, for values of α in the range 5-33°, as shown in Fig. 3. The following

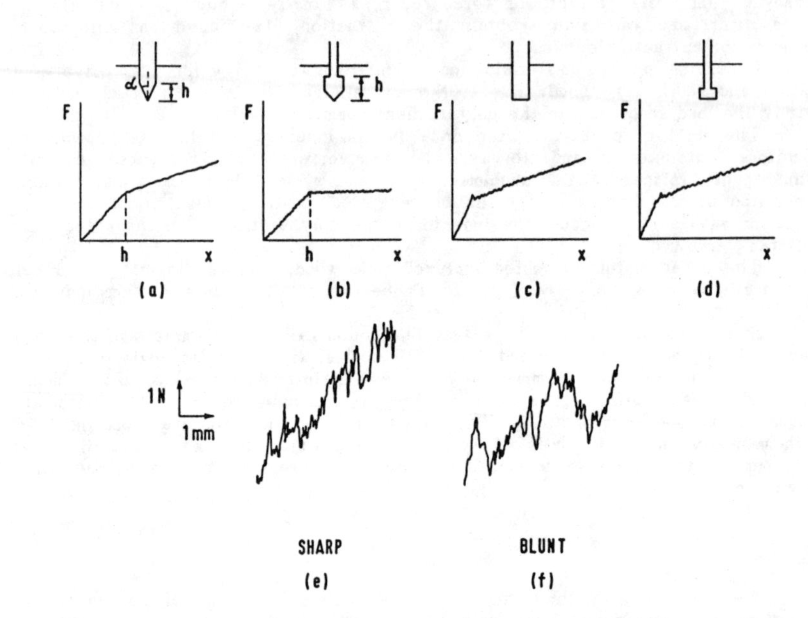

Fig.1-a-d) Load (F)-penetration depth (x) curves for various types of nails indicated on top, with sharp and blunt ends and with unreduced and reduced shank diameters. e-f) Load serrations at higher magnification for sharp (e) and blunt (f) ends.

values

$$\sigma_n = 0.29 \text{ MPa}$$
$$\mu = 0.48 \tag{4}$$

can be derived from Fig. 3. As will be shown latter,this value of μ is probably too high.

The force resisting penetration of the conical tip (eq. 2) can then be put in the form

$$F = \sigma_n \, tg\alpha \, (tg\alpha + \mu) \, \pi \, x^2 \tag{5}$$

or, in terms of the radius r at a distance x from the tip appex,

$$F = \sigma_n \, (1 + \mu \, ctg \, \alpha) \, \pi \, r^2 \tag{6}$$

After penetration of the tip, the force increases linearly with penetration depth x (Fig. 1a):

$$F = F_0 + \lambda \, x \tag{7}$$

where the origin for x is now at the beginning of shank penetration. If the shank diameter is reduced by machining, the friction term λx is absent, and the force of penetration is constant and equal to F_0, as in Fig.1b. The experimental values of λ shown in Fig. 2b indicate that this quantity is proportional to R. The continuum model predicts

$$\frac{\lambda}{2\pi R} = \mu \, \sigma_n \tag{8}$$

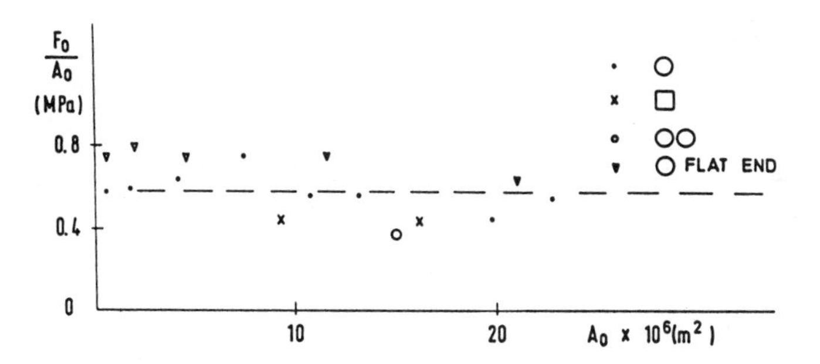

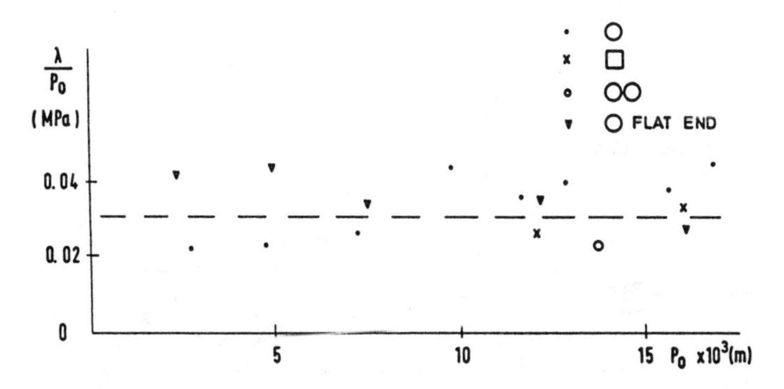

Fig.2 - Values of F_0/A_0 (a) and of λ/P_0 (b) for indenters of various types. F_0 is the load at the end of the initial transient and A_0 is the area of the cross-section of the indenter shank; λ is the slope of the curve region during shank penetration and P_0 is the perimeter of the cross-section. The sharp indenters all have apperture angles around 20°.

with the same μ and σ_n as above. The experimental values of $\lambda/2\pi R$ (Fig. 2b) are considerably smaller (by a factor of 4) than the one that can be derived from (4) and (8). This may indicate that the continuum model from which (3) and (8) were derived, is only a rough approximation. We can then write for the F(x) relation during shank penetration

$$\frac{F}{\pi R^2} = \frac{F_0}{\pi R^2} + \frac{2\lambda}{2\pi R}\frac{x}{R} \tag{9}$$

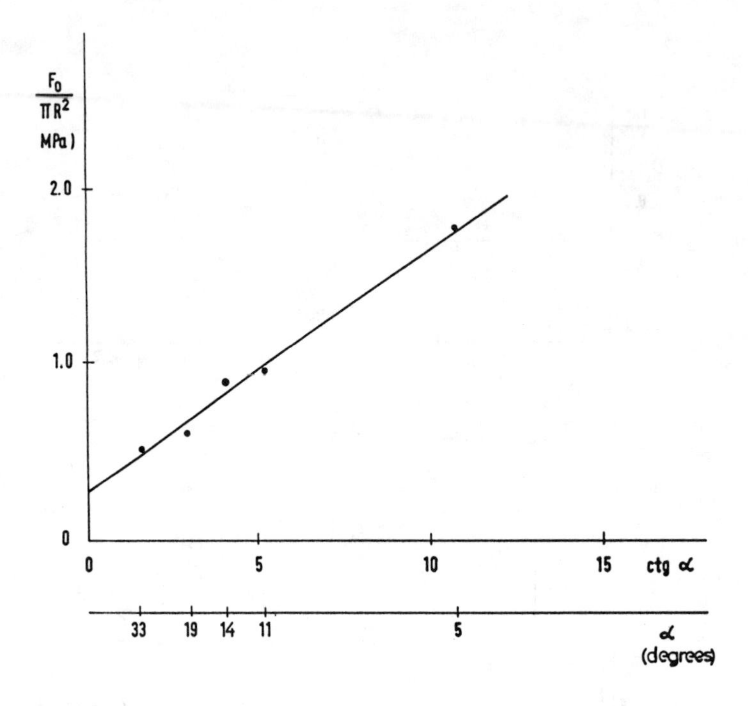

Fig.3 - Variation of $F_0/\pi R^2$ for round nails with the apperture α of the conical tip.

with $F_0/\pi R^2$ given by (3) and $\lambda/2\pi R$ by (8). It should be noted that $F_0/\pi R^2$ depends on the tip apperture, α (because of the friction term), but $\lambda/2\pi R$ is independent of α. The stress $F/\pi R^2$ is then a function of x/R only, for fixed α.

Experimental results obtained with square section nails and with two nails placed side by side are included in Fig. 2, and show that F_0 is proportional to the area of the cross section A_0, while the slope λ is proportional to the perimeter P_0.

Load serrations

The load serrations (i.e. load fluctuations) are quite irregular and it is difficult to measure their average amplitude and spacing (wave-length). There are fairly large load drops with intermediate smaller drops. The number of large drops per unit length traversed by the indenter is, on average, around 3 mm^{-1} and therefore correlates with the average number of intersections with cell walls (which is the reciprocal of the average cell diameter). The amplitude of these large load drops is variable, but for many of them the value is as much as 1.5N, between 3 and 30% of the load. The characteristics of the serrations do not seem to depend on the diameter of the nail nor on the apperture of its tip.

Introduction - removal cycles

Fig. 4a illustrates the variation of F with x in introduction-removal cycles of

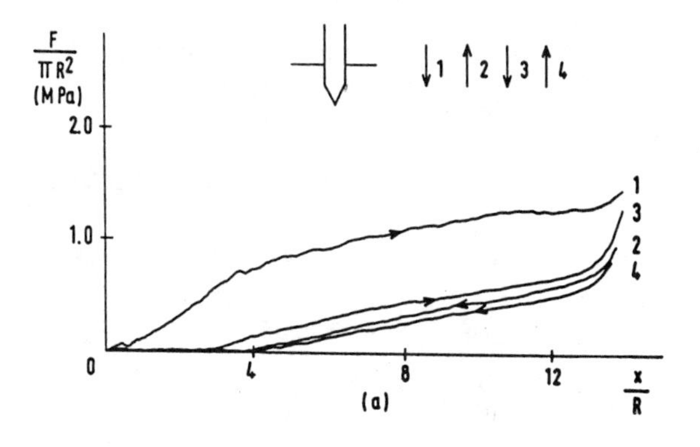

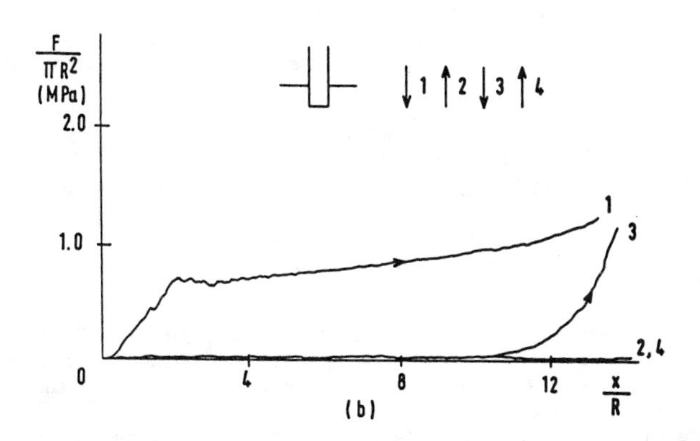

Fig.4 - Load (F) - penetration depth (x) curves for two penetration-removal cycles to the same final depth: a) sharp nail; b) flat end nail.

sharp indenters while in Table I are indicated some relevant quantities of the curves of Fig.4a. After indentation to a given depth x (curve 1, Fig. 4a) the removal of the indenter, at the same rate, gives a F(x) curve shown in Fig. 4a, curve 2. The value of F at the beginning of extraction is smaller than the value of F at the end of penetration, the difference ΔF_f (indicated in Table 1) being smaller than F_0 (the value of $F_0/\pi R^2$ is 0.66MPa). The slope of curve 2 is slightly smaller than that for curve 1. This may be due to stress relaxation in the material around the shank. The removal of the nail tip requires virtually no force. The second penetration gives curve 3. The introduction of the tip is accomplished with a very small force. The slope of the curve in the region of shank penetration is comparable to that for curve 1. The force rises rapidly when x approaches the final value indicating recovery of strain, but the final load is smaller than that at the end of the first penetration. Load serrations are almost absent in the second

TABLE I

Slopes, $\lambda/2\pi R$, and force $F/\pi R^2$ at maximum penetration (x=2.6 cm) for round sharp nails in successive penetration/removal cycles (values in MPa).

	$\lambda/2\pi R$	$F/\pi R^2$(x=2.6cm)	$\Delta F_f/\pi R^2$
1st penetration	0.06	1.45	
1st removal	0.05	0.92	0.53
	$\lambda/2\pi R$	$F/\pi R^2$(x=2.6cm)	$\Delta F_f/\pi R^2$
2nd penetration	0.06	1.31	
2nd removal	0.05	0.84	0.47

penetration curve. Curve 4 is for the second extraction.

If no stress relaxation occurred, the stress σ_n would be unchanged in an introduction-removal cycle. If the coefficient of friction μ is also unchanged, the difference ΔF_f between the values of F at the end of a penetration and at the beginning of the subsequent extraction should be given by

$$\frac{\Delta F_f}{\pi R^2} = \sigma_n \tag{10}$$

because the force of extraction is due to friction only. Combining with eq.3 yields

$$\frac{\Delta F_f}{\pi R^2} = \frac{F_0}{\pi R^2} (1+ \mu \text{ ctg } \alpha)^{-1} \tag{11}$$

For ctgα = 2.5 and taking μ=0.48 (eq. 4) we obtain $\Delta F_f/\pi R^2$=0.45 ($F_0/\pi R^2$). Using the experimental value $F_0/\pi R^2$=0.66MPa we get $\Delta F_f/\pi R^2$=0.30MPa which is below the experimental value of Table I. Again, the predictions of the continuum model should be regarded as rough.

When the nail is removed, the strain is partly recovered and the cavity left by the nail partly closes. In a subsequent penetration, the cavity is re-opened but this originates a smaller stress σ_n than in the previous introduction. There is work softening in the first penetration, in the sense that the stress to produce a given strain (or, more properly, a given displacement field) decreases due to prior deformation. This must be related to the viscoelastic behaviour of the cell wall material. It is this work softening that is responsable for the decrease in force (for a given penetration depth, including the final depth) in successive penetrations.

Rate effects

A strain rate sensitivity coefficient, m, was measured in strain rate change uniaxial compression tests of the polyurethane foam. The coefficient, m, is fairly independent of strain with an average value

$$m = \frac{d \ \ell n \ \sigma}{d \ \ell n \ \dot{\epsilon}} = 0.035$$

As illustrated in Fig. 5, it is possible, by suddenly changing the penetration rate, $\dot{x}$,

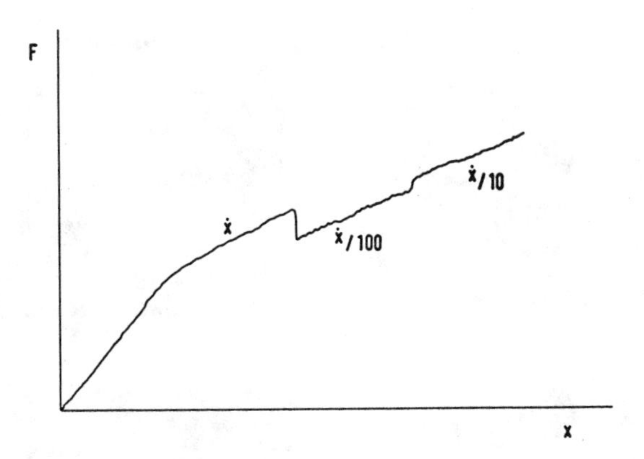

Fig.5 - Load (F) - penetration depth (x) curve with changes in the penetration rate.

in an indentation test, to measure the coefficient

$$m' = \frac{d\ \ell n\ F}{d\ \ell n\ \dot{x}}$$

The experimental values of m' are, within the experimental scatter, equal to those for m:

$$m' = m = 0.035$$

This result is not surprising because the compressive radial strain at a point in the axis, due to the passage of the tip, is achieved in time $\dot{x}/h$, where h is the tip length. The strain rate is then proportional to $\dot{x}$, while the compressive stress, σ_n, is proportional to F.

There is a very small effect of $\dot{x}$ in the slope of F(x): the slope tends to increase as $\dot{x}$ increases. This is probably due to the increase in σ_n (eq. 8), while μ is likely to change negligeably.

Microscopy

Fig. 6a,b are SEM micrographs of the material indented by a round sharp nail. In the axial section (Fig. 6a) it is apparent that the material around the indenter, including its tip, is little affected, except in a narrow region adjacent to the indentor surface up to a radius around 1.3R. This is also seen in the cross section through the shank (Fig. 6b).

Other materials

The F(x) curves for nailing of other cellular materials are similar to those for the rigid foam, but with values of $F_0/\pi R^2$ and $\lambda/2\pi R$ that depend, for a given nail, on the strenght of the indented material. The flexible foam is not easily penetrated, unless the tip is rather sharp; with tips of not too small radii of curvature at the appex, the material is depressed under the tip but is not penetrated.

Table II gives values of $F_0/\pi R^2$ and $\lambda/2\pi R$ for the different materials

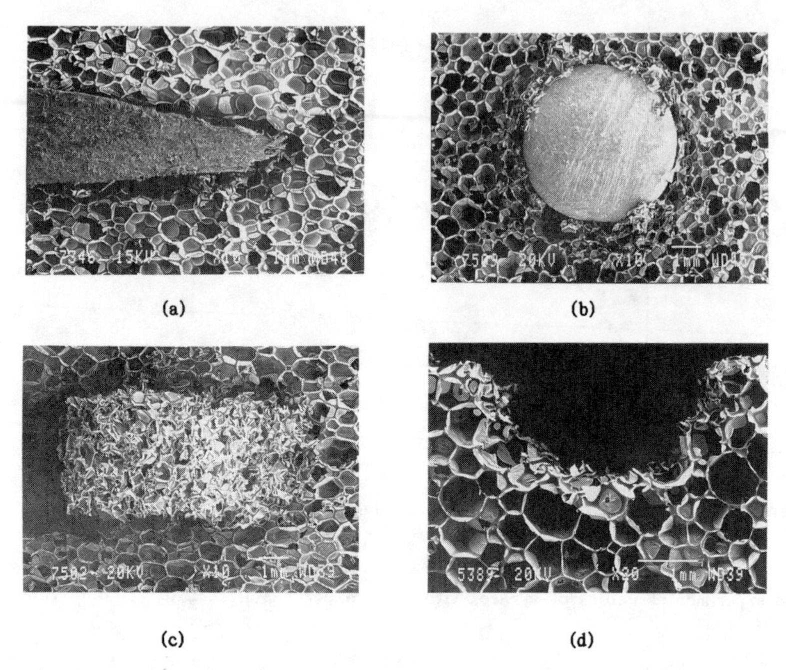

Fig.6 - SEM micrographs of sections in indented polyurethane foams: a) axial section with a sharp indenter; b) cross section through the shank of a sharp indenter; c) axial section through a blunt indenter; d) cross section of the hole left by the passage of a blunt indenter.

indented by round nails with a tip angle $\sim 20°$. Both values increase with increasing yield stress in uniaxial compression. It is also found that the ratio between $F_0/\pi R^2$ and $\lambda/2\pi R$ is fairly constant with a value between 14 and 24, with the exception of the potato which has a value of 59, possibly due to a lower friction coefficient.

Eqs. 3 and 8 predict for this ratio the value

$$\frac{F_0/\pi R^2}{\lambda/2\pi R} = \frac{1}{\mu} + \text{ctg}\ \sigma \tag{12}$$

Since α was approximately the same in all tests with ctg$\alpha \simeq 2.5$ it is concluded that the coefficient of friction is fairly constant (except perhaps for the potato) with values between 0.05 and 0.1. This is possibly a better estimate of μ than the one obtained by varying the apperture angle α (see eq. 4).

Table II also shows that $F_0/\pi R^2$ is roughly proportional to the yield stress in the materials for which this property could be determined.

Flat end nails

Indentations with blunt, flat ended round nails were made on the rigid polyurethane foam exclusively. Figs. 1c,d show F(x) curves, respectively for nails of uniform section and for nails with a reduced shank diameter. The two curves are identical.

The serrations are illustrated in Fig. 1f. They are not as fine as for sharp nails and are rare in the initial part of the curve; the amplitude of the largest load

TABLE II

Parameters F_0 and λ of the load-penetration depth curves for various materials

Material	$F_0/\pi R^2$ (MPa)	$\lambda/2\pi R$ (MPa)	$\frac{F_0/\pi R^2}{\lambda/2\pi R}$	Density (kg/m^3)	Young's modulus E (MPa)	Yield Stress σ_y (MPa)	$\frac{F_0/\pi R^2}{\sigma_y}$
Pine wood (perpendicular to grain)	35.0	1.5	23.8	550	600	6	5.8
Cork (axial direction)	5.6	0.27	20.7	164	8.2	0.85	6.6
Rigid polyurethane foam	0.58	0.035	16.9	34	3.0	0.2	2.9
Flexible polyurethane foam	0.20	0.030	14.0	66	0.024	—	—
Potato	2.25	0.038	59	106	3.71	—	—

drops is similar to that for sharp nails. Except for the serrations, the F(x) curves for blunt nails are similar to those for sharp nails, with two regions, the first, transient region having a higher slope than the second, nearly linear region. The value of F_0 at the end of the transient is proportional to R^2 and the slope of the second region proportional to R as for sharp nails, as the data included in Fig. 2 shows. It is noticeable that little or no difference occurs in the values of $F_0/\pi R^2$ and $\lambda/2\pi R$ between blunt and sharp nails (see Fig. 2).

While a sharp nail opens its way by pushing the material latterally, the

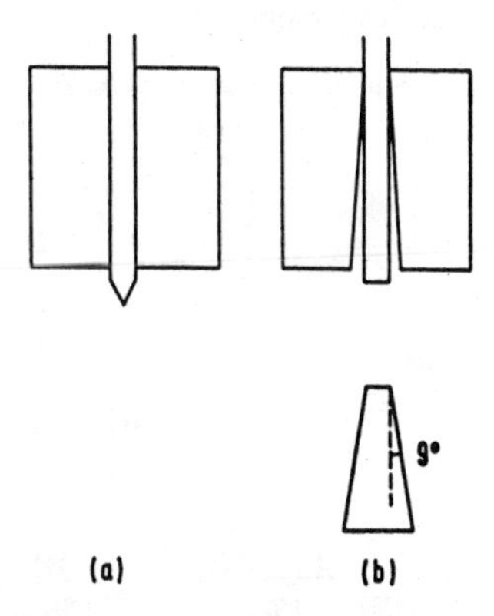

Fig.7 - (a) A sharp indenter does not remove material as it traverses a cellular material, but (b) a blunt indenter removes a cone of material, leaving a cavity of conical shape.

blunt nail compresses and pushes the material under it. If the nail traverses the entire thickness, it is found that virtually no material comes out from the exit face (Fig. 7a) when the nail is sharp, but for a blunt nail there is a "cylinder" of material emerging ahead of the indenter (Fig. 7b). This "cylinder" is in fact of conical shape as shown in Fig. 7b, the upper base having the radius of the indenter. When this cone is extended (by applying a small tension) its length becomes only slightly smaller than the height of the cavity left in the specimen, and the angle of the cone (Fig. 7b) is approximately 9°. The cavity left in the specimen by the indenter has the same shape. This geometry implies that the shank does not contact the material and there are no friction forces on the indenter as in the case of sharp indenters. This explains why it is indifferent to have a uniform or a reduced shank in a blunt indenter. The fact that the force increases linearly with penetration depth is more difficult to explain, and cannot be attributed to friction, as for sharp indenters.

The conical shape of the hole produced by the indenter also explains why the extraction force of a blunt indenter is very small, as shown in Fig. 4b.

The initial transient which ends at F_0 corresponds to a penetration depth, x_0, of the order of a few milimeters for the flat ended indenters used, the diameter of which ranged from 0.9 mm to 5.4 mm. There is experimental evidence that x_0 increases as the indenter radius increases, with $x_0 \simeq 1.5R$. The few load serrations in the initial region of the F(x) curve can be attributed to buckling and, eventually, fracture of cell walls under the indenter.

SEM observations made after the transient show that the material under the indenter is heavily compressed with buckled and fractured cell walls (Fig. 6c). The material around the indenter is affected only in a narrow region (Fig. 6d) as in the case of sharp nails.

MODELLING OF INDENTATION

Two main approaches can be envisaged for modelling indentation of cellular materials. In one, the material is treated as a continuum and average stresses at each point are considered over areas encompassing a large number of cells. A constitutive equation for deformation is used, which provides a macroscopic description of the material. In compression, the adopted constitutive relation is as shown in Fig. 8.b, with an initial linear region of slope E (elastic or viscoelastic bending of the cell walls) and a final linear region of slope E'>>E which corresponds to the crushing of the cell walls. The two regions are separated by a plateau at a stress σ_b.

In the other approach, the cellular structure is taken into account, and the

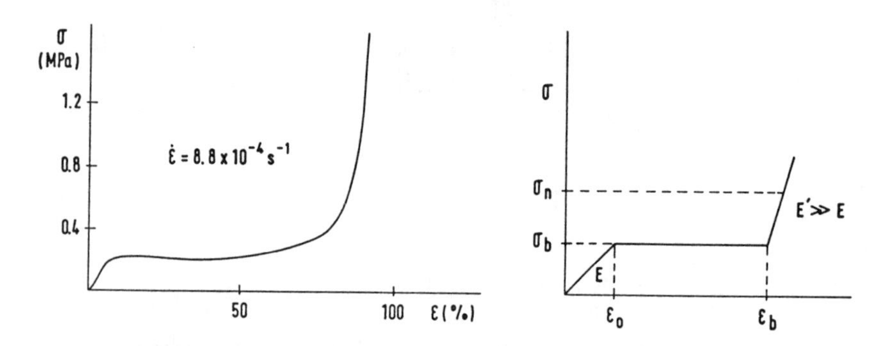

Fig.8 - (a) Stress-strain curve in uniaxial compression (perpendicular to the rise direction) of the rigid polyurethane foam at a strain rate $\dot{\epsilon}$=8.8x10^{-4}s^{-1}; (b) Curve for the ideal cellular material behaviour used in the calculations.

effects of the indenter on the cell walls or cell edges are analysed. This type of model was used by Gibson and Ashby [3] for the indentation of rigid open cell foams with a flat indenter. We shall apply this "discrete" model to the indentation by a sharp indenter. Sharp and blunt indentation will be considered separately, and examplified by a conical tip indenter and a flat end indenter, respectively.

Sharp indenters

For a nail with a conical tip, the cavity expansion treatments [8,9] indicate that the compressive normal stress on the tip surface is uniform, with a value σ_n, independent of geometry (Fig. 9), i.e. of R and α. For the tip region between the appex and a radius r, the resultant of these stresses is an axial force $\pi r^2 \sigma_n$.

The friction stresses are uniform with the value $\mu\sigma_n$; the resultant force, parallel to the indenter axis, is $\mu\sigma_n \pi r^2 \mathrm{tg}\alpha$. The net force resisting tip penetration under quasi-static conditions is then given by eq.6. The F(x) relation is eq.5 and the force F_0 at the end of penetration of the conical tip is given by eq.3.

The normal stress on the nail shank is also σ_n, by continuity reasons. The friction force in the shank is then $(2\pi R\mu\sigma_n)x$, where x is the shank length in contact with the indented material. The slope λ of the F(x) relation during shank penetration is therefore given by eq.8.

The stress σ_n is calculated in the Appendix for a material with a compression behaviour as in Fig. 8b. It is a stress in the final, high slope region of the compression curve, its value being given by eq. A12, independent of indenter size. The material around the shank, up to a radius r_b, is compressed to stresses in this high slope region. Outside r_b the material is elastically compressed. The value of r_b, given by eq. A11, agrees fairly well with the experimentally observed value ($r_b \sim 1.3R$).

For other geometries of the indenter, the pressure in a cross-section is not uniform, but the experimental results suggest that the average value, $\bar{\sigma}_n$, of the normal stress is independent of shape and size, with the value for round indenters. The average value is defined by

$$\bar{\sigma}_n = \frac{1}{P_0} \int \sigma_n \, ds \tag{13}$$

where ds indicates an element of arc of the cross-section where the normal stress is σ_n and P_0 is the perimeter of the cross-section.

We now take into account the cellular structure of the material. For a closed cell material, the tip of the indenter contacts successive cell faces, which can be identified with plates supported at their edges. In an attempt to model

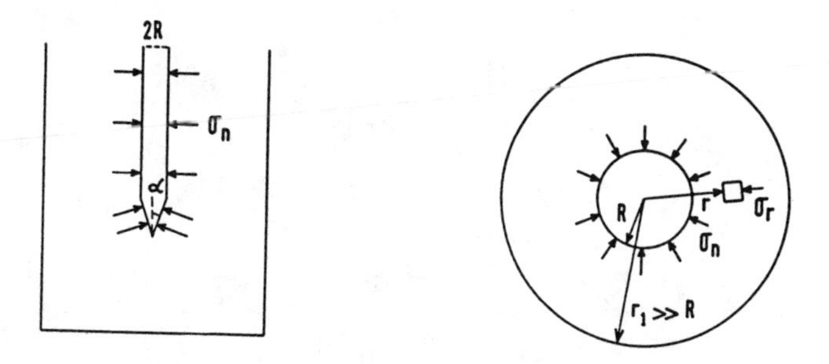

Fig.9 - The pressure, σ_n, on the indenter surface is assumed to be uniform (a).Cross-section (b) through the indented material, a cylinder of radius r_1; the radial compressive stress and strain are calculated as a function of r in the Appendix.

indentation of a cellular material we first consider a plate (for example, circular), supported or clamped at the periphery, which is indented at its centre by a conical tip (Fig. 10a). Simple experiments have been conducted with a sheet of paper, 2mm thick, clamped at the edge, which was indented by a sharp nail at the centre at a constant rate. The F(x) curve is shown in Fig. 10a. The plate initially bends under an increasing force. From linear elasticity theory, the slope dF/dx in bending is, for a circular plate of radius, a, and thickness, t, loaded at the centre [e.g.10]:

$$\frac{dF}{dx} = \frac{16\pi D}{a^2} \alpha \tag{14}$$

where D is the flexural rigidity of the plate (Young's modulus E_0; Poisson ratio, ν_0):

$$D = \frac{E_0 t^3}{12(1-\nu_0^2)} \tag{15}$$

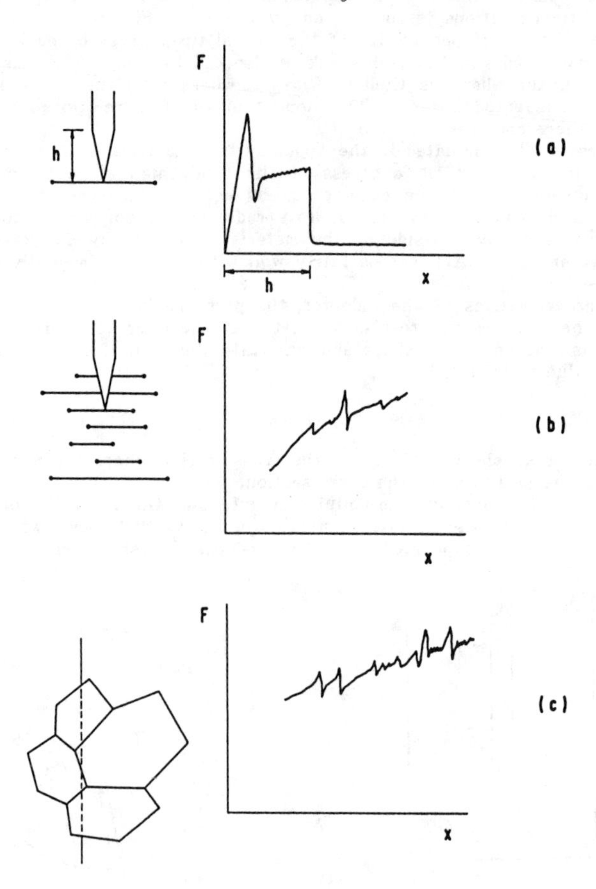

Fig.10 - Penetration of an indenter in a closed cell foam showing the expected load (F) - penetration depth (x) curves: (a) penetration through a single circular plate; (b) penetration through a series of plates of different sizes and spacings, contacted randomly; (c) penetration through the faces of a closed cell material.

and $\alpha=1$ for a clamped edge and $\alpha=(1+\nu_0)/(3+\nu_0)$ for a simply supported edge. Using $E_0=1.6$ GPa, $\nu_0=0.3$, $t=10$ μm and $a=200$ μm, eq. 14 gives the value (200α)N/m for the slope due to elastic bending of a wall in the polyurethane foam. Experimentally values as high as 9000 N/m are observed. These large values may be attributed to bending of faces loaded near their edges. The stress concentration under the appex eventually leads to fracture and the load drops. A small hole is formed which then expands as the conical tip penetrates. The force first rises quickly and then more slowly, with load serrations. The initial rise probably corresponds to the expansion of the original small hole. The force then increases more slightly with fluctuations. When the conical tip has fully crossed the plate, the force decreases to a low value which remains subsequently constant. This force is due to friction on the lips of the hole formed by the conical tip.

In a simple model of indentation of a cellular material we consider successive parallel plates of different sizes as in Fig. 10b, which are crossed by a conical tip in a direction perpendicular to the plates. Each plate is intersected at a point randomly located in the plate. The resulting F(x) curve would look like the one shown in Fig. 10b. The average number of plates intersected is proportional to x while the force due to a single plate indented by the tip increases with the length x of the part of the tip that has traversed the plate. If the increase in force is linear in x, the total force in a length x is proportional to x^2 as in eq 2. As the shank crosses a plate, the force remains constant. For a length x, the total force should then be proportional to x, as in eq. 7. It is also proportional to the perimeter of the shank.

In a real cellular material, the cell walls are inclined in relation to the indenter axis and are neither clamped nor simply supported at their edges (Fig. 10c). The plates adjacent to the one that is contacted by the appex of the tip also bend and may fracture. The fine details of the resulting F(x) are therefore complex. But it is likely that the large load drops are due to fracture of cell walls, while the intermediate smaller serrations may be due to the expansion of holes in successive walls.

Modelling of indentation by blunt indenters

Blunt indentation is rather more complex than sharp indentation, because the conditions under a blunt indenter change as it penetrates, with transport of material under the indenter. Fracture occurs at the edge of the indenter where both shear and tensile stresses are large. It is likely that fracture occurs by shear in planes at $\sim 9^\circ$ to the axis, where shear stresses are probably largest. This would explain the observed conical shape of the hole produced by the indenter. We are attempting a stress analysis of the cellular material around the indenter in order to explain that experimental observation.

The compressive strain under the indenter accommodates its displacement (Fig. 11). This strain, ϵ_x, changes both with x' and with r, where x' is the coordinate along the axis and r the distance to the axis. In a linear elastic material, a flat end indenter of radius R gives rise to a pressure distribution p(r) on the indenter end (i.e. the stress σ_n for x'=0 and $r \leq R$) which increases from the center to the edge [e.g., 11]:

$$p(r) = p_0 \, (R^2-r^2)^{-1/2} \qquad r \leq R \tag{16}$$

where p_0 is the pressure at the axis, r=0. The total force F is

$$F = 2\pi R^2 p_0 \tag{17}$$

and the displacement x of the punch is [e.g., 11]:

$$x = (1-\nu^2) \frac{F}{2RE} \tag{18}$$

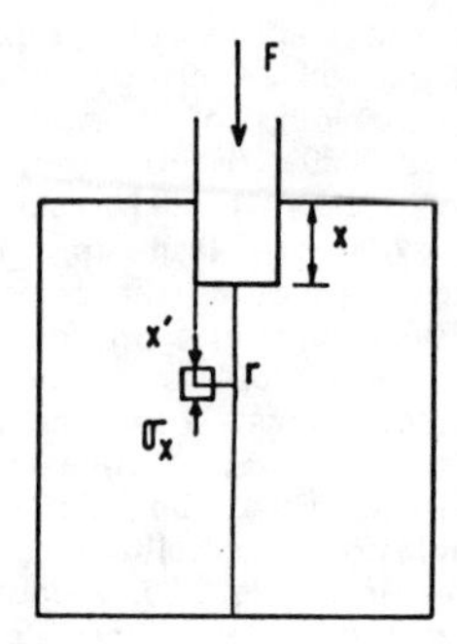

Fig.11 - A flat ended indenter in a material originates a non-uniform compressive stress σ_x. The displacement, x, is accommodated by the compressive strain ϵ_x under the indenter, which is a function of x' and r.

where E is the Young's modulus and ν is the Poisson ratio of the indented material (for the foam used, $\nu \simeq 0$). This gives for the initial slope of the F(x) curve the value

$$\frac{dF}{dx} = \frac{1-\nu^2}{2RE} \tag{19}$$

There is some uncertainly in the value of E for the rigid foam, because elasticity is not linear, but eq. 19 provides a good approximation for the slope at the origin. Subsequently the material under the indenter will buckle and F(x) will deviate from the linear relation, with a smaller force. When the shear stress reaches a critical value, fracture occurs at the edge and the second region of the F(x) curve begins. The value of x/R at this moment should be independent of R, as eq. 18 suggests. ($F/\pi R^2$ is a function of x/R independent of R).

In the second region of the F(x) curve, $F/\pi R^2$ is again expected to depend on x/R only, but the experimentally observed linear variation can only be explained upon a detailed calculation of the stress and strain distribution. In the region adjacent to the indenter end the material is in the buckling/fracture stage of compression while further away it is still in the initial elastic stage. SEM observations (Fig. 6c) indicate that the transition between the two regions is quite sharp.

CONCLUDING REMARKS

The behaviour under indentation depends strongly on how sharp the indenter is. Nevertheless, the force-penetration depth curves are very similar, comprising in all cases two regions, the first being of higher slope and associated with the penetration of the extremity of the indenter, i.e. the tip of a sharp indenter or the terminal part of a flat ended indenter. The other region has a nearly constant smaller slope. It is associated with penetration of the shank in sharp indenters and is due to the increasing friction force. With blunt indenters, the linearity of this region is somewhat unexpected and is not due to friction, but to the progressive compression of the material pushed under the end of the indenter. Other experimental findings that are worth noting are: 1) the stress F_0/A_0 at the end of the initial transient is fairly independent of tip shape, except for sharp indenters with very small apperture angles, for which $\mu \text{ctg}\alpha \geq 1$; 2) the slope λ/P_0 of the second region of the F(x) curve is about the same for sharp and blunt indenters.

An indenter can be classified as sharp if it advances by pushing the material laterally, and no material is expelled at the tip indenter exit. The hole produced by the indenter can seal off when it is removed, provided the strains are recovered. With a blunt indenter there is material expelled and a cavity is left in the specimen.

The frontier between sharp and blunt depends on the geometry of the indenter tip, more precisely, on the stress distribution under the indenter tip (Fig. 12). A flat indenter (Fig. 12a) concentrates the stress at its edge and fracture occurs there possibly by shear. In the other extreme of a very sharp indenter (Fig. 12c) the breaking stresses are larger at the indenter axis and fracture, possibly by tensile hoop stresses, occurs at or very close to the axis. As illustrated in Fig. 12 the transition between sharp and blunt has to do with the location of the maximum breaking stresses, which moves away from the axis as the indenter becomes blunter. A very thin indenter even if with a flat end behaves as a sharp indenter; and a tipped indenter with a large apperture angle α behaves as a blunt indenter (Fig. 12b).

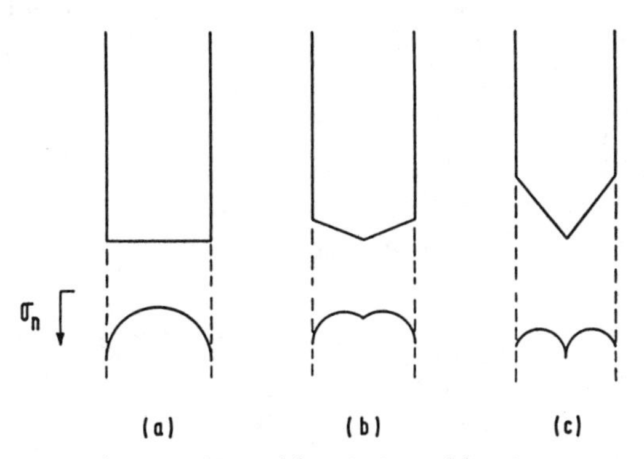

Fig.12 - The distinction between blunt (a) and sharp (c) indenters is related to the distribution of the stresses that may cause fracture. Indenter (b) behaves as a blunt indenter, because the stresses are largest at the edge, rather than at the axis.

APPENDIX

The normal stress on a nail shank

We shall attempt to calculate the normal stress σ_n on the shank of a sharp indenter, relating it to the compression properties of the nailed material that can be measured in simple, uniaxial compression tests. The stress-strain curve of the polyurethane foam used in the experiments is shown in Fig. 8a. It was obtained at a strain rate of $8.8\text{x}10^{-4}\text{s}^{-1}$. We shall use, for the calculations, the stress-strain relations shown schematically in Fig. 8b, which comprise three regimes. There is an initial, elastic regime, for which we take the constitutive equation:

$$\sigma = E\epsilon \qquad \sigma \leq \sigma_b \tag{A1}$$

assuming a linear elasticity. At the end of this regime ($\sigma=\sigma_b$ and $\epsilon=\epsilon_0$) a plateau regime starts which is characterized by the progressive buckling, eventually accompanied by fracture, of the cell walls. There is a buckling wave which propagates along the specimen. The deformation is not homogeneous as in the first regime. Finally there is a regime of high slope which starts when all cells have buckled. This regime starts at a strain ϵ_b and can be approximated by

$$\sigma = \sigma_b + E' (\epsilon - \epsilon_b) \qquad \sigma \geq \sigma_b \tag{A2}$$

where E' is much larger than E. We take compression stresses and compressive strains as positive.

The stress field around the circular nail (radius R) can be calculated using elementary elasticity theory. There is a region around the nail, of radius r_b , in which the material is in the high slope regime described by eq. A2. Outside this region and up to the outer surface, $r = r_1$, the constitutive eq. A1 applies. The stress field in polar coordinates is given by [e.g. 10]

$$\sigma_r = \frac{A}{r^2} + B$$
$$\sigma_\theta = -\frac{A}{r^2} + B \tag{A3}$$

where A and B are constants to be determined in each region. If the radial stress at the nail surface is σ_n and that at $r=r_b$ is σ_b we have, assuming that $\sigma_r=0$ at $r=r_1$, the following equations that fulfil these conditions on σ_r:

$$r_b \leq r \leq r_1 \qquad \sigma_r = \frac{\sigma_b r_b^2}{r_1^2 - r_b^2}\left[\left(\frac{r_1}{r}\right)^2 - 1\right]$$
$$\sigma_\theta = \frac{\sigma_b r_b^2}{r_1^2 - r_b^2}\left[\left(\frac{r_1}{r}\right)^2 + 1\right] \tag{A4}$$

which gives $(\sigma_r)_{r_b} = \sigma_b$ and $(\sigma_r)_{r_1}=0$; and

$$R \leq r \leq r_b \qquad \sigma_r = \frac{R^2 r_b^2}{r_b^2 - R^2}(\sigma_n - \sigma_b)\frac{1}{r^2} + \frac{\sigma_b r_b^2 - \sigma_n R^2}{r_b^2 - R^2}$$
$$\sigma_\theta = -\frac{R^2 r_b^2}{r_b^2 - R^2}(\sigma_n - \sigma_b)\frac{1}{r^2} + \frac{\sigma_b r_b^2 - \sigma_n R^2}{r_b^2 - R^2} \tag{A5}$$

which gives $(\sigma_r)_R = \sigma_n$ and $(\sigma_r)_{r_b} = \sigma_b$.

There are two conditions that have to be introduced to determine r_b and σ_n. The first is related to the continuity of the circumferential strain ϵ_θ. This implies the continuity of σ_θ, provided that the transverse Young's modulus in tension is the same in the buckled and unbuckled material, which we assume to be the case. The continuity of σ_θ for $r=r_b$ gives

$$\frac{r_1^2 + r_b^2}{r_1^2 - r_b^2} = \frac{-(R^2 + r_b^2) + \frac{2\sigma_n}{\sigma_b} R^2}{r_b^2 - R^2} \tag{A6}$$

The second condition is

$$R = \int_R^{r_1} \frac{\epsilon_r}{1-\epsilon_r}\, dr \tag{A7}$$

meaning that the material in the region occupied by the nail has been compressed radially with a radius change equal to R (the change in r_1 is neglected, i.e., $r_1 >> R$). The integrand in eq.A7 is the radial length change of an element dr, the radial strain in which is ϵ_r. Using the constitutive equations A1 and A2 relating ϵ_r to σ_r and eqs. A4 and A5 for σ_r , the integrand in equation A7 can be calculated and the resulting equation, combined with A6, can in principle be solved to calculate r_b/R and σ_n/r_b for a given R/r_1 and the material properties.

For $r_1 \to \infty$, the solution can easily be found. Eq. A6 gives•

$$\left(\frac{r_b}{R}\right)^2 = \frac{\sigma_n}{\sigma_b} \tag{A8}$$

while A7 leads to the following equation for $r_b/R=x$:

$$1=\frac{\sqrt{C}}{2} \times \ell n \frac{1+\sqrt{C}}{1-\sqrt{C}} + \frac{\epsilon_b - A}{1-\epsilon_b+A}(x-1) + \frac{kx}{2(1-\epsilon_b+A)} \ell n \frac{(1+kx)(1-k)}{(1-kx)(1+k)} \quad \text{(A9)}$$

with

$$C = \frac{\sigma_b}{E} \; ; \; A = \frac{\sigma_b}{E'} \; ; \; k^2 = \frac{A}{1-\epsilon_b+A} \quad \text{(A10)}$$

Eq.A9 can be solved numerically for x for given values of C, A and k^2 defined by A10.

Values of the material properties for the rigid polyurethane foam are

$$\frac{\sigma_b}{E} = 0.07 \qquad \frac{E'}{E} = 7 \qquad \epsilon_b = 0.7$$

Inserting in eq. A9 and solving leads to

$$\frac{r_b}{R} = 1.36 \quad \text{(A11)}$$

in reasonable agreement with the experimentally observed value. From eq. A8 we also get

$$\sigma_n = 1.8\, \sigma_b \quad \text{(A12)}$$

Since $\sigma_b \simeq 0.2$ MPa, the pressure on the indenter is $\sigma_n = 0.36$MPa, close to the value predicted by the continuum model (eq. 4). Combining with a friction coefficient μ=0.05-0.1 leads, through eqs. 3 and 8, respectively, to reasonable values for $F_0/\pi R^2$ and $\lambda/2\pi R$ comparable to the experimently determined values (Fig. 2).

REFERENCES

1. M. Wilsea, K. L. Johnson and M. F. Ashby, Int. J. Mech. Sci. 17, 457 (1975).
2. S.A.L. Salem, S.T.S. Al-Hassani and W. Johnson, Int. J. Mech. Sci. 17, 211 (1975).
3. L.J. Gibson and M.F.Ashby, Cellular Solids - Structure and Properties, 1st. ed. (Pergamon Press, 1988) ch.5, p.150.
4. E.G. Stern, Trans. ASME 72, 987 (1950).
5. M. Noguchi and H. Sugihara, Bull. Wood Res. Inst. 25, 1 (1961).
6. M. Noguchi and H. Sugihara, J. Jap. Wood Res. Soc. 6, 252 (1961).
7. T.A. McMahon and J.T. Bonner, On size and life (Scientific American Library, Freeman and Co. New York, 1983).
8. R. F. Bishop, R. Hill and N.F. Mott, Proc. Phy. Soc. 57, 147 part 3 (1945).
9. M.J. Forrestal, K. Okajima and V.K. Luk, J. Appli. Mech. 55, 755 (1988).
10. S.P. Timoshenko and K.W. Kriger, Theory of plates and shells, 2nd ed. (Mc Graw-Hill, 1989) ch.3.
11. K. L. Johnson, Contact Mechanics, 1st ed. (Cambridge University Press, 1985) ch.3.

ANISOTROPY AND SIMPLE MECHANICS OF THE FLESH OF APPLES.

JULIAN F. V. VINCENT
Biomechanics Group, Departments of Zoology and Engineering, The University, Whiteknights, PO Box 228, READING, RG6 2AJ, UK

ABSTRACT

Although its morphology is complex, the mechanical properties of the flesh of apples can be understood qualitatively in terms of the way the cells are assembled. There is no quantitative model for structures of turgescent cells which are not totally adherent.

Introduction

The flesh of apples is composed of thin-walled (1 micrometer or so) cells about 0.1 mm diameter (parenchyma) prestressed by an internal (turgor) pressure of about 0.5 MPa. Previous workers have all assumed that the cells are arranged isotropically and have therefore taken samples for mechanical tests with no regard for orientation and little regard for position within the fruit. None of them has found any correlation between density and mechanical properties, which is counter-intuitive. This study was undertaken in order to understand the nature of cellular materials in which the degree of adhesion between cells varies.

Initial observations on the morphology of apple flesh

A simple experiment (originally designed to enable the measurement of density by replacing the air in the spaces between the cells with water by floating slices of apple in isotonic sugar solution and applying a vaccum) showed that there are interconnecting air spaces radiating from the centre of the apple, but that the periphery of the apple contains isolated air spaces [1]. These simple observations were reinforced by direct SEM pictures of the spaces and measurements of the aspect ratio of the air spaces, which varies from unity in the peripheral zone to 20 or more towards the centre [2]. Thus the cells are arranged in two main ways; radially elongated with interconnecting spaces between, so that they appear as radial columns, and more spherical and arranged more randomly around isolated spaces.

Materials and methods

The stiffness of small cylinders of apple flesh, taken radially from the apple (with the air spaces running along the specimen), was measured in torsion by gluing the cylinder (using cyanoacrylate glue) between the platens of a modified

Deer rheometer and applying a sinusoidal strain of 0.1% at 0.1 Hz. Water was lost from the specimen during the experiment, so three determinations of G were made over a period of 5 minutes and extrapolated to zero time to find the value of G of the sample at full water content. There was no correlation between density and stiffness samples taken orthogonal to the radial air spaces. Several varieties of apple, early and late, were tested. In order to see the overall morphology and arrangement of cells in the flesh of the apple, sections 1 mm thick, cut on a bacon slicer, were photographed in an X-ray microscope. Overall density of apples was measured on entire fruits using Archimedes' principle.

Mechanical consequences of morphology

The density (measured by weighing cylinders of apple flesh 5 mm diameter and 10 mm long) of the radially orientated material is 0.8 to 0.95; adjacent cells therefore must have a significant area of the walls of adjacent cells adhering. The more random material has a density of 0.5 to 0.7 and occurs peripherally. The density of a regular close-packed array of spheres is 0.7405 so the peripheral material, since it is mechanically coherent, must have adjacent cells adhering in some controlled manner rather than just pressed together.

All varieties of apple are pollinated during a period of about two weeks in May or June (depending on the weather). Therefore "early" and "late" apples can spend very different amounts of time growing. Apples which mature early in the season (July in the UK) have been growing for 6-8 weeks. They have a preponderance of the less dense, more random, material and are less stiff (G = 1 - 2 MPa). Late maturing apples (harvested in November) have been growing six times longer, have much more of the well-orientated material and are much stiffer (G = 3 - 6 MPa), although a gradient of density remains. Therefore stiffness and density are associated with slower growth through differing cellular organisation. Stiffness increases as the square of density, so may be a function of the total area of mutual contact of adjacent cells. In some varieties of apple (e.g. Norfolk Beefing a late apple), stiffness and density are closely related irrespective of the morphology; in others (e.g. Cox a mid-season apple) the outer parenchyma is stiffer at a given density. This suggests that the cells in the outer parenchyma may have a greater proportion of their total surface area in contact.

An over-ripe apple swells and becomes 'mealy'; frequently it swells so much that the skin splits. Individual cells can easily be removed, giving a rough feel between the fingers. The dependence of stiffness on density is less: the cells swell and presumably lose mutual contact since the density of the apple is slightly reduced. The stiffness can fall by a factor of 2 or 3 whilst the turgor pressure within the cells remains constant or even increases [3]. This is most readily shown in early apples. The adhesive between the cells (pectin) tends to become more water-soluble.

Modelling plant parenchyma - some comments

None of this behaviour can yet be modelled convincingly, neither by the models produced for cellular materials in general nor by models based on turgescent cells stuck together into a tissue such as apple or potato parenchyma. Current theories of cellular (engineering) materials assume that the material is a continuum with holes in it. When biological materials are considered, this model seems to work for certain types of mammalian bone, cuttle bone and wood. It seems to work for turgid cell systems if the cells are maximally well stuck together [4]. However, it clearly will not work for natural systems composed of individual cells which are stuck together by less than the total cellular area available since there is currently no parameter either for a fraction of cell area nor for shear transfer between cells (analagous to stress transfer from matrix to fibre in a short-fibre composite). These systems are clearly also inherently unstable since they require a tensile epidermal layer around the outside of the assemblage of cells.

It is highly likely that biological cellular materials will have some useful lessons for us. At present we do not have the proper tools for their analysis.

REFERENCES

[1] J. F. V. Vincent, J. Sci. Food Agric. 47, 443 (1989).
[2] A. A. Khan, PhD Thesis, University of Reading (1989).
[3] E. Steudle & J. Wieneke, J. Amer. Soc. Hort. Sci., 110, 824 (1985).
[4] L. J. Gibson, M. F. Ashby & K. E. Easterling, J. Mater. Sci. 23, 3041 (1988).

FRACTURE PROPERTIES OF AN ANISOTROPIC BIOLOGICAL CELLULAR MATERIAL - APPLE FLESH

ALI A. KHAN* and JULIAN F.V. VINCENT
Biomechanics Group, Departments of Engineering and Zoology, University of Reading, Whiteknights, Reading, RG6 2AY.
*Unilever Research Fellow.

ABSTRACT

The texture of apple flesh is important in assessing its eating qualities. The texture in turn is related to the structure of the parenchyma. The parenchyma cells of the fruit are arranged in radial quasi-columnar form with radial spaces in between. This anisotropy has a marked effect on the fracture properties such that it is much easier to drive a crack between the columns (radially) than to drive it across them (tangentially). The fracture tests used were simple crack-opening tests under tension or using a wedge. This difference was also detected by a taste panel. The radial spaces ease the passage of cracks travelling along them, and act as crack stoppers for cracks travelling at right angles to them. They also allow the cells to deform more in one orientation more giving the structure ductility and making the apple tougher in that orientation. It is possible to increase this effect by controlled damage such as slow freezing which causes the intercellular spaces to expand increasing the crack-stopping mechanisms and increasing the ductility, therefore increasing the fracture toughness. Toughness first increases, then decreases with increasing damage. This effect can be mimicked with brittle paper: fracture toughness of tracing paper initially increases if holes are punched randomly in it.

INTRODUCTION

One of the most important components of texture of food as perceived in the mouth is the way in which it fractures during biting. The fracture properties of a material largely depend on its structure. Apple parenchyma was assumed to be isotropic and homogeneous with the cells evenly in contact with each other [1]. Mature apple parenchyma has a density less than 1, usually between 0.5 and 0.9 [2]. The parenchyma contains large intercellular spaces, up to 4000 μm in length and 100-200 μm in diameter, mostly filled with air, and clearly visible under a microscope [3]. The volume of the intercellular spaces has been estimated as about 20-25% [4] or 27% for Granny Smith [5].

Recent work carried out by [6] has reported the detailed structure of the apple cortex. The cells immediately underneath the surface of the fruit are small (50 μm) and radially flattened. Progressing towards the core, the cells gradually increase in size to a maximum of 200-300 μm in length (depending on the variety) at 5-10 mm from the surface. The inner cells become increasingly radially elongated and are organised into radial columns of cylindrical cells stuck end-to-end diverging from near the centre of the fruit towards the periphery. There are also radiating vascular strands [7]. Between columns of cells there are numerous radially elongated spaces. Near the surface of the apple the intercellular spaces are about 100-200 μm in diameter and have an Aspect Ratio (AR) of 1. Progressing inwards, the spaces, like the cells, elongate radially, lying in between adjacent columns of cells and the AR increases somewhat linearly to about 10 [8].

This arrangement of cells, intercellular spaces and vascular tissue makes the apple flesh anisotropic and heterogeneous. This paper deals with the influence of such structural anisotropy on the fracture properties of the flesh.

MATERIALS AND METHODS

Apples (*Malus sylvestris* Miller) varieties: Bramley, Cox's Orange Pippin, Gloster, Norfolk Beefing and Rock Pippin were obtained from the Institute of Horticultural Research, East Malling, and Royal Horticultural Society Gardens, Wisley. They were stored for no longer than four weeks at 4°C in sealed polythene bags until required.

Mechanical tests

Wedge penetration tests were done on rectangular blocks of flesh measuring 12 x 12 x 6 mm cut from the cortex. A 30° wedge mounted on an Instron was driven through these specimens at a constant speed of 1 mm/min until a reasonable sized free-running crack had developed ahead of the wedge tip. Specimens were cut so that the wedge cuts them either radially or tangentially. The energy required and the crack area were used to calculate the fracture toughness (R). Twelve tests were done on each variety of apple in each orientation. Wedge tests were also carried out in much smaller samples cut from the inner and outer parts of the cortex to measure differences in mechanical anisotropy between highly orientated inner cortex and slightly orientated outer cortex.

Tensile crack-opening tests were carried out on strips of apple flesh measuring 8 x 6 x 2 mm and stuck down lengthwise onto aluminium plates which were then gripped in the Instron. The specimen was stretched at 0.5 mm/min. until a crack starting at the notch transversed the entire specimen. Specimens were cut so that the crack either passes radially or tangentially. Fracture toughness (R) was measured as before.

RESULTS

As the wedge cuts and bend the two halves apart it stores strain energy until a crack is initiated at the tip. In a radial orientation the creack extends away in the samw direction as the wedge and the force falls.

In tangential orientation the wedge has to penetrate much further to initiate a crack indicating that a greater amount of energy has to be stored and hence greater fracture toughness. Usually the crack starts at the tip of the wedge but it is soon deflected at right angles. The wedge continues cutting and the force reading does not fall. Figure 1 shows the anisotropy in fracture properties of apple parenchyma. The differences are very apparent. Toughness is almost 40% higher tangentially than radially. It requires a greater degree of wedge penetration tangentially to store sufficient strain energy to initiate a crack.

Fracture properties determined by tensile tests give very similar numbers as the wedge penetration tests. A tangentially travelling crack takes a higher force and deformation of the specimen to propagate and the material has a significantly higher fracture toughness in that orientation.

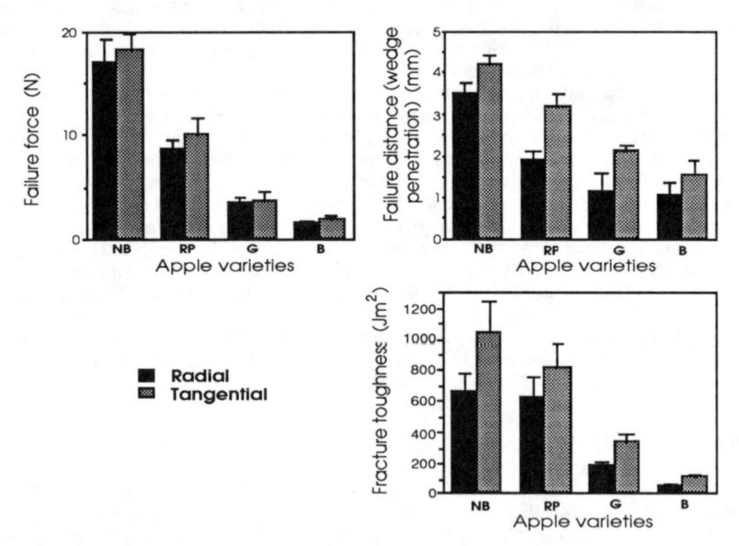

Figure 1: Fracture properties of the flesh of a variety of apples determined by wedge (30°) penetration tests (NB = Norfolk Beefing; RP = Rock Pippin; G = Gloster; B = Bramley). (n = 12; $p < 0.01$)

The inner cortex has a highly orientated structure than the outer cortex and hence the anisotropy in its toughness (R) is much pronounced (for Gloster apple: inner cortex: radial R = 263 Jm^{-3}, tangential R = 686 Jm^{-3}; outer cortex: radial R = 185 Jm^{-3}, tangential R = 294 Jm^{-3}).

Anisotropy in fracture properties is also detectable in the mouth. All panellists found that to bite a block of apple flesh tangentially requires a much greater force and energy and twice the tooth penetration. They did not initially know the outcome of the test.

Anisotropic differences also show up in the compressive properties of the flesh. Cylinders cut with their central axis along the radius of the fruit (radial) when compressed slowly show failure by an unstable collapse of one or two layers of cells at right angles to the direction of the force. Cylinders cut with the axis parallel to the core (tangential) show a stable shear failure at 45° to the direction of force. The stress and strain required to cause failure are also about 25% higher in this orientation.

DISCUSSION

The implications of anisotropy of apple flesh are far ranging. The distribution and arrangement of morphological structures contribute to the texture of the flesh [8]. The radial orientation of cells and spaces makes apple parenchyma mechanically different radially and tangentially and it is likely that this would affect the oral perception of apple texture. The orientation of the sample during a mechanical test (including biting) should be considered if the correct texture is to be realised.

In a mature apple the crack travels in between cells [8]. The shape and arrangement of the cells, the intercellular spaces and other structural components therefore dictate, to varying degrees, the crack path and fracture properties. The radially elongated intercellular spaces ease the passage of cracks travelling along them by acting as built-in notches and stress concentrators. They act as crack stoppers to the cracks travelling at right angles to them. When a crack tip runs into a spaces lying across its path it will have to start a new crack on the other side. Suffícent stress has to build up again at the blunt tip requiring extra strain energy and hence increased fracture toughness and failure strain. This confirmed by the fact that the highly orientated inner cortex is much more mechanically anisotropic than the less orientated outer cortex.

Cell and space orientation may also determine modes of failure under compression. The material compressed in the direction of the cell columns (radially) behaves as an array of straws compressed along their length, i.e. there is locallised unstable buckling of cell walls producing a compressive failure band at right angles to the applied force. If the same material is compressed at 90° to the cell columns, it fails as most materials do, by shear.

The effect of voids such as intercellular spaces in increasing the fracture toughness of a material can be further illustrated by measuring the changes in its toughness as the number and dimensions of the spaces changes. It can be expected that the more and the bigger the spaces, the tougher the material. This can be done by inducing controlled damage to the material by slowly freezing it so the as ice crystals expand in the voids, they push the cells apart increasing the dimensions of existing spaces and creating new ones. When thawed, the parenchyma has a much less dense structure with the volume fraction of cells falling drastically (for the inner cortex of Gloster apple, it falls from about 0.77 to about 0.65 when frozen to -10°C and thawed). The AR of the spaces increases from about 10 to as much as 25 [8]. Slow freezing keeps the cells intact, hence the cellular properties of the material are largely unchanged. This increase in space dimensions causes a corresponding increase in the fracture toughness in the initial stages of cooling and mainly in the tangential orientation where the effect of the expanded size of the spaces is much more pronounced (Figure 2). Thus by increasing the void volume, the structure is made tougher. There is a subsequent decrease proportional to the reduction in volume fraction and load bearing area.

Various explanation for this increase and decrease in toughness were put forward. It can be either due to increase crack-stopping mechanisms. The crack initiation energy being greater than crack propagation energy, the more times a new crack has to start, the greater the energy requirement increasing the overall toughness. There can be increased ductility and plastic deformation acting as an energy dissipating mechanism. These effects can be effectively mimicked with paper. Brittle paper like tracing paper with holes randomly punched in it to represent the spaces in the apple flesh and subjected to a tensile test shows an increase in fracture toughness at first with increasing number of holes. There is a subsequent decrease corresponding to the reduction of material. Ductile paper like tissue paper shows no initial increase in toughness with increasing number of holes (Figure 2). If crack initiation energy is solely responsible for the increasing toughness, both tracing and tissue paper will show an initial increase as this applies to both cases. The difference between the behaviour of the two papers is that tracing paper under tension develops locallised zones of plastic deformation around the holes (appear opaque) and it is this that absorbs extra energy and increases the toughness. Tissue paper shows only global elastic deformation and hence most of the energy is released into the fracture. The more the voids, the more the energy dissipators in tracing paper and the higher the energy requirement. Eventually as the density of voids becomes very high these flow-zones start merging together and have a reduced effect. The decreasing load bearing area then causes toughness to fall.

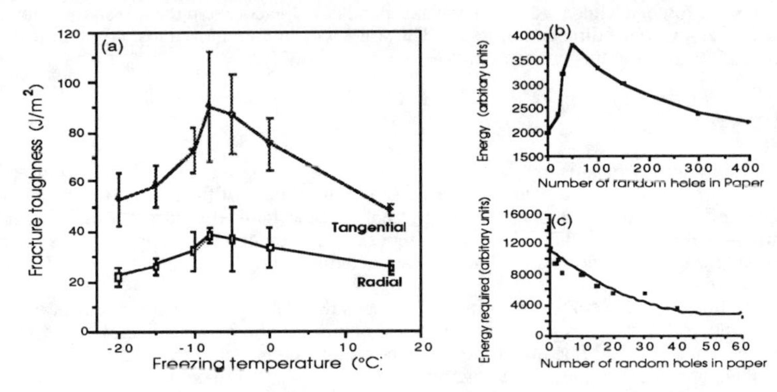

Figure 2: (a) Effect of freezing and thawing on the fracture toughness of Cox determined by wedge tests (n = 10; $p < 0.01$). Effect of increasing number of holes on the toughness of (b) tracing paper and (c) tissue paper.

As apple flesh shows a considerable degree of flow [8], it is very likely that such energy dissipating mechanisms are operating in apple flesh and flow-zones develop around the intercellular spaces.

Only the effect of voids is considered in any detail here. They have been shown to dictate to a large extent the crack-path, strength and fracture toughness of the material. The anisotropic orientation of these spaces make the material behave very differently according to the direction it splits. Other morphological factors may have an equally important role in influencing the mechanical properties of the parenchyma. These systems may well exhist in all fruit and vegetables and any structural orientation of these components makes the material mechanically anisotropic, a property very important in determining the accurate texture of the material. In turn this means that it should be possible to correlate texture more closely with the morphology and in turn to breed for texture by selecting for certain morphology/density relationships. We are currently exploring some of these ideas.

References

[1] T. T. Lin and R. E. Pitt, J. Tex. Studies 17, 291 (1986).

[2] J. F. V. Vincent, J. Sci. Food Agric. 47, 443 (1989).

[3] R. M. Reeve, Food Res. 18, 604 (1953).

[4] C. Sterling, Recent Advances in Food Sci 3, Biochemistry and Biophysics in Food Research, ed R. M. Leitch and D. N. Rhodes, Butterworths, London, 259 (1963).

[5] J. M. Bain and R. N. Robertson, Aust. J. Sci. Res. B4, 75 (1951).

[6] A. A. Khan and J. F. V. Vincent, J. Sci. Food Agric. 52, 455 (1990).

[7] U. Tetley, J. Pomol. Hort. Sci. 9, 278 (1931).

[8] A. A. Khan, PhD Thesis, University of Reading (1989).

PART II

Microstructural Influences on Elasticity and Fracture

HIGH STRENGTH, POROUS, BRITTLE MATERIALS

J.S. Haggerty*, A. Lightfoot*, J.E. Ritter** and S.V. Nair**
*Massachusetts Institute of Technology, Cambridge, MA 02139
**University of Massachusetts, Amherst, MA 01003

ABSTRACT

Contrary to existing models, strengths need not be a strong function of porosity for intermediate density, brittle materials. Flaw sizes can remain small (<50μm) if the void space is distributed uniformly in minimum dimension pores. For RBSN, fracture toughness decreases linearly with porosity for 0< porosity <40%. Strains to failure and specific strengths of these materials are higher than fully dense counterparts.

INTRODUCTION

Several expressions have been proposed for describing the relationships between mechanical properties and porosity in brittle materials. They apply to either high or low porosity levels; none have been derived specifically for intermediate levels of porosity.

The most familiar of the expressions generally applied to nearly fully dense materials (P<10%) was first proposed by Ryshkewitch [1]; it relates strength to porosity through a simple exponential relationship. Similar expressions are used to relate porosity to modulus, fracture toughness and hardness of less than fully dense brittle materials [2]. Although poorly justified theoretically, the empirically observed exponential dependence on porosity shows why fractional density has been maximized for applications requiring high strengths. Some of the models for mechanical properties of highly porous materials (P>85%) have a better theoretical basis. These apply elastic thin-beam theory to the idealized 2- and 3-dimensional structures and predict exponential or power law relationships between the fractional density and geometrical features of the cell structure in the porous material. The best known of the high porosity level models originate from Ashby's group [3]. Other models have been proposed for limited ranges of porosity levels by Nielsen [4], Gent & Thomas [5], Rice [6], Woignier [7], and Green [8].

All of the expressions predict extremely sensitive relationships between properties and porosity levels. However, none apply over a wide density range, at best the constants are poorly related to microstructural features, and all of the constants change with relatively minor variations in the materials. In general, the modulus and hardness data are better behaved than strength, fracture toughness and fracture surface energy data which exhibit a great deal of variation about predicted values.

We [9] and others have demonstrated that the strengths of intermediate density (40%>P>20%) brittle materials are described by the same fracture criteria as fully dense materials. Tensile strength (σ) is:

$$\sigma = \frac{Z}{Y} \frac{K_{IC}}{\sqrt{a}} \qquad (1)$$

where K_{IC} is the fracture toughness, **a** is the half-flaw size and Z and Y are constants relating to the location and shape of the strength controlling flaw. Although this expression does not predict strength directly, it does provide a basis for making improvements through the effects that processing variables and resulting microstructural features have on K_{IC} and **a**. It is important to note from equation 1 that strength is not directly related to porosity.

Our research has reduced the flaw dimensions and explored the variation in K_{IC} through controlled processing variables. There appeared to be no fundamental reason why porous materials must contain large flaws if even high levels of porosity were distributed uniformly as very small pores. Also, the observed variation in K_{IC} values suggested opportunities for improving the toughnesses of intermediate density porous materials once the underlying factors are understood. It was obvious that single-parameter, porosity-level models are inadequate descriptions of both strength and toughness.

We have studied reaction bonded Si_3N_4 (RBSN) because it could be used advantageously in many demanding applications if strength and oxidation resistance were improved. Intrinsically, the material must remain porous until the end of the nitriding process so N_2 can penetrate into interior regions. Thus, the material is an excellent model material for exploring opportunities to improve the mechanical properties of intermediate density porous brittle materials.

EXPERIMENTAL

Reaction bonded silicon nitride samples were made from SiH_4 derived Si powders having unusual characteristics. The spherical powders are small (~0.2μm), uniform in size (geometric standard deviation ~ 1.4), pure (total impurities < 50ppm), and free of agglomerates [10]. Additionally, the surfaces exhibit strong Si-H bonding which prevents spontaneous oxidation and permits nitriding to be completed in short, low-temperature cycles. Samples were made by colloidally pressing disks from powders dispersed in alcohol. The nitriding schedules (8-20h for heating between 1200-1400°C) are much less severe than those typically required for RBSN, which may extend up to 150h at T > 1500°C. If as-synthesized purities are maintained, samples will fully nitride in schedules like 1h at 1150°C or 10min at 1250°C [11,12]; with pre-nucleation they will nitride to completion in 1h at 1050°C. Typical characteristics of the samples and the techniques used to make the determinations are summarized in Table 1.

The properties of the RBSN samples were defined as follows. Fracture toughnesses were measured by the indentation technique, by identifying the fracture origin and using fracture mechanics to define K_{IC} from the fracture stress at the site, and by analysis of mirror constants. Biaxial strengths were measured using the ball-on-ring technique. Hardnesses were measured by the Vickers indentation technique. Elastic moduli were calculated from the hardness measurements.

Table 1. Summary of typical MIT-RBSN microstructural characteristics

- Density: ~ 67 and ~ 75%
- 100% converted to Si_3N_4 (XRD, TGA)
- α/β ratio: 2 -14 (XRD)
- Grain Size ~ 30nm (XRD, TEM)
- Structure:
    - Solid Phase: 0.2-0.7 μm (BET, SEM)
    - Interparticle Pores: 0.1-0.2 μm (SEM, Hg porosimetry)
    - Pore Channels: 0.01-0.06 μm (Hg porosimetry)
    - Isolated Pores: 5-70 μm (SEM)

Initial samples made from these high purity Si powders exhibited unusually rapid nitriding kinetics but did not achieve superior strengths. We systematically identified the strength controlling defects [13] and eliminated them through improved processing techniques. The defects we identified originated both during sample fabrication and during surface finishing. Careful lapping and removal of surface damage induced by prior coarser grits was required before strengths became limited by processing flaws. The developed surface finishing procedures reduced surface flaw dimensions from ~275μm to ~20μm. Processing flaws originated from several sources listed in Table 2. By employing the indicated modified procedures, flaw dimensions were reduced from values that approached millimeters to tens of microns. Occasionally, samples broke from flaws that were less than 10μm in diameter. Once these improved processing techniques were employed, we systematically investigated the effects modified process variables had on properties.

Table 2. Sources and means by which flaws were eliminated from MIT-RBSN

Defect	Source	Modified Procedure
Cracks	Colloidal Pressing	Reduce pressure application rate Reduce pressing pressure Reduce pressure release rate
	Die-Release	Cold isostatic pressing (CIP'ing)
	Handling	No green-part measurements Strip membranes without gloves Eliminate vacuum exposures
	Nitriding Gradients	Modify nitriding schedule
Laminations	Colloidal Pressing	Use unidirectional flow
Pores	Ultrasonic Cavities	Substitute ultrasonic bath for probe
Inclusions & Contaminates	Die (Cu, Zn)	Improve finish & technique
	Sonicator (Fe, Ni, Cr)	Substitute ultrasonic bath for tip
	Filter (Na, Ca, K)	Avoid scraping filter when removing powder

Table 3. Mechanical Properties of RBSN and RBSN plus SiC Composite Samples

Sample	Strength (MPa)	Toughness Indentation ($MPa.m^{1/2}$)	Toughness Fractography ($MPa.m^{1/2}$)	Mirror Constant ($MPa.m^{1/2}$)	Hardness (GPa)	Modulus (GPa)	Flaw Size a (μm)
RBSN							
Methanol (with CIP)							
Optimization Gp.	441 ± 129	2.4 ± 0.3	2.6 ± 0.3		8.9 ± 0.5	156 ± 23	34 ± 29
Baseline Gp.	401 ± 107	2.3 ± 0.1	1.9 ± 0.1	3.8 ± 0.2	9.1 ± 0.4	160 ± 23	26 ± 19
Octanol (without CIP)	592 ± 148	2.5 ± 0.1	2.7 ± 0.1	4.6 ± 0.6	10.2 ± 0.4	182 ± 13	23 ± 11
RBSN+12% Si rich SiC	438 ± 54	2.4 ± 0.1	1.7 ± 0.02	3.2 ± 0.1	7.8 ± 0.5	150 ± 10	33 ± 10
RBSN+10% C rich SiC	517 ± 130	2.4 ± 0.2	2.3 ± 0.03	4.1 ± 0.3	9.8 ± 0.7	166 ± 14	35 ± 10
RBSN+20% C rich SiC	313 ± 66	2.2 ± 0.1	2.0 ± 0.1	3.4 ± 0.2	7.4 ± 0.4	115 ± 10	29 ± 9

RESULTS

The mechanical properties for several types of RBSN samples and for a group of RBSN + SiC particulate composites are summarized in Table 3. The strengths of the RBSN and RBSN plus 10-12% SiC are both unusually high compared with conventional RBSN, which typically exhibits strengths in the 75-150 MPa range. Two points about the strength results in Table 3 should be emphasized. First, the strengths of the octanol processed RBSN are significantly higher than have been achieved previously. The average strength of this set of samples (592 MPa) is >150 MPa higher than any average strength we have observed for any previous batch of samples; also, the maximum strength exhibited by this set (860 MPa) is more than 160 MPa higher than we have ever measured. Second, the high strengths of the RBSN plus 10-12% SiC samples indicate that highly perfect microstructures were achieved in green bodies made from very dissimilar powders.

The strengths of the MIT-RBSN are plotted in Figure 1 as a function of porosity along with the results of three literature surveys of observed strength-density relationships for Si_3N_4. This figure shows that MIT-RBSN's are 3-5 times stronger than average values reported for the same density levels. They also show that the 75% dense material achieves strength levels normally associated only with fully dense Si_3N_4's that typically have much higher toughnesses. The 67% dense RBSN was produced before systematically eliminating processing defects; strength degradation resulted more from the effect of larger flaws than a reduction of K_{IC}. It should be entirely feasible to fabricate samples having these lower densities with the smaller flaw dimensions achieved in the higher density samples. This should permit nearly equal strengths at lower densities, contrary to the strength-density dependence implied in Figure 1.

The flaw dimensions summarized in Table 3 show that all of these sample sets (porosity ≈ 25%) are free of large processing flaws, and that both the average flaw sizes and the standard deviations are approximately the same for all sample types. This result is encouraging, especially for the RBSN/SiC composites. It is difficult to achieve uniform, defect free green bodies from slips containing such different powders, particularly since >95% of the particles are SiC because of their small size.

Table 3 also reports the results of the fracture toughness measurements made by two techniques as well as the mirror constant which is proportional to fracture toughness [14]. We have observed that the fracture toughnesses defined by the crack lengths emitted from indentations are essentially constant for all of the samples studied. In contrast, the toughnesses indicated both by the fractography and by mirror constant analyses vary considerably for the samples investigated; analyses show an excellent correlation between the fractography and mirror constant values. These results lead us to conclude that the indentation technique is not sensitive to differences for porous materials. We have characterized these samples extensively but have not identified any correlation between microstructural properties and fracture toughness. Although toughnesses have been quite uniform within a specific batch of samples, they have exhibited batch to batch variation. For instance, a batch of samples made to duplicate the Octanol group shown in Table 3 exhibited a low toughness of 1.9 $MPa.m^{1/2}$ by both fractographic analyses and by its mirror constant. These results demonstrate that porosity level alone provides an inadequate prediction of toughness.

Comparison of the maximum K_{IC} values we observed with the value Rice and coworkers [2] projected for fully dense, high purity Si_3N_4 gives one measure of porosity's effect. Rice showed that most of the toughness improvements in fully dense Si_3N_4 result from residual stresses between the glassy grain boundary phase and the crystalline Si_3N_4, an improvement that intrinsically degrades high temperature properties. They reported a value of 3.9 $MPa{\cdot}m^{1/2}$ for fully dense high purity Si_3N_4. As shown in Figure 2, our best K_{IC} results at 67 and 75% density levels give a linear, rather than exponential, relationship with respect to Rice's value.

The combination of the high strengths and low specific gravities (~2.4) gives RBSN two important advantages relative to other structural materials. With the strengths we have achieved, specific strengths (up to 3.6×10^5 m^2/sec^2) equal or exceed those of the best metallic and intermetallic alternatives [15] while having substantially higher temperature capability. The reduced modulus, resulting from the low specific gravity, combined with the high strengths we achieved, gives a strain to failure (~0.3%) that is much higher than conventional RBSN (~0.07%) or even state-of-the-art, fully-dense Si_3N_4 (~0.2%). This material illustrates the enormous potential that less than fully dense materials have for structural applications, as monoliths and as matrices of ceramic matrix composites.

We have also studied the high temperature oxidation rates and the effect of oxidizing exposures on the room temperature strength of the MIT-RBSN [16]. These results show that at both 1000 and 1400° C, the SiH_4-originating MIT-RBSN exhibited average oxidation rates between 5-20 times lower than more dense RBSN samples made from conventional Si powders and also ~10 times lower than is observed with fully dense HPSN. Oxidation had no significant effect on the strength controlling flaw population in the high purity MIT-RBSN.

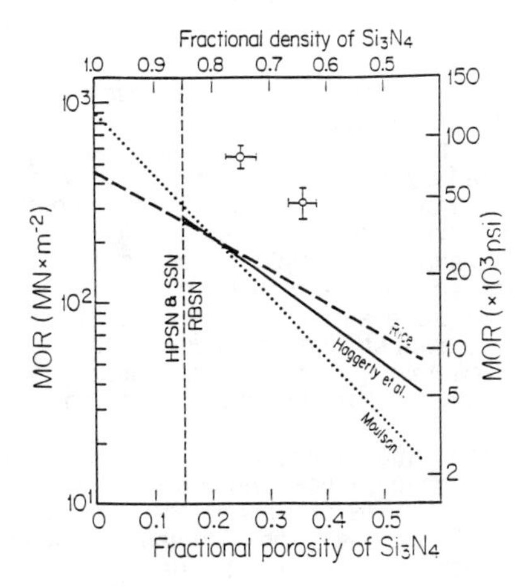

Figure 1. Strength of Si_3N_4 as a function of porosity. The MIT-RBSN strengths are 3-5 times stronger than values shown for three literature surveys.

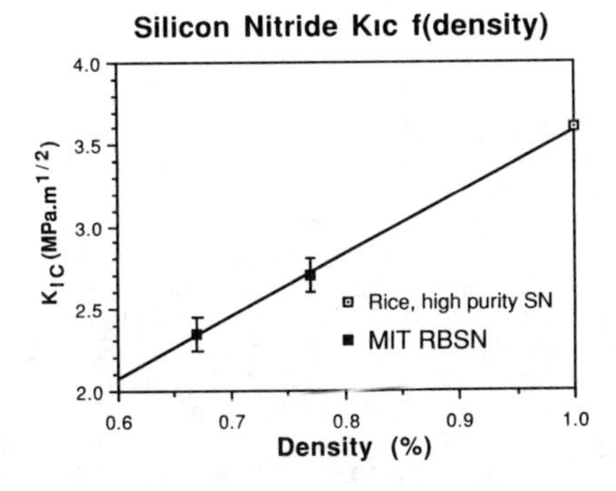

Figure 2. Fracture toughness vs. fractional density for high purity Si_3N_4. The results exhibit a linear rather than an exponential relationship.

ACKNOWLEDGEMENTS

Many students and staff cited in the references have contributed to this research. Currently the research is supported by ARO (DAAL-03-88-K0099), NASA Lewis Research Laboratory (NAG-3-845) and DOE/BES(DE-FG02-87ER45313). All contributions are gratefully acknowledged.

CONCLUSIONS

The RBSN results have confirmed our hypothesis that the properties of porous brittle materials could be improved substantially from traditional values. Our strength and oxidation results with the high purity RBSN have shown that both properties can equal or exceed those of fully dense Si_3N_4 made by conventional techniques. In combination, they indicate that the high purity RBSN can probably be used in structural applications in oxidizing environments up to its dissociation temperature, ~1800°C. These results show that simple porosity-level models for properties are inadequate descriptions, and that the models obscure the potential that can be realized through superior process control. These property improvements were achieved by making parts with small channel diameters, high purities, and small defect dimensions, while retaining reasonable fracture toughness values. If porosity is distributed uniformly in small diameter pores, it need not have any adverse effect on properties.

Fracture toughness is more complex. The results revealed variations in fracture toughnesses that have not yet been correlated with specific microstructural features or process variables. The maximum observed values show that the decrease in toughness with increasing porosity can be much more gradual than exponential models predict. There is almost certainly an opportunity for improving the toughnesses of the porous materials when the microstructural basis is understood.

Rather than degrading properties, porosity can be seen as improving the materials by raising specific strengths, lowering the modulus, and raising the strain to failure. These results appear applicable to other porous brittle materials if the required microstructural perfection is achieved. Porous brittle materials can indeed be included in the high performance category.

REFERENCES

1. E. Ryshkewitch, J. AM. CERAM. SOC. 36 (2), 65-68 (1953).
2. R.W. Rice, K.R. McKinney, C. Cm. Wu, S.W. Freiman, and W.J.M. Donough, J. MAT. SCI. 20, 1392-1406 (1985).
3. M.F. Ashby, METAL. TRANS. A, Volume 14A, pp. 1755-69, (1983).
4. L.F. Nielsen, J. AM. CERAM. SOC. 73 (9), 2684-89 (1990).
5. A.N. Gent and A.G. Thomas, RUBBER CHEM. AND TECH. 36 (1), 596-610 (1963).
6. R.W. Rice, J. AM. CERAM. SOC. 59 (11-12), 536-7 (1976).
7. T. Woignier and J. Phalippou, J. NON-CRYSTALLINE SOLIDS 100, 404-8 (1988).
8. D.J. Green, J. AM. CER. SOC. 66 (4), 288-292 (1983).
9. J.S. Haggerty, A. Lightfoot, J.E. Ritter, S.V. Nair and P. Gennari, Ceramic Engineering and Science Proceedings 9 (7-8), 1073-77 (1988).
10. J.H. Flint and J.S. Haggerty, AEROSOL SCI. AND TECH. 13, 72-84 (1990).
11. B. W. Sheldon, The Formation of Reaction Bonded Silicon Nitride From Silane Derived Silicon Powders, Sc.D. Thesis, Department of Materials Science and Engineering, Massachusetts Institute of Technology, Cambridge, MA, January 1989.
12. B.W. Sheldon and J.S. Haggerty, Ceramic Engineering and Science Proceedings 10 (7-8), 784-793 (1989).
13. J.E. Ritter, S.V. Nair, P.A. Gennari, W.A. Dunlay, J.S. Haggerty, and G.J. Garvey, ADV. CERAM. MAT. 3 (4), 415-417 (1988).
14. A. Lightfoot, H.L. Ker, J.S. Haggerty, and J.E. Ritter, Ceramic Engineering and Science Proceedings 11 (7-8), 842-856 (1990).
15. J.-M. Yang, W.H. Kao, and C.T. Liu, MATLS. SCIENCE AND ENGG. A107, 81-91 (1989).
16. J.S. Haggerty, A. Lightfoot, J.E. Ritter, P.A. Gennari, and S.V. Nair, J. AM. CER. SOC. 72 (9), 1675-79 (1989).

FRACTURE AND YIELD OF A POROUS COMPOSITE OF TRINITROTOLUENE AND CYCLOTRIMETHYLENE TRINITRAMINE.

Donald A. Wiegand and J. Pinto, U.S. Army Armament Research, Development, and Engineering Center, Picatinny Arsenal, NJ 07806-5000.

ABSTRACT

Composites of trinitrotoluene and cyclotrimethylene trinitramine have been studied in uniaxial compression and in triaxial confined compression. Failure by crack propagation and fracture was observed in all uniaxial studies while failure by yield and plastic flow was found for all triaxial confined compression work. The yield strength is approximately a factor of 2 greater than the uniaxial compressive fracture strength. The relationships between the fracture and yield strengths and Young's modulus indicate that all three decrease with increasing porosity. Based on this porosity interpretation, the compressive strength, the yield strength and Young's Modulus may be increased significantly by reducing the porosity.

INTRODUCTION

This paper is concerned with mechanical failure of a composite of trinitrotoluene (TNT) and cyclotrimethylene trinitramine (RDX).[1,2,3] This composite usually also contains 1% wax and is commonly known as Composition B (Comp B). TNT is the matrix material and both TNT and RDX are molecular organic (poly)crystalline solids. In this paper two modifications of the composite are also considered and the fracture and yield strengths are compared to those of the unmodified composite (Comp B). There are minor differences in compositions of the three composites which are not thought to be significant for the results reported here and are discussed elsewhere.[4] Changes in processing conditions which were made in attempts to obtain good quality casts appear to be predominantly responsible for the differences between the three composites.

Comp B is prepared by adding particulate RDX (and wax) to molten TNT and casting from the melt. During the preparation process defects such as cracks, porosity and larger voids and strain are introduced. The fracture and yield strengths and Young's modulus can be significantly affected by these defects.

EXPERIMENTAL

The equipment and the data handling and reduction procedures are described elsewhere.[1,2,3] Samples were in the form of right circular cylinders and two modes of compression were used. Samples were subjected to uniaxial compression parallel to the cylinder axis or they were confined in a tight fitting thick walled steel cylinder so that the radial strain was negligible during the applied axial compression (triaxial stress). All results presented here were obtained using a strain rate of approximately 2 s^{-1} and typical curves of strain rate versus time are given in references (1) and (2). Fracture in the uniaxial case or yield in the triaxial case occurred in 3 to 5 ms for these strain rates. All of the uniaxial data presented here were taken at 23°C but in a few cases were corrected to 35°C to allow direct comparison with the triaxial data which were taken at 35°C.

All samples were obtained from material cast in a split mold approximately 4 inches in diameter and 10 inches long.[3] The composites contain approximately 40% TNT and 60% RDX and two of the composites also contained approximately 1% wax. The two modifications of Comp B, Comp B M1 and Comp B M2, were cast using procedures similar to those for Comp B. In addition, vacuum was applied to the melts of these materials for about 12 minutes and the filled mold of Comp B M2 was vibrated for 12 seconds. All casts were radiographed and acceptance or rejection was based on the quality as determined from the radiographs.[3]

The acceptable casts from the split mold were cut into sections perpendicular to the cast axis and these sections were then further cut and machined into cylindrical samples.[3] The samples were approximately 1.5 inches long with ends flat and parallel to ± 0.001 inches. The sample diameters were 0.7520 inches or slightly less and were uniform to 0.0005 inches. All samples were radiographed after final machining and samples with cracks and/or excessive porosity were discarded.

RESULTS

The three composites have been studied in uniaxial compression and the results indicate that the compressive strength and Young's Modulus increase on going from Comp B to Comp B M1 and Comp B M2. The average compressive strengths of Comp B M1 and Comp B M2 are about 30% greater than that of Comp B. In addition, the average moduli for Comp B M1 and Comp B M2 are about 35% greater than the modulus for Comp B[3]

The compressive strengths and moduli for Comp B M1 and Comp B M2 are functions of the position in the cast from which the samples were taken and the data indicate that both increase approximately linearly on going from the top to the bottom of the casts. In contrast, the compressive strength and modulus of Comp B are independent of the position within the accuracy of the data.[3] If the uniaxial results for Comp B M1, Comp B M2 and Comp B are plotted as compressive strength versus modulus as shown in Figure 1 the results clearly indicate a correlation between the moduli and the compressive strengths. The results suggest that the correlation is between the compressive strength and the modulus with position as a parameter. The straight lines of this Figure and Figure 2 were obtained by least squares fits to the data points assuming zero intercepts (see below).

Because the yield strength is approximately a factor of two greater than the compressive fracture strength for these materials, yield is only observed when fracture is inhibited. Therefore, yield has been studied using a thick walled tight fitting steel confinement cylinder to minimize radial strain and inhibit fracture.[2,3] The average yield strengths of Comp B M1 and Comp B M2 are approximately 50% greater than the value for Comp B. Thus, the yield strength changes in the same manner as the compressive fracture strength.

Results (not shown) indicate that the yield strength for Comp B is independent of position within the accuracy of the data.[3] Only very limited triaxial data were taken as a function of position for Comp B M1 and triaxial data was not obtained as a function of position for Comp B M2. In Figure 2 the yield strength is given versus the modulus for Comp B, Comp B M1 and Comp B M2. The data of this Figure indicate a correlation between the yield strength and the modulus which is similar to the correlation between the compressive strength and the modulus of Figure 1. Because of the limited data as a function of position a correlation can only be made between the yield strength and the modulus and not between these two and position.

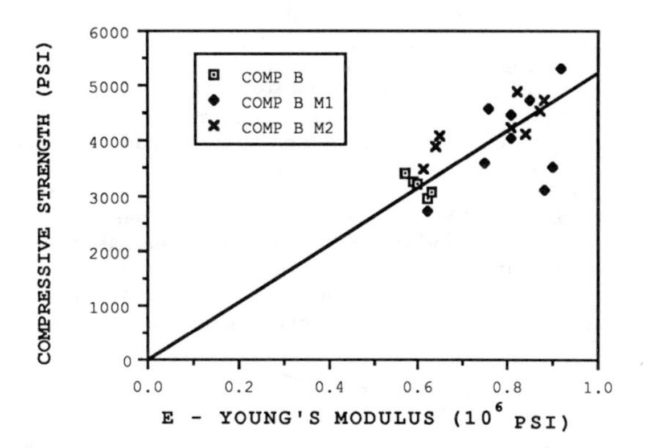

Figure 1. Compressive strength versus Young's modulus for Comp B, Comp B M1, and Comp B M2. Temperature 23°C.

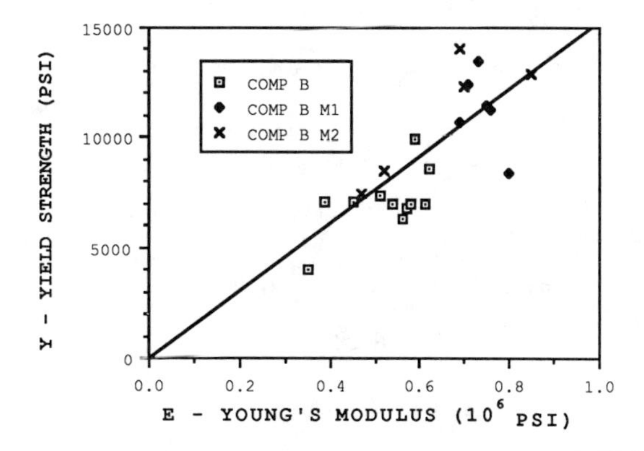

Figure 2. Yield strength versus Young's modulus for Comp B, Comp B M1 and Comp B M2. Temperature 35°C.

DISCUSSION

In order to develop an understanding of the relative strengths of the three composites for which data has been presented it is necessary to consider the effects of inhomogeneities such as porosity and microcracking. The uniaxial (fracture) results and the triaxial (yield) results are considered separately.

Many factors can influence the fracture strength of brittle solids, including porosity, microcracking, crack size, crack path considerations, the modulus and the surface energy.[5] Because there appears to be a correlation between the modulus and the compressive strength it is of interest to consider factors which also can influence the modulus. These include porosity and microcracking. Because porosity has a direct effect on the strength and the modulus and the relationship between them, it is considered in some detail. Microcracking is considered elsewhere.[4]

Densities were measured and porosities calculated for many but not all of the samples used in this work. However, for a number of reasons the porosities are not considered very reliable and are not presented.[4] Instead, emphasis is placed on the relationship between the strength and the modulus when porosity is a major factor in this relationship. The relative porosities are considered more reliable, however, and are referred to below. Unfortunately, the importance of porosity was not anticipated when the measurements were made and the density measurements were made only for routine characterization purposes.

The applied stress, σ_a, for failure has been calculated as a function of porosity for one particular pore structure and a relationship of the form

$$\sigma_a = \sigma_o e^{-b_\sigma P} \tag{1}$$

has been obtained. σ_o is the zero porosity failure stress, b_σ is a constant and P is the porosity.[6] The form of this equation has been found to describe the experimental relationship between fracture strength and porosity for many brittle materials.[5] While equation (1) was obtained for tensile failure, a similar relationship has been found to apply for compressive failure.[5] More recently the apparent modulus, E, has been calculated as a function of porosity for the same pore structure and a relationship of the form

$$E = E_o e^{-b_E P - cP^2} \tag{2}$$

has been obtained where E_o is the zero porosity modulus and b_E and c are constants.[7] Many materials can be described by this type of relationship.[5]

While these two equations are strictly only valid for a particular pore structure, they have been found to have more general applicability.[5] Thus, it is reasonable to apply them here although the pore geometry is unknown. By combining equations (1) and (2)

$$\sigma_a = E \frac{\sigma_o}{E_o} e^{-(b_\sigma - b_E)P} \tag{3}$$

$$\sigma_a \approx E \frac{\sigma_o}{E_o}\left[1 - (b_\sigma - b_E)P\right] \tag{4}$$

for

$$(b_\sigma - b_E)\, P \ll 1 \tag{5}$$

where the quadratic term has been omitted from the expression for E, equation (2). This omission and (5) are valid for sufficiently small values of P. Equations (3) and (4) predict a linear relationship between σ_a and E for sufficiently small values of porosity. The observed relationship as shown in Figure 1 is approximately linear in agreement with this model of porosity and the hypothesis that the relationship between strength and the modulus is determined by porosity. The points of Figure 1 for Comp B M1 and Comp B M2 are for samples taken from different distances for the top of the cast and these results (not shown) indicate that the porosity decreases as this distance increases as observed elsewhere.[3,8] Variations with this distance were not found for Comp B indicating a uniform cast. In addition, the relative average porosities decrease on going from Comp B to Comp B M1 to Comp B M2. The results indicate that (5) is valid.

These relative differences in average porosities can be attributed to differences in the processing conditions. A vacuum was applied to the melts of Comp B M1 and Comp B M2 for twelve minutes but not to the melt of Comp B. This treatment is expected to decrease the porosities of the former relative to the latter in agreement with observations. In addition, the filled mold of Comp B M2 was vibrated for twelve seconds while the filled molds of Comp B and Comp B M1 did not receive this treatment. The average porosity of Comp B M2 was found to be lower than the average porosities of the other two as expected for this difference in treatment. Thus, the relative porosities and so compressive fracture strengths and moduli of Comp B, Comp B M1 and Comp B M2 can be attributed to these differences in processing conditions.

Microcracking can also cause changes in both the fracture strength and the modulus. However, the relationship between the fracture strength and the modulus is not clear in this case.[9] Therefore, it is not possible to so clearly relate the observed correlation between the fracture strength and the modulus to microcracking.

Because the triaxial data and so yield strength results are not nearly so extensive as the uniaxial data the discussion of the triaxial results is more tentative than the discussion of the uniaxial results. While triaxial data is available for Comp B as a function of position in the cast, triaxial data is available for Comp B M1 only for the bottom portion of the cast and is not available as a function of position for Comp B M2. All triaxial data for Comp B M2 are for samples taken from the bottom section of the cast. The following is concerned largely with the relationship between the yield strengths and moduli of Comp B, Comp B M1 and Comp B M2 and the effects of porosity and cracking are addressed.

The general approach to the dependence of the fracture strength on porosity which is based in part on the ideas that the local stress is increased relative to the applied stress when porosity is present because of a decrease of the load bearing area and/or stress concentration effects are also applicable to yield strength considerations. Thus, to a first approximation the yield strength, Y, should be given as a function of porosity by an expression of the same form as equation (1), i.e.

$$Y = Y_o e^{-b_Y P} \tag{6}$$

and the relationship between Y and E is

$$Y = E \frac{Y_o}{E_o} e^{-(b_Y - b_E)P} \tag{7}$$

$$Y \approx E \frac{Y_o}{E_o}\left[1 - (b_Y - b_E)P\right] \tag{8}$$

for

$$(b_Y - b_E)\, P \ll 1 \tag{9}$$

using equations (2) and (6). The quadratic term in equation (2) has been omitted in obtaining (7).

The data for Comp B, Comp B M1 and Comp B M2 as plotted in Figure 2 suggest an approximately linear relationship between Y and E as predicted by equation (8) when (9) is satisfied. The data indicate that the latter is valid. The results thus suggest that the relative yield strengths of the three forms of Comp B are determined by porosities. Thus, the yield strength results can be interpreted in the same way as the compressive fracture strength results.

The yield strength may also be influenced by the presence of cracks and the reduction of the yield strength and the modulus by a particular arrangement of cracks has been calculated.[10] Cracks are taken as all aligned in one direction and all cracks have the same lengths, widths and separations. With loading perpendicular to this direction there are significant reductions of the modulus and the yield strength which increase with increases in the crack lengths. The predicted (and observed) relationship between the yield strength and the modulus with increasing crack length is not linear but could easily be used to describe the data of Figure 2 because of the scatter of the data points in this Figure. The porosity is directly proportional to the crack length. The calculated yield strength is obtained by elastic strain energy considerations for the material containing cracks. However, the model does include the effects of porosity as represented by the cracks. While the model is highly idealized and most probably does not in any way represent the conditions in the samples used in this study, it does indicate that the observed relationship between Y and E may be obtained by a consideration of the effect of cracks by the general approach used by Litewka.

SUMMARY

The relationship between the compressive fracture strength and the moduli for the three forms of Comp B can be attributed to differences in porosity and the relative differences in the average porosities are consistent with the differences in processing conditions used in preparing the casts. The changes in compressive fracture strength and modulus with position in the cast are then also due to porosity changes. The latter are consistent with the expected changes in porosity. Microcracking could also contribute to the observed changes in fracture strength and modulus.

The limited triaxial data which indicates a relationship between the yield strength and the modulus for the three forms of Comp B can also be interpreted on the bases of porosity being the primary cause of this relationship and the differences in these quantities from sample to sample and from one form of Comp B to another. The idealized crack model used by Litewka also predicts a relationship between yield strength and modulus which is compatible with the available data.

Based on the porosity interpretation, the compressive fracture strength, the yield strength and the modulus of Comp B may be significantly increased by reducing the porosity.

REFERENCES

1. Wiegand, D. A., Pinto, J. and Nicolaides, N., J. Energetic Materials, In Press.
2. Pinto, J. and Wiegand, D. A., J. Energetic Materials, In Press.
3. Pinto, J., Nicolaides, S. and Wiegand, D. A., Technical Report ARAED-TR-85004 (1985).
4. Wiegand, D. A. and Pinto, J., To Be Published.
5. For example see: Rice, R. W., in "Treatise on Materials Science and Technology Vol 11, Properties and Microstructure", Ed. MacCrone, R. K., Acedemic Press, New York (1977).
6. Knudsen, F. P., J. Am. Ceramic Soc., 42, 376 (1959).
7. Wang, J. C., J. Mat. Sci., 19, 801 (1984).
8. M. Joyce, Unpublished Results.
9. For example see: Montagut, E. and Kachanov, M., J. Fracture, 37, R55 (1988).
10. Litewka, A., Engr. Fracture 25, 637, 1986.

REPRESENTATION OF THE MICROSTRUCTURAL DEPENDENCE OF THE ANISOTROPIC ELASTIC CONSTANTS OF TEXTURED MATERIALS

STEPHEN C. COWIN
Department of Mechanical Engineering, City College of the City University of New York, New York, N. Y. 10031

ABSTRACT

This paper addresses the question of representing the dependence of the elastic coefficients in the anisotropic form of Hooke's law upon the microstructure of a material. The concern is with textured material symmetries, that is to say materials such as natural and man-made composites whose material symmetry is determined by microstructural organization. The approach is to relate the anisotropic elastic coefficients to local geometric or stereological measures of the microstructure. The predictions of micromechanical models and continuum mechanical models are compared and are found to be consistent with each other.

INTRODUCTION

The elastic properties of anisotropic porous materials are dependent both on the solid volume fraction of the material and the geometrical organization of the material microstructure. A relationship between the fourth rank tensor of elastic constants of a porous, anisotropic, linear elastic material and stereological parameters characterizing the anisotropy of the microstructure of the material was presented by Cowin [1]. The stereological parameter characterizing the anisotropy of the microstructure is called the *fabric ellipsoid or fabric tensor* of the material. The concept of a fabric tensor applies to any phase of a distributed, multiphase mixture, relative to the other phases. It has been suggested that the fabric tensor is the second measure of microstructure in a porous material after porosity (or relative volume fraction). The exact definition of the fabric tensor varies with the material being considered and with the investigator. Some of the definitions of the fabric tensors for various porous geological and biological materials are described in the following section.

In developing this relationship between elastic constants and fabric, Cowin [1], it was assumed that the matrix material of the porous elastic solid was isotropic and that the anisotropy of the porous elastic solid was completely determined by the fabric ellipsoid or tensor representing the architecture of the material microstructure. It was then shown that the material symmetries of orthotropy, transverse isotropy and isotropy correspond to the cases of three, two and one distinct eigenvalues of the fabric tensor, respectively. In terms of the fabric ellipsoid, these cases correspond to cases of an ellipsoid with three, two and no unequal axes. The fabric measure is described in the next section and the stress-strain-fabric relationship, the extension of the anisotropic form of Hooke's law to include a dependence of the elastic coefficients upon the fabric tensor, is described in the section that follows. The stress-strain-fabric relation has been employed as a model of the highly porous bone cancellous tissue by Cowin [2]. The paper of Turner, *et al.* [3] reports on the experimental determination of the coefficients in the stress-strain-fabric relationship in the case of cancellous bone. Specific formulas for the dependence of the orthotropic elastic constants upon fabric are developed in the section that follows the stress-strain-fabric relationship.

The formulas for the dependence of the orthotropic elastic constants upon fabric for the class of micromechanical models of open cell foam considered by Huber and Gibson [4] are summarized and converted into the notation of this paper in the next to last section. It is then shown that the relationship between ratios of the orthotropic elastic constants and ratios of the mean intercept lengths presented by Huber and Gibson [4], and obtained from dimensional arguments applied to a specific class of structural models of an open celled foam, are special cases of the formulas given in [1] and [5] and based on a general, not cell wall bending specific, model of an anisotropic porous material. This coincidence reinforces both the general approach of Cowin [1] and the cell wall bending model approach of Huber and Gibson [4]. This matter is considered further in the final discussion.

FABRIC

It is recognized that porosity or solid volume fraction is the primary measure of local material structure in a porous material. Porosity does not, however, reflect any directionality of the specimen's microstructure. What then is the second best measure of local structure after porosity? This question was posed in the context of granular materials by Cowin [6]. Now there appears to be general agreement that a fabric ellipsoid is the best second measure of local material microstructure in many porous materials. In this paper the term fabric ellipsoid indicates any ellipsoid (i.e. any positive definite second-rank tensor in three dimensions) that characterizes the local anisotropy of the material's microstructure. The fabric ellipsoid is a point property (even though its measurement requires a finite test volume) and is therefore considered to be a continuous function of position in the material. Methods exist to measure fabric ellipsoids in cellular materials, rocks and granular materials and are described below.

In cellular materials, foams and cancellous bone, a fabric ellipsoid can be associated with the directional variation of a lineal measure called the mean intercept length. The *mean intercept length* is the average distance between two solid/void interfaces in a given direction. Measurements are made on a plane of the material which is prepared to show contrast between the solid matrix and the pores. A grid of parallel test lines is laid over a specimen in a given direction, as illustrated in Figure 1a, and the length of the test line is divided by the number of intercepts to give mean intercept length. Measurements are repeated for test lines orientated in several directions. The experimental procedure for this type of measurement is described by Whitehouse [7], [8], Harrigan and Mann [9] and Turner, *et al.* [3]. Whitehouse showed that when the mean intercept lengths measured in cancellous bone were plotted in a polar diagram as a function of the direction (i.e.,of the slope of the line along which they were measured) the polar diagram produced ellipses. If the test lines are rotated through several values of θ and the corresponding values of mean intercept length

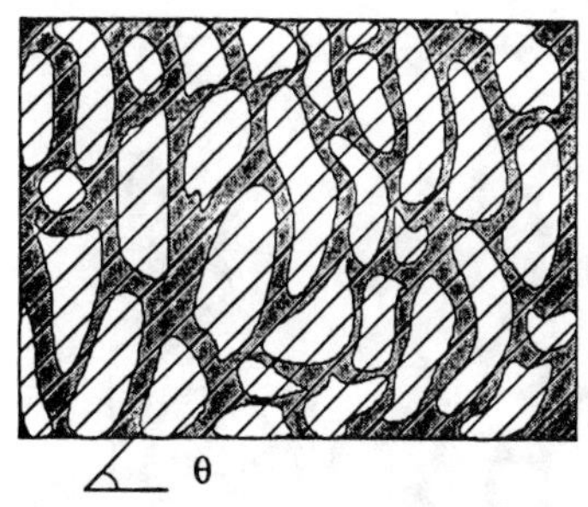

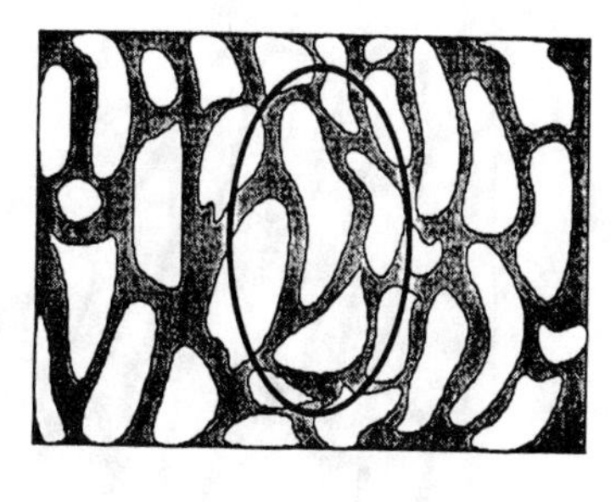

Figure 1. Left (a) Test lines superimposed on a porous material specimen. The test lines are oriented at the angle θ. The mean intercept length at this angle is denoted by L(θ). Right (b) The fabric ellipse is illustrated superimposed on the porous material specimen from which it was determined.

L(θ) are measured, the data are found to fit the equation for an ellipse very closely,

$$\frac{1}{L^2(\theta)} = M_{11}\cos^2\theta + M_{22}\sin^2\theta + 2M_{12}\cos\theta\sin\theta\ , \qquad (1)$$

where M_{11}, M_{22} and M_{12} are constants when the reference line from which the angle θ is measured is constant. The subscripts 1 and 2 indicate the axes of the x_1, x_2 coordinate system to which the measurements are referred. A fabric ellipse is shown superimposed on the porous structure it represents in Figure 1b. The result of the application of this procedure to the cancellous bone from a horse is illustrated in Figure 2.[10]

Harrigan and Mann [9] observed that in three dimensions the mean intercept length would be represented by ellipsoids and would therefore be equivalent to a positive definite second rank tensor. The constants M_{11}, M_{22} and M_{12} introduced in the foregoing are then the components in a matrix representing the tensor **M** which are related to the mean intercept length L(**n**), where **n** is a unit vector in the direction of the test line, by

$$\frac{1}{L^2(\mathbf{n})} = \mathbf{n}\cdot\mathbf{M}\mathbf{n}\ . \qquad (2)$$

The work of Whitehouse [7], [8], Harrigan and Mann [9] and Turner, *et al.* [3] has shown that the mean intercept length tensor is a good measure of the structural anisotropy in cancellous bone tissue. Cowin [1], noting the similarity to the structural ellipsoid/tensor employed by Oda [11] in a soil mechanics study, called an algebraically related tensor the fabric tensor of cancellous bone. The fabric tensor is denoted by **H** and is related to the mean intercept length tensor **M** by

$$\mathbf{H} = \mathbf{M}^{-1/2}\ . \qquad (3)$$

The positive square root of the inverse of **M** is well defined because **M**, which represents an ellipsoid, is a positive definite symmetric tensor. A measure of fabric that depends inversely on the mean intercept length is introduced so that the Young's modulus of the material will increase with increasing values of the measure.

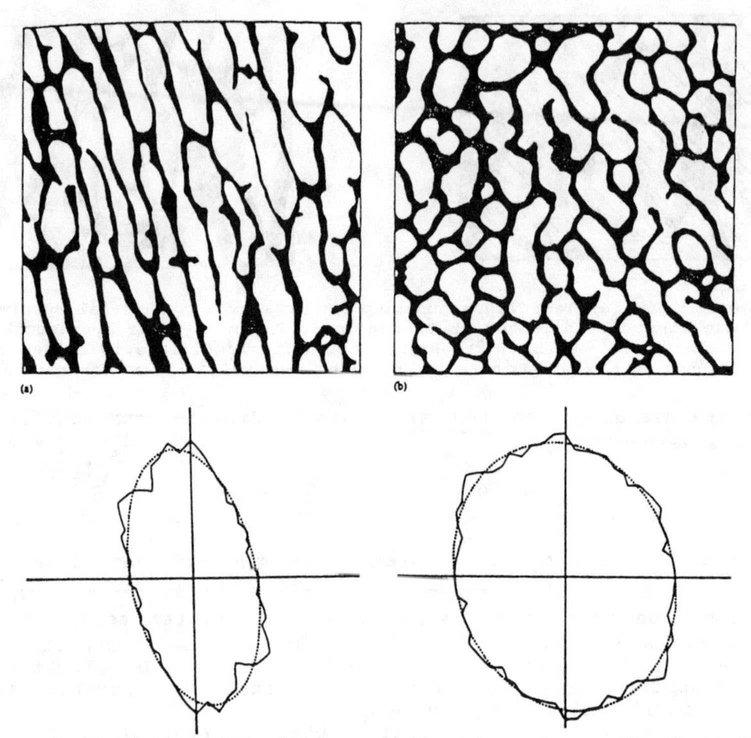

Figure 2. The fabric ellipses for cancellous horse bone. The faces of two cubes of cancellous horse bone, (a) and (b) above

The results presented in this work do not depend upon the precise details of the definition of the fabric tensor. It is required only that the fabric tensor be a positive definite tensor that is a quantitative stereological measure of the microstructural architecture, a measure whose principal axes are coincident with the principal microstructural directions and whose eigenvalues are proportional to the massiveness of the microstructure in the associated principal direction.

The methodology of making measurements is easily adapted to an automated computational system as shown by Harrigan and Mann [9] and Turner, *et al.* [3] The fabric ellipsoid is constructed by measuring mean intercept length as a function of direction in three orthogonal planes. In each of the three planes the data are fitted to the equation of an ellipse as described above. The three orthogonal ellipses that are formed are projections of an ellipsoid which is the fabric ellipsoid.

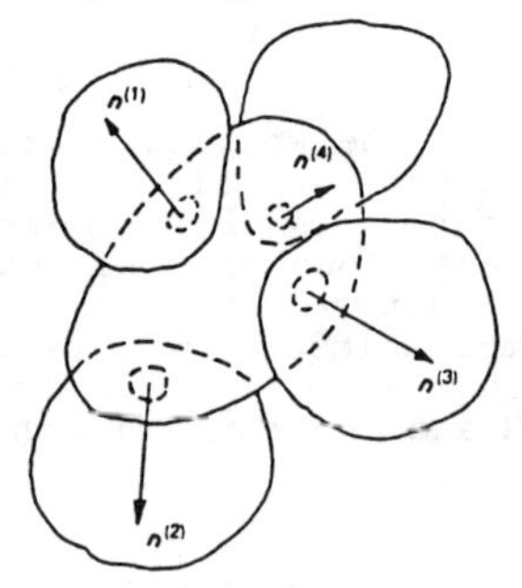

Figure 3. The contact normals associated with a granular particle in a granular material.

A different fabric measure is employed for granular materials. Oda [11], Oda et al. [12], and Satake [13] suggest that the best indicator of fabric in granular materials is the probability density function of the distribution of the orientation of contact normals, i.e. the normal at the point of contact between two granular particles, see Figure 3. This distribution is found to be periodic with respect to direction and can be represented by an ellipsoid. The number of contact normals in a given direction is measured on a section of a granular material taken from a load test apparatus. An illustration of data of this type is shown in Figure 4. A polar plot of the contact normals measured in a triaxial compression test of glass beads is shown on the left panel of Figure 4. Since this data has only to do with the geometry of the particulate architecture, it is labeled as geometric. On the right panel of Figure 4 the data is modified to account for the magnitude of the force associated with the contact normals it is therefore labeled kinetic. Figure 4 The data is from Ishibashi et al. [14] Data taken in a number of directions can be fitted to the equation of an ellipse. And the data from three orthogonal planes form the fabric ellipsoid.

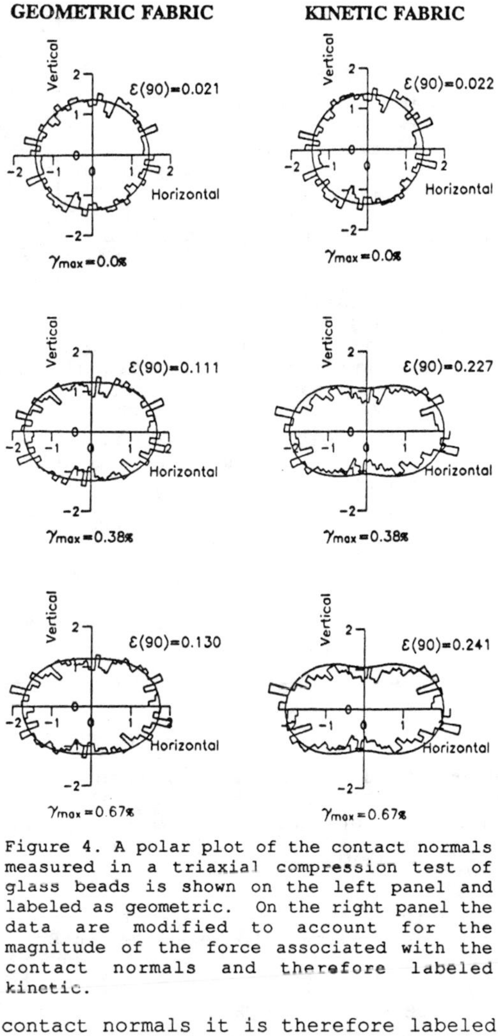

Figure 4. A polar plot of the contact normals measured in a triaxial compression test of glass beads is shown on the left panel and labeled as geometric. On the right panel the data are modified to account for the magnitude of the force associated with the contact normals and therefore labeled kinetic.

An automated computational system has been developed to quantify fabric from electron micro-graphs of soil by Tovey [15]. This system examines the gradients of the intensity of an image at each spatial point. The distribution of gradient vectors with respect to direction plotted on a polar plot tends to correlate to an ellipse.

In the analysis of rock mechanics, the fabric is best determined by the orientation of cracks within the rock. A fabric ellipsoid can be formed from mean intercept length measurements taken from three orthogonal planes of a rock. This definition of fabric is similar to those proposed by Satake [13] and Harrigan and Mann [9]. Figure 5 illustrates the measurement of mean intercept length in rock mechanics using a grid of parallel test lines superimposed on the surface of a rock. Oda [16] and Oda *et al.* [17] proposed a fabric ellipsoid for rock which incorporated the concentration of cracks as well as the directionality of cracks. In this definition the volume of the fabric ellipsoid is proportional to the concentration of cracks.

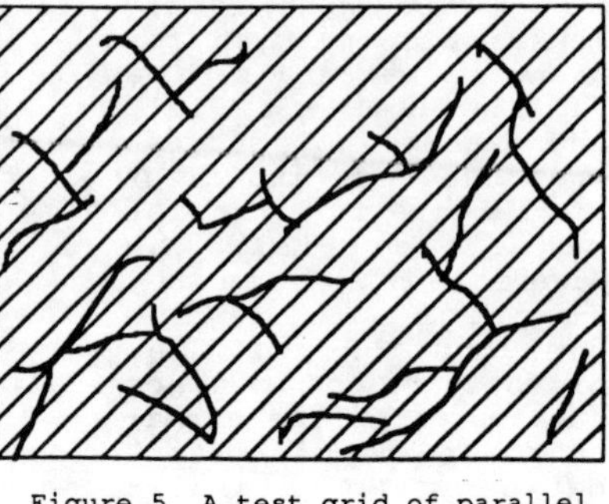

Figure 5. A test grid of parallel lines superimposed upon a surface of a rock specimen.

Let H_i, $i = 1,2,3$, denote the lengths of the three axes of the fabric ellipsoid or, equivalently, the eigenvalues of the fabric tensor **H**. In the principal coordinate system of **H**, the matrix of tensor components has the form

$$\mathbf{H} = \begin{bmatrix} H_1 & 0 & 0 \\ 0 & H_2 & 0 \\ 0 & 0 & H_3 \end{bmatrix} . \tag{4}$$

The parameter H_i represents the mean intercept length in the coordinate direction x_i, where x_i is a principal direction of **H**. Each of the H_i satisfies the cubic equation

$$H_i^3 - I\, H_i^2 + II\, H_i - III = 0, \qquad i = 1,2,3, \tag{5}$$

where I, II, and III represent the invariants of the fabric tensor **H**. The invariants are related to the principal components H_i by the formulas

$$I = H_1 + H_2 + H_3, \qquad II = H_1H_2 + H_1H_3 + H_3H_2, \qquad III = H_1H_2H_3. \tag{6}$$

These invariants are related to the traces of **H**, $\mathbf{H}^2$ and $\mathbf{H}^3$ by the formulas obtained, for example, in Ericksen [18]:

$$\mathrm{tr}H = H_1 + H_2 + H_3 = I, \qquad \mathrm{tr}H^2 = H_1^2 + H_2^2 + H_3^2 = I^2 - 2II,$$
$$\mathrm{tr}H^3 = H_1^3 + H_2^3 + H_3^3 = I^3 - 3I(II) + 3III. \tag{7}$$

The general problem of the quantification of structural anisotropy in materials with distinctive microstructure like geological and biological materials is considered in a series of papers by Kanatani [19],[20], [21], [22], and [23]. The problem of the determination of structural anisotropy by the methods of quantitative stereology was considered by Philofsky and Hilliard.[24].

STRAIN-STRESS-FABRIC RELATIONS

The anisotropic form of Hooke's law can be written as

$$E_{ij} = S_{ijkm}T_{km}, \tag{8}$$

where the S_{ijkm} are the components of the compliance tensor, and T_{km} and E_{ij} are the components of the stress and strain tensors, respectively. The components of the compliance tensor S_{ijkm} can be expressed as the components of a six by six matrix by the standard rules (Voigt, [25]; Love, [26]; Hearmon, [27]; Lekhnitskii, [28]; Fedorov, [29]) for representing a fourth-rank tensor of compliance coefficients as a six-by-six matrix; thus (8) can be rewritten as the matrix equation

$$\begin{bmatrix} E_{11} \\ E_{22} \\ E_{33} \\ 2E_{23} \\ 2E_{13} \\ 2E_{12} \end{bmatrix} = \begin{bmatrix} s_{11} & s_{12} & s_{13} & s_{14} & s_{15} & s_{16} \\ s_{12} & s_{22} & s_{23} & s_{24} & s_{25} & s_{26} \\ s_{13} & s_{23} & s_{33} & s_{34} & s_{35} & s_{36} \\ s_{14} & s_{24} & s_{34} & s_{44} & s_{45} & s_{46} \\ s_{15} & s_{25} & s_{35} & s_{45} & s_{55} & s_{56} \\ s_{16} & s_{26} & s_{36} & s_{46} & s_{56} & s_{66} \end{bmatrix} \begin{bmatrix} T_{11} \\ T_{22} \\ T_{33} \\ T_{23} \\ T_{13} \\ T_{12} \end{bmatrix} . \tag{9}$$

If it is assumed that fourth-rank elastic compliance tensor is a function only of the fabric tensor and some scalar parameters, then, using tensor algebra, it can be shown, Cowin [1], that the relationship between the fourth-rank elastic compliance tensor and the fabric tensor is

$$\begin{aligned} S_{ijkm} = {} & b_1\delta_{ij}\delta_{km} + b_2(\delta_{ki}\delta_{mj} + \delta_{mi}\delta_{kj}) + b_3(H_{ik}\delta_{jm} + \delta_{im}H_{kj} + H_{im}\delta_{kj} + \\ & \delta_{ik}H_{jm}) + b_4(H_{ij}\delta_{km} + \delta_{ij}H_{km}) + b_5(\delta_{mj}H_{iq}H_{qk} + \delta_{im}H_{kq}H_{qj} + \\ & \delta_{kj}H_{iq}H_{qm} + \delta_{ki}H_{mq}H_{qj}) + b_6(\delta_{ij}H_{kq}H_{qm} + \delta_{km}H_{iq}H_{qj}) + b_7H_{ij}H_{km} + \\ & b_8(H_{ij}H_{kq}H_{qm} + H_{km}H_{iq}H_{qj}) + b_9H_{is}H_{sj}H_{kq}H_{qm} , \end{aligned} \tag{10}$$

where the coefficients b_a, $a = 1,\ldots,9$, are functions of I, II and III defined by (6) and any scalar parameters introduced. The least elastic material symmetry for which the representation holds is orthotropy. It therefore holds for transverse isotropy and isotropy as well as orthotropy. The symmetry of the material is orthotropy if the three eigenvalues of **H** are distinct, transverse isotropy if only two are distinct and isotropy if all three eigenvalues are equal.

When (10) is substituted into (8) and the result rewritten in the matrix notation, we obtain the following representation for the strain-stress-fabric relationship:

$$\begin{aligned} \mathbf{E} = {} & b_1(\mathrm{tr}\mathbf{T})\mathbf{1} + b_2\mathbf{T} + b_3(\mathbf{HT} + \mathbf{TH}) + b_4(\mathbf{1}(\mathrm{tr}\mathbf{HT}) + (\mathrm{tr}\mathbf{T})\mathbf{H}) + \\ & b_5(\mathbf{H}^2\mathbf{T} + \mathbf{TH}^2) + b_6(\mathbf{1}(\mathrm{tr}\mathbf{H}^2\mathbf{T}) + (\mathrm{tr}\mathbf{T})\mathbf{H}^2) + \\ & b_7(\mathrm{tr}\mathbf{HT})\mathbf{H} + b_8((\mathrm{tr}\mathbf{H}^2\mathbf{T})\mathbf{H} + (\mathrm{tr}\mathbf{HT})\mathbf{H}^2) + b_9(\mathrm{tr}\mathbf{H}^2\mathbf{T})\mathbf{H}^2 . \end{aligned} \tag{11}$$

The formal inversion of the strain-stress-fabric relations (11) is given by stress-strain-fabric relationship

$$\mathbf{T} = a_1(\mathrm{tr}\mathbf{E})\mathbf{1} + a_2\mathbf{E} + a_3(\mathbf{HE} + \mathbf{EH}) + a_4(\mathbf{1}(\mathrm{tr}\mathbf{HE}) + (\mathrm{tr}\mathbf{E})\mathbf{H}) + a_5(\mathbf{H}^2\mathbf{E} + \mathbf{EH}^2) + a_6(\mathbf{1}(\mathrm{tr}\mathbf{H}^2\mathbf{E}) + (\mathrm{tr}\mathbf{E})\mathbf{H}^2) + a_7(\mathrm{tr}\mathbf{HE})\mathbf{H} + a_8((\mathrm{tr}\mathbf{H}^2\mathbf{E})\mathbf{H} + (\mathrm{tr}\mathbf{HE})\mathbf{H}^2) + a_9(\mathrm{tr}\mathbf{H}^2\mathbf{E})\mathbf{H}^2. \tag{12}$$

It was shown in Cowin [2] that either the principal axes of **T**, **E** and **H** are all coincident or they are all distinct. Since the eigenvectors of **H** are the axes of material symmetry for orthotropy, this coincidence of the principal axes of **T**, **E** and **H** is equivalent to the observation that the principal axes of **T** and **E** coincide in an orthotropic material only when the axes of **T** and **E** are coincident with the axes of material symmetry. In order to express this result algebraically, note that the eigenvectors of two matrices coincide if the multiplication of the two matrices is commutable; thus either each of the relations,

$$\mathbf{TH} = \mathbf{HT}, \quad \mathbf{EH} = \mathbf{HE}, \quad \mathbf{TE} = \mathbf{ET} \tag{13}$$

holds, or none of them does.

THE DEPENDENCE OF THE ORTHOTROPIC ELASTIC CONSTANTS UPON FABRIC

In the special case of orthotropic symmetry there exists a unique coordinate system in which the six-by-six matrix in (9) takes a special form characterized by the vanishing of twenty-four of its elements (only twelve of the twenty-four are distinct), thus

$$\begin{bmatrix} E_{11} \\ E_{22} \\ E_{33} \\ 2E_{23} \\ 2E_{13} \\ 2E_{12} \end{bmatrix} = \begin{bmatrix} s_{11} & s_{12} & s_{13} & 0 & 0 & 0 \\ s_{12} & s_{22} & s_{23} & 0 & 0 & 0 \\ s_{13} & s_{23} & s_{33} & 0 & 0 & 0 \\ 0 & 0 & 0 & s_{44} & 0 & 0 \\ 0 & 0 & 0 & 0 & s_{55} & 0 \\ 0 & 0 & 0 & 0 & 0 & s_{66} \end{bmatrix} \begin{bmatrix} T_{11} \\ T_{22} \\ T_{33} \\ T_{23} \\ T_{13} \\ T_{12} \end{bmatrix}. \tag{14}$$

The components of the matrix representation (14) can be written in terms of the technical elastic constants which consist of three Young's moduli, three shear moduli and six Poisson's ratios (only three of which are independent). The Young's moduli are denoted by E_i, $i = 1,2,3$; the shear moduli by G_{ij}; and the Poisson's ratios by ν_{ij} where $i, j = 1,2,3$ and $i \neq j$; thus

$$\begin{bmatrix} E_{11} \\ E_{22} \\ E_{33} \\ 2E_{23} \\ 2E_{13} \\ 2E_{12} \end{bmatrix} = \begin{bmatrix} E_1^{-1} & -\nu_{12}E_1^{-1} & -\nu_{13}E_1^{-1} & 0 & 0 & 0 \\ -\nu_{21}E_2^{-1} & E_2^{-1} & -\nu_{23}E_2^{-1} & 0 & 0 & 0 \\ -\nu_{31}E_3^{-1} & -\nu_{32}E_3^{-1} & E_3^{-1} & 0 & 0 & 0 \\ 0 & 0 & 0 & G_{23}^{-1} & 0 & 0 \\ 0 & 0 & 0 & 0 & G_{13}^{-1} & 0 \\ 0 & 0 & 0 & 0 & 0 & G_{12}^{-1} \end{bmatrix} \begin{bmatrix} T_{11} \\ T_{22} \\ T_{33} \\ T_{23} \\ T_{13} \\ T_{12} \end{bmatrix}. \tag{15}$$

The interrelationships between the Young's moduli and Poisson's ratios are given by

$$E_j \nu_{ij} = E_i \nu_{ji}, \quad i \neq j, \text{ no sum on } i,j. \tag{16}$$

It was shown in [1] (see also [5]) that these elastic constants are related to the fabric components by

$$\frac{1}{G_{ij}} = b_2 + b_3(H_i + H_j) + b_5(H_i^2 + H_j^2), \quad i \neq j, \tag{17}$$

$$\frac{1}{E_i} = b_1 + 2b_2 + d_1 + 2(2b_3 + b_4 + d_2)H_i + (4b_5 + 2b_6 + b_7 + d_3)H_i^2, \tag{18}$$

and

$$-\frac{\nu_{ij}}{E_i} = -\frac{\nu_{ji}}{E_j} = b_1 + b_4(H_i + H_j) + b_6(H_i^2 + H_j^2) + b_7 H_i H_j + b_8 H_i H_j(H_i + H_j) + b_9 H_i^2 H_j^2 \quad , \quad i \neq j,\ i,\ j = 1,2,3, \tag{19}$$

where the b_i and d_i, $i = 1,2,3$, are functions of I, II and III and some measure of the porosity of the material. The coefficients d_i, $i = 1,2,3$, are functions of the coefficients b_8, b_9 and the invariants I, II, and III,

$$d_1 = (2b_8 + b_9 I)III, \quad 2d_2 = b_9(III - I(II)) - 2b_8 II, \quad d_3 = 2b_8 I + b_9(I^2 - II). \tag{20}$$

In the representation for E_i the characteristic equation (5) has been used to express the third- and fourth-order terms in H_i given in the formulas in [1] and [5] in terms of the second- and first-order terms; the coefficients d_i arise from this algebraic manipulation. The points considered in this paper do not concern the porosity and other scalar parameter dependence of the coefficients b_i directly; this dependence is important and always permitted, however.

THE ANIOSTROPY OF FOAMS

In Huber and Gibson [4] the elements of mechanics of materials were applied to a class of cell-wall-bending models of an elastically orthotropic foam to construct the algebraic relationship between ratios of the three Young's moduli, the ratios of the three shear moduli, and the principal mean intercept lengths. The unit cells in the structural model were connected to adjacent cells at the cell midpoint to develop a bending response of the structure. Examples of these results are the ratios G_{12}/G_{23} and E_3/E_1 as functions of the mean intercept lengths H_i, $i = 1,2,3$,

$$\frac{G_{12}}{G_{23}} = \frac{(H_2 + H_3)}{(H_2 + H_1)}, \tag{21}$$

and

$$\frac{E_3}{E_1} = \frac{H_3^2\left(1 + \frac{H_2^3}{H_1^3}\right)}{H_1^2\left(1 + \frac{H_2^3}{H_3^3}\right)}, \tag{22}$$

respectively. The notation employed in Huber and Gibson [4] for these results is slightly different: the ratio R_{ij}, where $i, j = 1,2,3$ and $i \neq j$, is used in place of H_i/H_j. Straightforward algebraic manipulations, combined with the use of the notation introduced in the third equation of (7), namely $\mathrm{tr}\mathbf{H}^3$ for $(H_1)^3+(H_2)^3+(H_3)^3$, permits (22) to be rewritten in the form

$$\frac{\frac{1}{E_1}}{\frac{1}{E_3}} = \frac{\frac{1}{H_1^5(\mathrm{tr}\mathbf{H}^3 - H_1^3)}}{\frac{1}{H_3^5(\mathrm{tr}\mathbf{H}^3 - H_3^3)}}. \tag{23}$$

These results suggest that the reciprocals of G_{ij} and E_i may be represented by

$$\frac{1}{G_{ij}} = G(H_i + H_j), \quad i \neq j, \tag{24}$$

and

$$\frac{1}{E_i} = \frac{E}{H_i^5(\mathrm{tr}\mathbf{H}^3 - H_i^3)}, \tag{25}$$

where G and E represent arbitrary functions of I, II, III and the porosity.

The representations (24) and (25) for G_{ij} and E_i are special cases of the representations given above by (17) and (18), respectively. It is easy to see that (24) is a special case of (17) because the two formulas coincide if one sets b_2 and b_5 equal to zero and b_3 equal to G in the formula (17). A casual inspection suggests that it might not be possible to obtain coincidence between (25) and (18), but such is not the case. The coincidence can be seen in the following way: formally divide the denominator into the numerator of (25). The result is an infinite series in H_i, the nth term being of the form $(H_i)^n$ times a scalar-valued function of I, II and III. (Recall from (7) that $\mathrm{tr}\mathbf{H}^3$ can be expressed in terms of I, II and III). All terms of order higher than $(H_i)^2$ can then be eliminated from the infinite series by repeated use of (5). The result is then a specific form of (18), and it has been established that the results (24) and (25) are special cases of (17) and (18), respectively.

DISCUSSION

There are several interesting points concerning the simple algebraic observations described above. The observations represent a positive and reinforcing interaction between the development of a general theory and the development of specific structural models for the anisotropy of highly porous materials. The results that were common to both the general model and the cell wall bending model are the following:

A. The orthotropic elastic Young's moduli E_i depend directly upon H_i and the elementary symmetric functions I, II and III of H_1, H_2, H_3. This means that the influence of the components of **H** other than H_i only occurs through the elementary symmetric functions I, II and III.

B. The orthotropic shear moduli G_{ij} depend directly upon H_i, H_j, I, II and III. This means that the influence of the component of **H** other than H_i and H_j occurs only through I, II and III. Both approaches predict a dependence of the inverse of G_{ij} upon the sum $H_i + H_j$.

The notations employed in (17), (18) and (19) for the relationships between the orthotropic elastic constants and the mean intercept length measures present these results in a relatively concise form, emphasizing the appropriate symmetries in the functional dependencies. In order to obtain the corresponding results for transversely isotropic symmetry one has only to set two of the H_i equal. The number of elastic coefficients then reduces from nine to five. Note that, if one takes H_1 and H_2 to be equal, then E_1 and E_2, G_{13} and G_{23}, ν_{12} and ν_{21}, ν_{13} and ν_{23}, and ν_{31} and ν_{32} are all equal, G_{12} equals E_1 divided by $2(1 + \nu_{12})$, and $E_3\nu_{13}$ equals $E_1\nu_{31}$.

In the development above of the correspondence between the general approach of Cowin [1] and the cell-wall-bending model approach of Huber and Gibson [4], it is assumed that the spatial distribution of the directed mean intercept lengths of the cell-wall-bending model approach will form a second rank tensor or, equivalently, an ellipsoid. The arguments of Huber and Gibson [4] apply to a class of models, a class that includes geometrically regular, rectangular, structural cell models of the type illustrated in Huber and Gibson [4], as well as less regular models. The consideration of the mean intercept length as a function of spatial direction for idealized linear plane geometric structures such as those reported by Tozeren [30] and Luo *et al.*[31] shows that the directed mean intercept lengths of the geometrically very regular model types considered in Huber and Gibson [4], (e. g., the rectangular structural cell model type) will not form an ellipsoid, but rather a polyhedron with three orthogonal axes of symmetry. Along each orthogonal axis of symmetry the half-dimension of the polyhedron will be equal to one of the mean intercept length, and the eight octants of the polyhedron will be identical. The exterior boundary of each octant will be a mosaic of polygons. It is also suggested by the results reported in Tozeren [30] and Luo *et al.*[31] that, as the precise regularity of the cell model tends to the irregularities of natural porous materials, the polyhedron representing the spatial distribution of the directed mean intercept lengths will approach more closely an ellipsoid. In the analysis presented above it is assumed that even the polyhedron representing the spatial distribution of the directed mean intercept lengths of the geometrically very regular model types considered in Huber and Gibson [4], (e. g., the rectangular structural cell model type) can be approximated by an ellipsoid. This approximation may be accomplished, for example, by a least squares fit of an ellipsoid to the actual polyhedron representing the spatial distribution of the directed mean intercept lengths.

ACKNOWLEDGEMENT

This investigation was supported by NSF Grant No. BSC-8822401

REFERENCES

1. S. C. Cowin, *Mech. Mater.* 4, 137 (1985).
2. S. C. Cowin, *ASME J. Biomech. Eng.* 108, 83 (1986).
3. C. H. Turner, S. C. Cowin, J. Y. Rho, R. B. Ashman, J. C. Rice, *J. Biomech.* 23, 549 (1990).
4. A. T. Huber and L. J. Gibson, *J. Mater. Sci.* 23, 3031 (1988).
5. C. H. Turner and S. C. Cowin, *J. Mater. Sci.* 22, 3178 (1987).
6. S. C. Cowin, in US-Japan Seminar on Continuum Mechanical and Statistical Approaches in the Mechanics of Granular Materials, edited by S. C. Cowin and M. Satake, (Gakujutsu Bunken Fukyu-kai, Sendai, 1978), pp. 162-170.
7. W. J. Whitehouse, *J. Microscopy* 101, 153 (1974).
8. W. J. Whitehouse, E. D. Dyson, *J. Anatomy* 118, 417 (1974).
9. T. P. Harrigan, R. W. Mann, *J. Mater. Sci.* 19, 761 (1984).
10. R. Hodgskinson and J. D. Currey, *Proc. Instn. Mech. Engrs.* 204, 101 (1990).
11. M. Oda, *Soils Found.* 12, 17 (1972).
12. M. Oda, J. Konishi, S. Nemat-Nasser, *Geotechnique* 30, 497 (1980).
13. M. Satake, *Theor. Appl. Mech.* 26, 257 (1978).
14. I. Ishibashi, T. Agarwal, S. A. Ashraf, in Proc. 1st US Conf. on Discrete Element Methods, (1989)
15. N. K. Tovey, *J. Microscopy* 120, 303 (1980).
16. M. Oda, *Mech. Mater.* 2, 163 (1983).
17. M. Oda, K. Suzuki, T. Maeshibu, *Soils Found.* 24, 27 (1984).
18. J. L. Ericksen, in *Encyclopedia of Physics* edited by C. A. Truesdell, (Springer, Berlin, 1960), pp. 794-858.
19. K. Kanatani, *J. Jpn. Soil Mech. Found Eng.* 23, 171 (1983).
20. K. Kanatani, *Int. J. Eng. Sci.* 22, 149(1984).
21. K. Kanatani, *Int. J. Eng. Sci.* 22, 531 (1984).
22. K. Kanatani, *Soils Found.* 25, 77 (1985).
23. K. Kanatani, *Int. J. Eng. Sci.* 23, 587 (1985).
24. E. M. Philofsky, J. E. Hillard, *Quart. Appl. Math.* 27, 79 (1969)
25. W. Voigt, Lehrbuch der Kristallphysik Leipzig, 1910).
26. A. E. H. Love, Elasticity (Dover, New York, 1927).
27. R. F. S. Hearmon, An Introduction to Applied Anisotropic Elasticity (Oxford University Press, Oxford, 1961).
28. S. G. Lekhnitskii, *Theory of Elasticity of an Anisotropic Elastic Body* (Holden Day, San Francisco, 1963).
29. F. I. Fedorov, Theory of Elastic Waves in Crystals (Plenum Press, New York, 1968)
30. A. Tozeren, R. Skalak, *J. Mater. Sci.* 24, 1700 (1989).
31. G. M. Luo, A. M. Sadegh, S. C. Cowin, *J. Mater. Sci.* (1991, in press).

DIGITAL-IMAGE-BASED STUDY OF CIRCULAR HOLES IN AN ELASTIC MATRIX

A.R. DAY[*], M.F. THORPE[**], K.A. SNYDER[***], AND E.J. GARBOCZI[***]
[*]Marquette University, Dept. of Physics, Milwaukee, WI 53233
[**]Michigan State University, Dept. of Physics and Astronomy, East Lansing, MI 48824
[***]National Institute of Standards and Technology, Building Materials Division, Bldg. 226, Room B348, Gaithersburg, MD 20899

ABSTRACT

Using a digital-image-based representation of a continuum composite, we apply computer simulation techniques to obtain the elastic moduli of a matrix containing randomly-centered circular voids. As the area fraction of the voids increases, the elastic moduli of the composite decrease until they eventually vanish at the percolation threshold. We compare our results with an effective medium theory, which predicts that Poisson ratio tends to a fixed value as the percolation threshold is approached, independent of the values of the elastic moduli in the pure system. Our results are also compared with recent experimental results.

INTRODUCTION

Calculating the elastic properties of a material that is made up of randomly distributed phases that have different elastic properties is a difficult problem that has received much attention [1]. Materials of interest range from semi-conductor alloys, where the random mixing of elastic properties is at atomic length scales, to fiber-reinforced materials, like SiC-whisker reinforced ceramics or steel-fiber-reinforced concrete, where the mixing occurs on macroscopic length scales. Atomic-scale composite problems are treated with lattice calculations, while macroscopic problems are usually treated with continuum mechanics.

This paper describes a merging of these two methods, using a new, simple finite-element-type scheme realized on a digital-image-based model of a two-dimensional two-phase material, where one of the phases has zero elastic moduli. This second phase is generated by punching out circular holes in a sheet. The holes are randomly- centered, so that they may freely overlap. The remaining area fraction of the matrix is denoted by p. When the area fraction (1-p) of circular holes reaches about 0.68, the holes percolate [2] and the elastic moduli of the sheet go to zero in a percolation transition.

DIGITAL-IMAGE-BASED MODEL

Representing images with pixels on a lattice is a technique that has been used for many years for quantitative image analysis [3]. Usually a square lattice of equal-sized square pixels is superimposed over an image, with the pixels being assigned to different phases of the picture according to whether or not the center of the pixel falls inside a phase boundary. Once the original continuum picture is thus represented digitally, all mathematical operations of interest can now be carried out on the underlying lattice, greatly simplifying their application.

The idea of breaking up a continuum into a lattice, and then carrying out mathematical operations on the lattice instead of the continuum is of course the basic idea behind finite-difference [4] and finite-element [5] methods for solving partial differential equations, and has been used for many years. The approach is to start out with continuum partial differential equations, and convert them into linear algebraic equations by approximating derivatives by quotients of differences. This can of course be done for the partial differential equations of elasticity as well as any other set of equations.

In this paper we take a different approach. We represent a continuum model by pixels, and then use the underlying pixel lattice and material properties to define a random spring network having force constants that model the elastic properties of the original continuum material. Fig. 1 shows a portion of a 210 x 210 pixel triangular lattice of pixels, in which 11-pixel diameter circles have been introduced. The circles are all centered on pixels, so that their shapes are all exactly the same. The elastic network is defined by connecting nearest-neighbor pixels with the appropriate force constant spring, as will be discussed below. A space-filling triangular network of hexagonal pixels is used instead of a square network of square pixels, because the square lattice is unstable with only nearest-neighbor central-force springs [6,7].

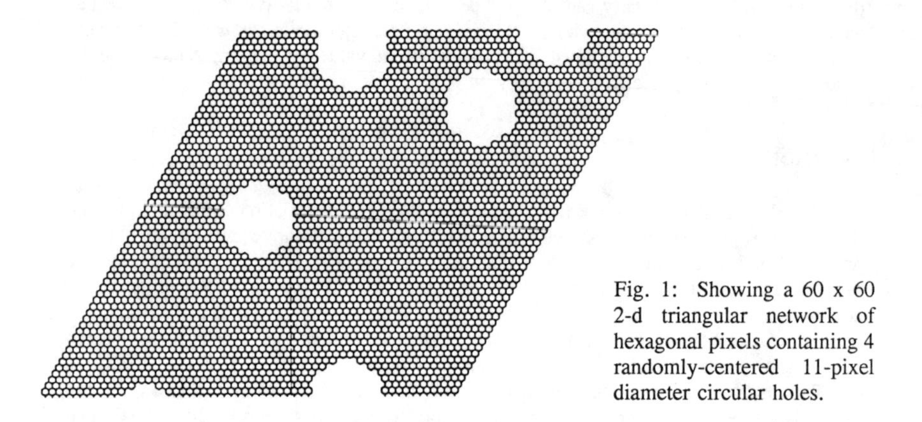

Fig. 1: Showing a 60 x 60 2-d triangular network of hexagonal pixels containing 4 randomly-centered 11-pixel diameter circular holes.

NUMERICAL METHOD

Once an image is defined, as in Fig. 1, with a certain number of holes, a random spring network is created by connecting each pixel with its six nearest neighbors with springs of the appropriate force constant. Since the perfect triangular network with all the same nearest-neighbor springs obeys Cauchy's relation, it is necessary to use more than one kind of spring in order to have a variable Poisson ratio [7]. Fig. 2 shows how three different kinds of springs, α, β, and γ, are defined in the triangular spring lattice, so as to preserve isotropy, but allow for different values of Poisson ratio in the perfect lattice before any holes are introduced. For the perfect lattice, the area bulk modulus, K_o, and the shear modulus, μ_o, are given by:

$$K_o = \frac{1}{\sqrt{12}} (\alpha + \beta + \gamma) \qquad (1)$$

$$\mu_o = \sqrt{\frac{27}{16}}\left(\frac{1}{\alpha} + \frac{1}{\beta} + \frac{1}{\gamma}\right)^{-1} \qquad (2)$$

The perfect lattice is isotropic because the α, β, γ springs form three interlocking honeycomb lattices, which are elastically isotropic. In 2-d, the Young's modulus E and area Poisson ratio σ can be found from the area bulk modulus and the shear modulus μ through [8]

$$E = \frac{4k\mu}{k + \mu} \qquad (3)$$

$$\sigma = \frac{k - \mu}{k + \mu} \qquad (4)$$

For the perfect lattice, using eqs. (1)-(4), the Young's modulus E_o and Poisson ratio σ_o are given by:

$$\sigma = 1 - \frac{2}{1 + \frac{2}{9}(\alpha + \beta + \gamma)\left(\frac{1}{\alpha} + \frac{1}{\beta} + \frac{1}{\gamma}\right)} \qquad (5)$$

$$E_o = \frac{2\sqrt{3}\,(\alpha + \beta + \gamma)}{3\left[1 + \frac{2}{9}(\alpha + \beta + \gamma)\left(\frac{1}{\alpha} + \frac{1}{\beta} + \frac{1}{\gamma}\right)\right]} \qquad (6)$$

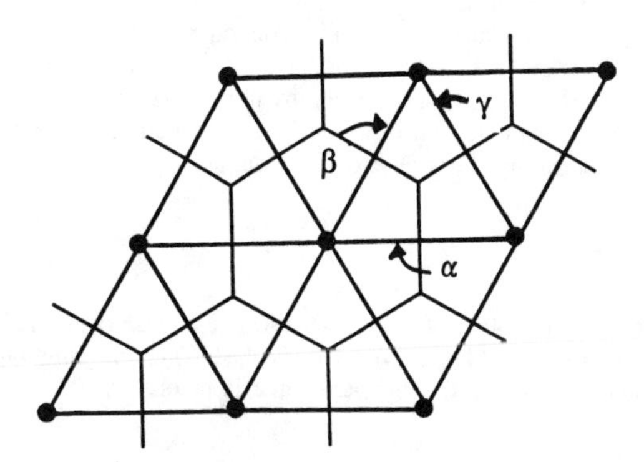

Fig. 2: Showing a piece of a triangular grid of hexagonal pixels. The three kinds of bonds, which connect centers of nearest-neighbor hexagonal pixels, have spring constants α, β, and γ.

Eq. (5) for σ_o allows values between 1/3 and 1 to be realized. It is convenient to use three, rather than two, parameters, because of the symmetry of the triangular network. The extra free parameter can be used as a consistency check.

When holes are introduced, springs with zero force constant are connected between pairs of hole pixels and across the hole boundary, while α, β, and γ springs connect the matrix pixels.

Once the spring network has been set up, a uniaxial strain is applied, typically ~ 10^{-3}. The nodes of the networks and the length of the unit cell perpendicular to the applied strain are simultaneously relaxed using a specialized conjugate gradient algorithm [9]. When the network is relaxed, the Young's modulus E can be immediately obtained from the total elastic energy per unit area, and Poisson ratio σ from the new length of the unit cell perpendicular to the applied strain. Equal tensile and compressive strains are used to average out any non-linearities, which were found to be small for strains ~ 10^{-3}. Non-linearities are present because the (harmonic) springs in the network are not colinear.

EFFECTIVE MEDIUM THEORY

The problem of a single circular hole can be solved exactly, and this single-inclusion result forms the basis for effective medium theories (EMT). The simplest EMT for circular holes [9] in a 2-d sheet leads to the results:

$$\frac{E}{E_o} = 3p - 2 \tag{7}$$

$$\sigma - \frac{1}{3} = (\sigma_o - \frac{1}{3})\,(3p - 2) \tag{8}$$

The other elastic constants can be obtained by inverting eqs. (3) and (4). The results (7) and (8) are remarkably simple, and are exact for small (1-p), where it can be seen that E/E_o is (unexpectedly) independent of σ_o. Indeed, eq. (7) predicts that E/E_o is independent of σ_o for all p. However, (7) and (8) incorrectly predict the percolation threshold p_c to be 2/3. In reality, percolation occurs $p_c \simeq 0.32$ [2]. At percolation, the critical Poisson ratio reaches a universal value of 1/3, independent of the Poisson ratio of the matrix material σ_o. Similar results have been found in lattice systems. The result can be understood by noting that any external stress will produce similar uniaxial strains in the narrow necks between adjacent circles at percolation.

The area fraction of material remaining, p, is related to the number of holes per unit area, n, via

$$p = e^{-na} \tag{9}$$

where a is the area of a single circular hole. This is a statistical relationship that holds if a sufficiently large number of holes are present [10]. We have checked (9) in analysing our data, since p may be evaluated directly by counting the number of matrix pixels remaining.

RESULTS

We have studied the dependence of the Young's modulus E and the Poisson ratio σ on p, the area fraction of the matrix material remaining, for three different choices of α, β, and γ. These are (α, β, γ) = (1,1,1), (1,1,4), and (1,6,7).

Two checks can be made on the accuracy of this digital-image approach. The first check is on whether we are really simulating a continuum system or not. The results for given values of

E_o and σ_o should not depend on the specific choice of lattice force constants, if we are truly carrying out a continuum simulation. Table I shows the results for E/E_o and σ at p = 0.7377, averaged over the same ten geometrical configurations. The first row is for $\alpha=\beta=1$ and $\gamma=4$, the second row interchanges α and γ, and the third row is for a completely different choice of force constants having the same values of E_o and σ_o. The results on these three different systems all agree, to within less than 1%.

Table I: Continuum check on elastic moduli

α	β	γ	E/E_o	σ
1	1	4	0.2899	0.3880
4	1	1	0.2898	0.3886
2	3.3	0.69	0.2916	0.3879

The second check is on how accurately the single-hole exact result is reproduced. Eqs. (7) and (8) give the exact initial slope for the one-defect problem when (1-p) is small:

$$\delta\sigma = \sigma - \sigma_o = (1 - 3\sigma_o)(1 - p) \qquad (10)$$

$$\delta E = \frac{E}{E_o} - 1 = -3(1 - p) \qquad (11)$$

Table II shows the percent difference from the exact result for δE and $\delta\sigma$, for three different hole diameters d, and (α,β,γ) = (1,1,4). As expected, the results for E/E_o become better as

Table II: One-hole results

(1-p)	d	$\Delta(\delta E)$%	$\Delta(\delta\sigma)$%
0.00247	11	10.2	0.02
0.01077	23	3.2	0.03
0.02220	33	0.2	0.01

more pixels are used to define the circular hole. The diameter of 11 pixels used in this paper results in a 10% error in the change in E caused by the introduction of one circular hole. The percent error in E itself is, of course, very much smaller, less than a tenth of a percent. It is interesting to compare with the equivalent electrical problem. We have recently calculated the effect of one elliptical hole on the conductivity of a conducting sheet, using a square lattice digital-image scheme, and found that the error in the initial slope decreased linearly with the number of pixels used per unit length to represent the hole [11]. Table II implies that this error decreases more quickly for the elastic problem, as doubling the number of pixels per unit length caused the error to decrease by more than a factor of three, and tripling the number of pixels per unit length caused the error to decrease by more than a factor of 10. The surprising result in Table II is the fact that, at least for diameters greater than or equal to 11 pixels, the error in the initial slope for the Poisson ratio seems to be independent of the number of pixels per unit length, and is essentially computer round-off error.

Figs. 3 and 4 present the results for E/E_o and σ vs. p, averaged over 10 configurations, and computed for the three choices of the force constants given above. The points are the simulation results, and the dashed lines are the effective medium results (7) and (8). The solid line is Fig. 3 is an interpolation formula fit to the simulation data points, and is discussed below. In Fig. 3 it is interesting to note that it appears that E/E_o is independent of σ_o for all values of p, and not just for p close to 1. We have independently computed K/K_o and μ/μ_o, and have seen that they are clearly not independent of σ_o. Fig. 4 shows that σ flows to a fixed point of about 1/3, as p decreases, as was predicted by EMT. Interestingly, the flow to the fixed point is essentially

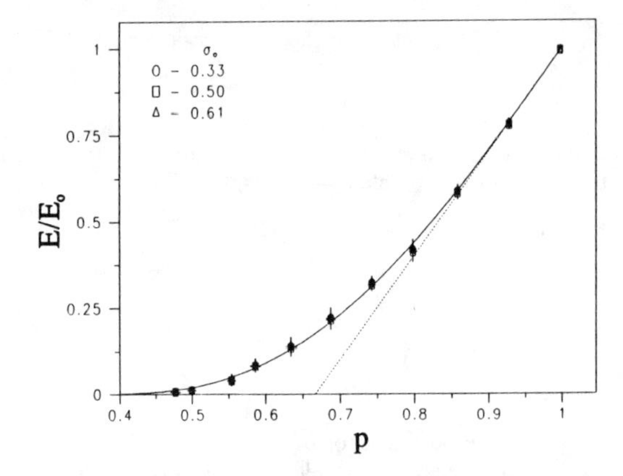

Fig. 3: Showing simulation data for the Young's modulus E/E_o for various values of the Poisson ratio σ_o of the matrix material. The dashed lines are the EMT results (7) and (8). The solid line is the interpolation formula (12), with m = 3.4.

complete by p = 2/3, the EMT percolation threshold, which is well above p = 0.32, the true percolation threshold [2]. The error bars on the Poisson ratio are large for $p \lesssim 0.6$. These are the statistical error bars from the 10 configuration average. The error bars of the Young's

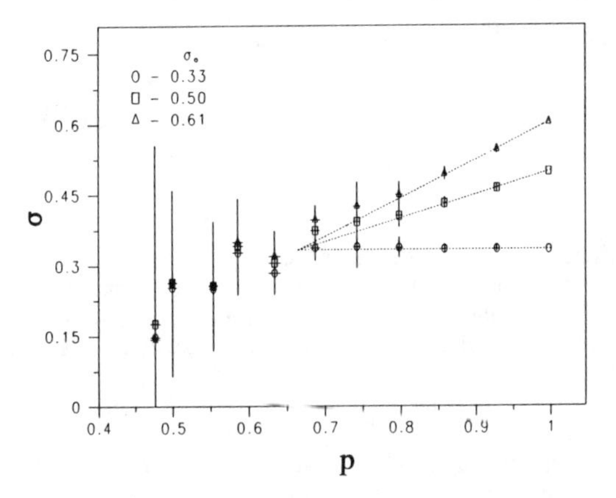

Fig. 4: Showing Poisson ratio σ vs. p for various values of the Poisson ratio σ_o of the matrix material. The dashed lines are the EMT results (7) and (8).

modulus are much smaller as can be seen from Fig. 3. Although the Poisson ratio is computed directly, it can be regarded as the ratio of two elastic moduli, both of which become very small for $p \lesssim 0.6$. Nevertheless, the error bars in this region are disappointingly large.

For d = 11 pixel circles in a 210 by 210 lattice, we have directly computed the percolation threshold, using a lattice "burning algorithm" [12], and found it to be $p_c = 0.34 \pm 0.03$, in good agreement with the known result that was obtained using a true continuum simulation [2].

SUMMARY

We have shown that this simplest of finite element algorithms, a scheme based on linear Hookean springs, can give good results in 2-d composites. Such calculations are now feasible with the use of modern supercomputers. The calculations reported in this paper took ~ 150 hours of CPU time on a CYBER 205 supercomputer. It is important to emphasize that the present simple scheme, combined with averages over many samples to give good statistics, is best suited for the calculation of the elastic behavior of random composites. More sophisticated grids and algorithms have the capability of doing better on any particular realization, but that is not what is required here.

In the past, various EMTs have been compared with one another and checked against various exact bounds on the elastic constants [2]. This is rather unsatisfactory. The present work opens up the possibility of producing high quality numerical results on well-controlled composite geometries, in order to evaluate which, if any, EMTs are valid away from the dilute limit. A direct comparison between EMTs and real experiments is always dangerous because of the uncertainties present in the characterization of the geometry of real composites. Thus computer simulation can form an important bridge between approximate EMTs and experiments.

Recent experiments [13,14] have measured the elastic moduli of square sheets, containing randomly-centered circular holes, under uniaxial loading. Because of the clamping arrangement at the ends, a modulus intermediate between Young's modulus E (completely free sides) and $C_{11} = E/(1 - \sigma^2)$ (clamped sides). For simplicity, we compare our results for E/E_o with the experiments in Fig. 5. This should be a good approximation, as our numerical results show that if $\sigma_o = 1/3$, then $\sigma \sim 1/3$ for all p. The materials used in the experiments [13,14] would be

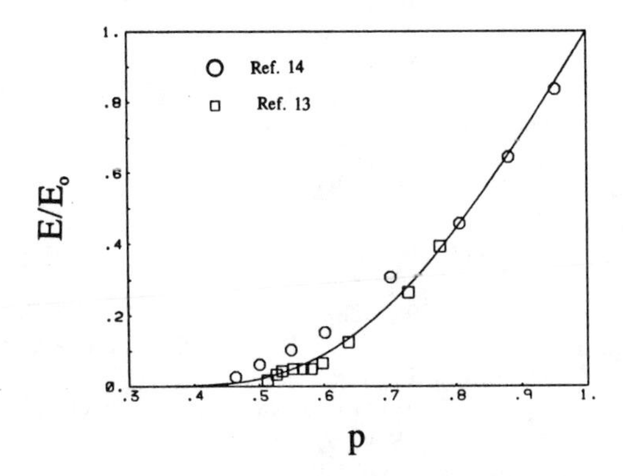

Fig. 5: A comparison of the simulation results for the Young's modulus E/E_o, as represented by the interpolation formula (12), shown as a solid line, with the experiments of Ref. 14 (circles) and Ref. 13 (squares).

expected to have a Poisson ratio close to 1/3. To facilitate comparison between our simulations and the experiments, we have fitted our results to the empirical form

$$\frac{E}{E_o} = \left[1 - \frac{(1-p)}{m(1-p_I)} - \frac{[m(1-p_I)-(1-p_c)](1-p)^2}{m(1-p_I)(1-p_c)^2}\right]^m \qquad (12)$$

where we have set the initial slope parameter $p_I = 2/3$; the percolation threshold $p_c = 0.34$, to match our numerical results; and the exponent $m = 3.4$ is obtained by a least-squares-fit to the simulation data. The experimental data points have been obtained from the published graphs using image analysis.

The exponent m is not meant to be a critical exponent, as our data is quite far from the critical region near p_c. Rather, m is merely a free parameter in the interpolation formula (12). It is very difficult to obtain good results in the critical region as E/E_o is so small. On theoretical grounds, the true critical exponent f is expected to be about 5.1 [13-15].

The interpolation formula (12) fits the data in Fig. 3 well. Fig. 5 shows that the interpolation formula, with the same exponent m, also fits the data from Ref. 13. For reasons that are not clear to us, the data from Ref. 14 is systematically quite far above our results.

ACKNOWLEDGEMENTS

We would like to thank P. Duxbury and I. Jasuik for useful discussions. Two of us (MFT and ARD) would like to thank NIST for its hospitality during extended visits. One of us (MFT) acknowledges partial support from the State of Michigan Research Excellence Fund.

REFERENCES

[1] For a review, see for example J.P. Watt, G.F. Davies, and R.J. O'Connell, Rev. Geoph. Space Phys. 14, 541 (1976).
[2] E.T. Gawlinski and H.E. Stanley, J. Phys. A: Math. Gen. 14, L291 (1981).
[3] K.R. Castleman, Digital Image Processing (Prentice-Hall, Englewood Cliffs, 1979).
[4] W.H. Press, B.P. Flannery, S.A. Teukolsky, and W.T. Vetterling, Numerical Recipes (Cambridge University Press, Cambridge, England, 1989).
[5] R.D. Cook, D.S. Malkus, and M.E. Plesha, Concepts and Applications of Finite Element Analysis, Third Edition (John Wiley and Sons, New York, 1989).
[6] S. Feng and P.N. Sen, Phys. Rev. Letts. (1984).
[7] S. Feng, M.F. Thorpe, and E.J. Garboczi, Phys. Rev. B31, 276 (1985).
[8] M.F. Thorpe and P.N. Sen, J. Acoust. Soc. Am. 77, 1674 (1985).
[9] This algorithm was written by and can be obtained from A.R. Day, at the address given above.
[10] W. Xia and M.F. Thorpe, Phys. Rev. A38, 2650 (1988).
[11] E.J. Garboczi, M.F. Thorpe, M. deVries, and A.R. Day, "Universal Conductance Curve for a Plane Containing Random Holes", submitted to Phys. Rev. A.
[12] Dietrich Stauffer, Introduction to Percolation Theory (Taylor and Francis, London, 1985).
[13] J. Sofo, J. Lorenzana, and E.N. Martinez, Phys. Rev. B36, 3960 (1987).
[14] C.J. Lobb and M.G. Forrester, Phys. Rev. B35, 1899 (1987).
[15] B.J. Halperin, S. Feng, and P.N. Sen, Phys. Rev. Lett. 54, 2391 (1985).

Rigidity of Layered Random Alloys

Wei Jin

Materials Science Division
and Science and Technology Center for Superconductivity
Argonne National Laboratory, Argonne, IL 60439

S.D. Mahanti and M.F. Thorpe

Department of Physics and Astronomy
and Center for Fundamental Materials Research
Michigan State University, East Lansing, MI 48824

Abstract

Randomly intercalated layered alloys are excellent models for two-dimensional microporous systems. We have studied the nonlinear gallery expansion and the gallery height fluctuations by constructing a double layer model that describes the layer rigidity and the size and stiffness of the intercalant species. Exact solutions, simulations and an effective-medium theory (EMT) results are compared. Applications of the results to ternary intercalation compounds are discussed.

Introduction

There is a large variety of ternary layered alloys whose chemical composition can be written as $A_{1-x}B_xL$, where L represents the host layer such as graphite, dichalcogenide and layered sheet silicate (vermiculite).[1] Two distinct atoms (ions) A and B are intercalated into the galleries between the host layers. All these alloys show a composition dependence of the the average interlayer spacing $\langle h \rangle$ which increase with the concentration (x) of the largest constituent.[1] The linear variation of $\langle h \rangle$ with x is the well known Vegard's law,[2] although most alloys exhibit a complex nonlinear (superlinear, sublinear, sigmoidal) behavior.[1,3] The physical origin of this nonlinear behavior is a subject of considerable theoretical interest.[3-8] Furthermore these layered alloys, particularly those associated with the sheet silicates are excellent models for two-dimensional microporous systems[9] which are potentially useful for solid state catalysis.[10]

In this paper, we discuss our recent theoretical work[7] on the structural properties of these layered alloys. We will introduce a harmonic model that describes both the layer rigidity and the size and stiffness of the intercalant ions in these alloys. We give an exact solution for the case when A and B have the same stiffness. We will also give an EMT solution and compare the results with numerical simulations. Finally we apply the results to experiments in ternary intercalated graphite and in layered sheet silicates.

The Model

The basic idea of our microscopic model is that the x-dependence of $\langle h \rangle$ depends on the competition between local and global energies associated with forming a solid solution.[4,7] These energies depend upon the relative size and compressibility of the different atomic species and overall rigidity of the system.[4,7,8] The host layer-intercalant interaction is approximated by a harmonic spring of strength $K_i(= K_A$ or $K_B)$, characterizing the compressibilities of the intercalant atoms (ions),[4] and equilibrium height

$h_i(= h_A^0$ or $h_B^0)$. The intercalants are assumed to occupy randomly a set of well defined lattice sites (Fig. 1). The energy associated with the host layer deformation has two types of contributions. The first one, proportional to K_T, is the *transverse* layer rigidity,[4,7] and the second one, proportional to K_F is the *bending* layer rigidity.[3,11] Therefore the total elastic energy of the layer-intercalant system can be written as

$$E = \frac{1}{2}\sum_i K_i(h_i - h_i^0)^2 + \frac{1}{2}K_T\sum_i\sum_{<\delta>}(h_i - h_{i+\delta})^2 + \frac{1}{2}K_F\sum_i\left[\sum_\delta(h_i - h_{i+\delta})\right]^2, \quad (1)$$

where h_i is the gallery height at the site i where an intercalant (either A or B) sits. If we define a local dimensionless height d_i such that $h_i = h_A^0 + d_i(h_B^0 - h_A^0)$, then the heights and the energy are scaled by $(h_B^0 - h_A^0)$ and $(h_B^0 - h_A^0)^2$, respectively.[7]

We minimize the energy E given in Eq. (1) with respect to the heights $\{h_i\}$ for a given realization of the random variables $K_i(= K_A, K_B)$ and $h_i^0(= h_A^0, h_B^0)$ to determine the stable structure of the random alloys. We calculate the average heights $\langle d\rangle$, $\langle d_A\rangle, \langle d_B\rangle$, the average energy per site e, the fluctuations in height $\langle(d - \langle d\rangle)^2\rangle$, $\langle(d_A - \langle d_A\rangle)^2\rangle$ etc., as functions of $x, K_A, K_B, h_A^0, h_B^0, K_T, K_F$ and the structure of the host lattice by averaging over different configurations.

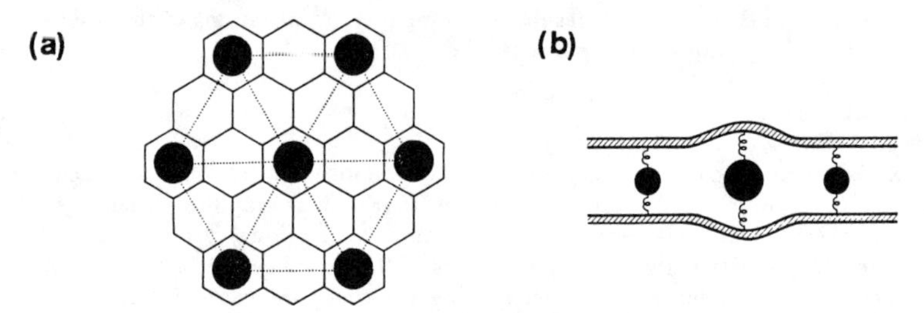

Fig. 1 (a) Top view of a hexagonal lattice host layer where the intercalants form a triangular lattice. (b) Side view of the intercalants.

Exact Solution

When the layers are either completely floppy or perfectly rigid, the model can be solved exactly.[4,7] If $K_T = K_F = 0$, we have $\langle d\rangle = x$; the usual Vegard's law[2] with $\langle d_A\rangle = 0$ and $\langle d_B\rangle = 1$. The corresponding fluctuations are given by $\langle(\Delta d)^2\rangle = \langle d^2\rangle - \langle d\rangle^2 = x(1-x)$, and the site-specific fluctuations are zero. On the opposite end, in the limit of infinite layer rigidity[4] the energy E is minimized by having all the d_i equal ($d_A = d_B = d$), i.e.,

$$\langle d\rangle = \langle d_A\rangle = \langle d_B\rangle = \frac{x}{x + (1-x)K_A/K_B}, \quad (2)$$

and fluctuations in all the three heights vanish in this limit.

When the stiffness of the A and B ions are equal ($K_A = K_B = K$), the model can also be solved exactly. The non-trivial exact solution in this case can be expressed in terms of Watson integrals $W(K) \equiv W_0(K), W_1(K)$ and $W_2(K)$, which are related to the rigidities of the layers.[7] We write

$$\langle d_B \rangle = 1 - (1-x)[1 - W(K)], \quad \langle d_A \rangle = x[1 - W(K)], \tag{3}$$

$$\langle (d_B - \langle d_B \rangle)^2 \rangle = \langle (d_A - \langle d_A \rangle)^2 \rangle = x(1-x)\left[W_1(K) - W(K)^2\right], \tag{4}$$

$$\langle (d - \langle d \rangle)^2 \rangle = x(1-x) W_1(K), \tag{5}$$

$$e = \frac{1}{2} K x (1-x)[1 - W(K)], \tag{6}$$

where the Watson integrals are defined by

$$W_n(K) = \int \frac{d\mathbf{q}}{(2\pi)^D} \left(\frac{K}{\lambda_\mathbf{q}}\right)^{n+1}, \tag{7}$$

where D is the dimension ($D = 1$ for linear chain, $D = 2$ for square and triangular lattices), and the factor $\lambda_\mathbf{q}$ is given by

$$\lambda_\mathbf{q} = K + K_T z(1 - \gamma_\mathbf{q}) + K_F[z(1 - \gamma_\mathbf{q})]^2, \tag{8}$$

where $\gamma_\mathbf{q} = \frac{1}{z}\sum_\mathbf{r} e^{i\mathbf{q}\cdot\mathbf{r}}$, and z is the number of nearest neighbors ($z = 2$ for linear chain, $z = 4$ for the square lattice, $z = 6$ for the triangular net). Clearly the geometry of the lattice will determine $\lambda_\mathbf{q}$ and hence the layer distortion characteristics. In Fig. 2 we plot $W(K)$ and $W_1(K)$ for various lattices as a function of $K/(zK_T)$ for $K_F = 0$. The average heights $\langle d \rangle, \langle d_A \rangle, \langle d_B \rangle$ show straight line behavior; with $\langle d \rangle$ obeying Vegard's law. The fluctuations and average energy are symmetric about $x = 0.5$. The difference $\langle d_B \rangle - \langle d_A \rangle = W(K)$ is independent of x. It depends only on the transverse layer rigidity parameters and decreases as the rigidity of the layer increases.[7]

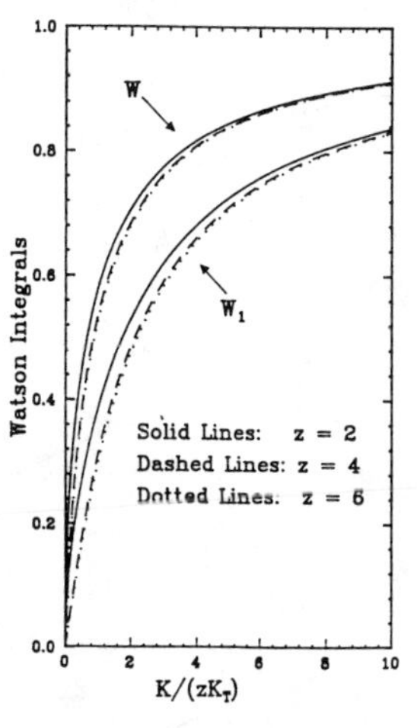

Fig. 2 The Watson integrals $W_n(K)$ are show for various lattices as a function of $K/(zK_T)$ and for $K_F = 0$.

Effective Medium Theory

We have developed an EMT to handle the case of unequal stiffness ($K_A \neq K_B$). In this approximation,[7,12] the effect of the layer on the intercalant ions is contained within an effective local spring constant K_e. This can be found by applying a force $\mathbf{F}$ to a single site in the non-random system where $K = K_A = K_B$ as shown in Fig. 3(a), where the effective spring constant for this kind of displacement K_e is given by

$$K_e = \frac{K}{W(K)}. \tag{9}$$

The whole system is now replaced by a single spring K_e as shown in Fig. 3(b). The problem is now reduced to just two springs in parallel as shown in Fig. 3(c). One of these springs is K_α, where α can be either A or B with probability $1-x$ or x respectively. The other spring is $K_e' = K_e - K$ formed by removing the spring K. From variational procedures, we obtain the self-consistency condition that determines the effective spring constant K

$$x\frac{K - K_B}{K_e' + K_B} + (1-x)\frac{K - K_A}{K_e' + K_A} = 0. \tag{10}$$

Various solutions for K from Eqs. (10) and (11) are shown in Fig. 4 for the triangular net (Fig. 1) and compared to the virtual crystal result $K_v = xK_B + (1-x)K_A$. The averages can also be obtained from the same variational procedures.[7] We write

$$\langle d \rangle = x + x(1-x)\,\frac{K_e K_e'(K_B - K_A)}{K(K_e' + K_A)(K_e' + K_B)}, \tag{11}$$

$$\langle d_B \rangle = 1 - (1-x)\frac{K_e K_e' K_A}{K(K_e' + K_A)(K_e' + K_B)}, \quad \langle d_A \rangle = x\,\frac{K_e K_e' K_B}{K(K_e' + K_A)(K_e' + K_B)}, \tag{12}$$

$$e = x(1-x)\frac{K_e' K_e K_A K_B}{2K(K_e' + K_A)(K_e' + K_B)}. \tag{13}$$

We note that when $K_B > K_A$ the average height shows superlinear behavior and the partial heights $(\langle d_A \rangle, \langle d_B \rangle)$ are no longer linear. The fluctuations can be obtained using the Feynman-Hellman theorem,[13]

$$\langle (d_A - \langle d_A \rangle)^2 \rangle = x(1-x)\,\frac{\left[\dfrac{K_A K_B K_e^2}{K(K_e' + K_A)(K_e' + K_B)(K_e' + K_\alpha)}\right]^2}{\dfrac{W(K)^2}{W_1(K) - W(K)^2} + \dfrac{(K - K_A)(K - K_B)}{(K_e' + K_A)(K_e' + K_B)}}, \tag{14}$$

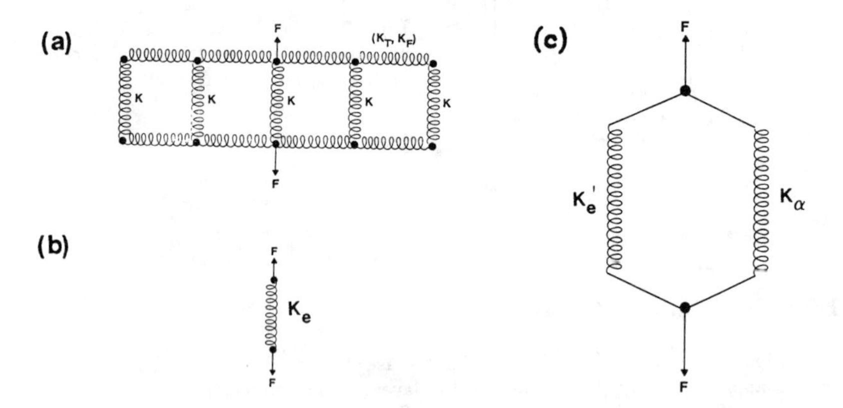

Fig. 3 Illustrating how (a) the equivalent spring K_e, (b) the local rigidity, and (c) the effective coupling constant K are determined within the EMT.

$$\langle (d - \langle d \rangle)^2 \rangle = x(1-x)\, \frac{\dfrac{W_1(K)}{W_1(K) - W(K)^2} \left[\dfrac{K_A K_B K_e}{K(K_e' + K_A)(K_e' + K_B)} \right]^2}{\dfrac{W(K)^2}{W_1(K) - W(K)^2} + \dfrac{(K - K_A)(K - K_B)}{(K_e' + K_A)(K_e' + K_B)}}. \tag{15}$$

These EMT results contain the previous exact results as special cases.

Numerical Simulation

We test the above EMT results via a direct numerical simulation method. We have performed extensive simulations for the triangular lattice where typical lattice sizes used were $N = 50 \times 50 = 2500$ nodes with periodic boundary conditions. The simulations were carried out by first generating an initial configuration of random system and then relaxing it by use of the conjugate gradient total energy minimization method.[7] In Fig. 5 we show a typical configuration of a relaxed layer obtained from the simulation at $x = 0.2$ and $K_A \neq K_B$. One sees that the relaxed surface is not flat, but rather have the appearance of "waves on the ocean". However, in the limit of very large K_T or K_F, the layers do become flat, similar to the "flat phase" of polymerized membranes.[14] The simulation results of the average heights, average fluctuations for both the A and B intercalants, and the average energy are shown as solid dots in Fig. 6 and compared with

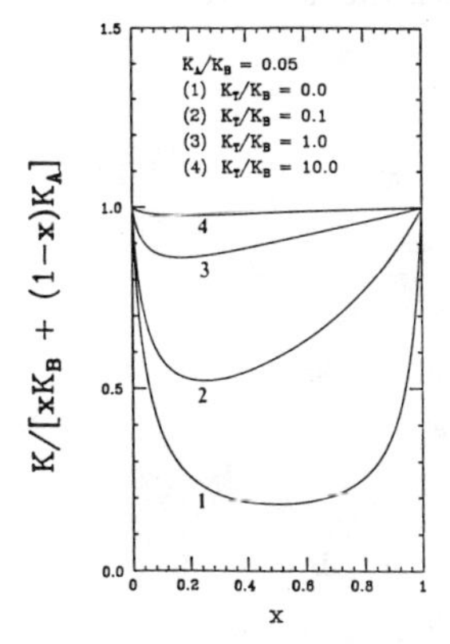

Fig. 4 The effective spring constant K for a triangular lattice.

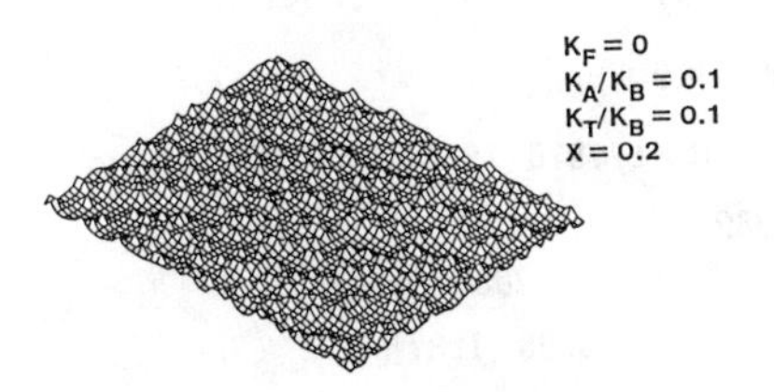

Fig. 5 A configuration of a relaxed triangular layer.

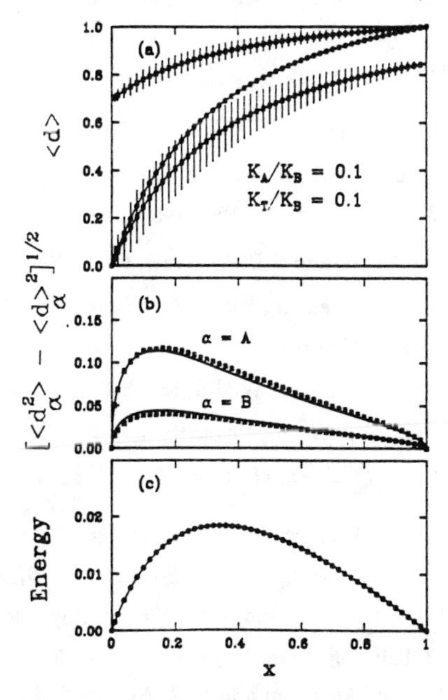

Fig. 6 Showing (a) average height, (b) fluctuations in height and (c) the average energy. The solid lines are EMT results and the solid dots are simulation results.

the EMT results. We see that the EMT gives very good agreement with the simulation results.

The two ternary graphite intercalation systems whose average c-axis separation can be understood within our model are $V_{1-x}Li_xC_6$ (V is a vacancy) and $K_{1-x}Rb_xC_8$ (Ref. 3, 15). For the lithium ternary, where the gallery expands from 3.36Å for $x = 0$ to 3.78Å for $x = 1$, the A atom is actually a vacancy so that $K_B/K_A >> 1$. A choice of $K_A/K_B = 0.1$, and $K_T/K_B = 0.1$ can semi-quantitatively fit the experimental data excepting for $x = 1$ where anharmonicity effects may be important. The potassium-rubidium ternary data, where the gallery expands from 5.47Å for $x = 0$ to 5.68Å for $x = 1$, shows nearly a Vegard's law behaviour.[15] This can be understood if we assume that $K_A/K_B \approx 1$ which seems physically reasonable.

In summary, we have set up an elastic model that describes the structural properties of random layered alloys. This model incorporates both the layer rigidity and the compressibilities of the intercalants. Vegard's law is only obtained when $K_A = K_B$ (Ref. 16).

Acknowledgments

This work was supported by the NSF Science and Technology Center for Superconductivity under DMR-88-09854 and by NSF grants DMR-85-14154 and DMR-87-14865.

References

1 S.A. Solin and H. Zabel, Adv. Phys. **37**, 87 (1988).

2 L. Vegard, Z. Phys. **5**,17 (1921).

3 H. Kim, *et. al.*, Phys. Rev. Lett. **60**, 877 (1988).

4 W. Jin and S.D. Mahanti, Phys. Rev. B **37**, 8647 (1988).

5 S. Lee, *et. al.*, Phys. Rev. Lett. **62**, 3066 (1989).

6 M.F. Thorpe, Phys. Rev. B **39**, 10370 (1989).

7 M.F. Thorpe, W. Jin, and S.D. Mahanti, Phys. Rev. B **40**, 10294 (1989).

8 Y. Cai, *et. al.*, Phys. Rev. B **42**, 8827 (1990).

9 Z.X. Cai, *et. al.*, Phys. Rev. B **42**, 6636 (1990).

10 T.J. Pinnavaia, Science **220**, 365 (1983).

11 K. Komatsu, J. Phys. Soc. Japan, **6**, 438 (1951).

12 G. Garboczi and M.F. Thorpe, Phys. Rev. B **31**, 7276 (1985); **33**, 3289 (1986).

13 R.P. Feynman, Phys. Rev. **56**, 340 (1939).

14 F.F. Abraham and D.R. Nelson, J. Phys. France **51**, 2653 (1990).

15 J.E. Fischer and H.J. Kim, Phys. Rev. B **35**, 3295 (1987); P.C. Chow and H. Zabel, Phys. Rev. B **38**, 12837 (1988).

16 M.F. Thorpe, W. Jin, and S.D. Mahanti, in *Proceedings of Sir Roger Elliott Symposium*, (to be published by Oxford University Press).

DENSIFICATION MODELS FOR PLASMA SPRAYED METAL MATRIX COMPOSITE FOILS

D.M. Elzey, J.M. Kunze, J.M. Duva and H.N.G. Wadley
University of Virginia, Charlottesville, VA 22901

ABSTRACT

The development of a constitutive model describing the deformation response of plasma sprayed metal matrix composite (MMC) foils during hot isostatic pressing (HIP) is described. A representative volume element of the composite is chosen whose constitutive behavior, including inelastic compressibility, is analyzed for the particular case of constrained uniaxial compression. Results of this preliminary model for the case of a SCS–6/Ti–14wt%Al–21wt%Nb intermetallic matrix composite are presented in the form of a density–pressure HIP map.

INTRODUCTION

Fiber–reinforced intermetallic matrix composites (IMC's) are being developed to satisfy the design requirements of future hypersonic vehicles and propulsion systems. Processing intermetallic matrix materials by conventional methods is difficult and a number of novel processing approaches have recently emerged.[1] Among these is the induction–coupled plasma deposition (ICPD) technique, wherein plasma melted droplets of intermetallic matrix material are sprayed onto a fiber wound mandrel.[2] Hot isostatic pressing (HIP) is then used for fabricating components from the plasma sprayed foils. Idealy, this near net shape processing route leads to perfect interlaminar bonding, zero internal porosity and no incidental fiber damage. Here, we seek to develop a model to predict the densification kinetics and shape changes of plasma sprayed IMC foils during HIP consolidation processing.

MATERIAL AND LOADING GEOMETRY

A typical cross section of a plasma sprayed composite foil is shown in Figure 1. The ICPD process produces a monotape with one rough (plasma sprayed) and one smooth (mandrel) side. Due to the surface roughness, foil stacking leads to a high degree of interlaminar porosity in addition to the isolated porosity within the matrix. Foils are typically about 250 μm thick with a fiber volume fraction of between 35 and 45%.

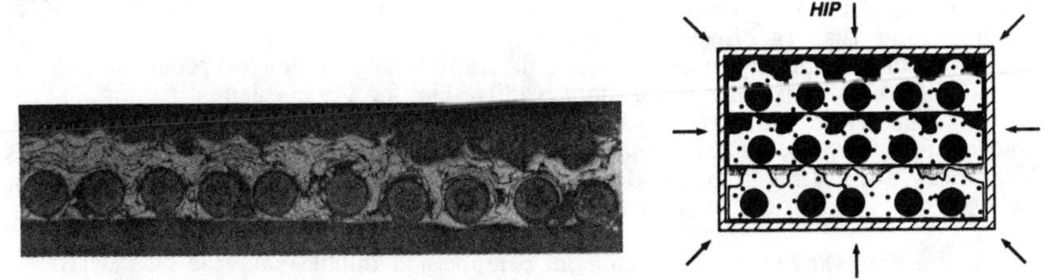

Fig. 1 Plasma sprayed SCS–6/Ti–14Al–21Nb foil.

Fig. 2 Schematic representation of IMC foils during HIP consolidation.

Hot isostatic pressing is accomplished by first placing a stack of foils cut to the desired size and shape in a preformed metal tool or cannister (Figure 2), which is then evacuated and sealed, followed by exposure to an optimal pressure and temperature cycle.

UNIT CELL IDENTIFICATION

To simplify analysis, we seek the smallest portion of the porous composite whose constitutive behavior is the same as the overall response of the entire body exposed to the same average state of stress. This volume element, referred to here as the unit cell, represents the averaged behavior of the entire composite. Referring to Figure 3, we have chosen to split the foil into two sub–laminae; the 'R'–layer is smooth on all sides and contains the fiber reinforcement while the 'P'–layer is isotropic and contains the surface roughness of the original foil. A unit cell is identified within each of the R– and P–laminae. These sub–cells are then combined as shown to construct a composite unit cell. As densification proceeds, the P–element (in contact with an adjacent foil) is gradually transformed from a discrete contact to a porous continuum. Therefore, adopting the approach already taken in modeling the behavior of spherical powders, we apply models for porous continua to describe the behavior of the P–element for densities higher than about 90%, the density at which pores are no longer interconnected.[3-5] These behavioral regimes are referred to as stages I and II.

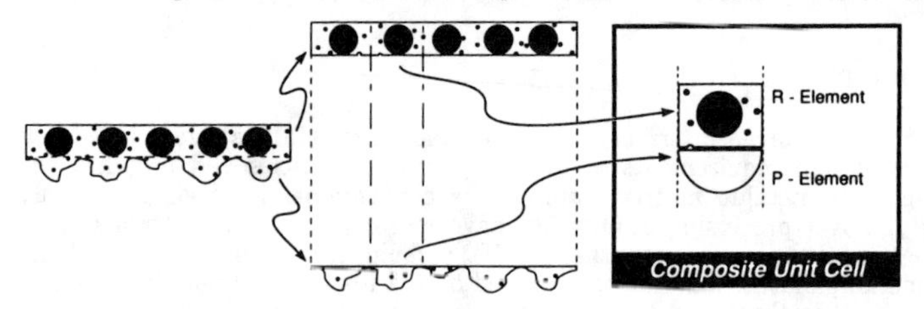

Fig. 3 The representative volume element used to predict the overall behavior.

STRESS STATE DURING HIP'ING

In order to demonstrate the validity of the chosen composite unit cell and of the associated constitutive models it is important to compare the unit cell prediction with results of experiments for which a nearly homogeneous stress state exists. The isostatic pressing of a thin, flat plate represents a simple geometry for which a macroscopically uniform state of stress is expected, (except near the foil edges).

For thin plate–like samples subjected to an external hydrostatic stress, the initial state of stress within the densifying matrix corresponds closely to that of constrained uniaxial compression, i.e., the in–plane strains are taken to be zero. This can be visualized by noting that (1) the fibers are likely to support most of the load along their direction, so that the stress–assisted densification in the direction of the fibers will be much smaller than in directions perpendicular to the fibers, (2) owing to the high concentration of porosity between the foils, initial deformation perpendicular to the plane of the plate will occur much more readily than for the in–plane directions. Thus the in–plane deformations are expected to be small in comparison with the out–of–plane deformation. Finally, for the particular case of a thin, flat plate sample, load shielding by the can walls is much greater for the two in–plane directions, and thus the driving force for lateral densification is lower.

We note that constrained uniaxial compression implies that the lateral stresses are just great enough to prevent the lateral expansion (Poisson effect) that would otherwise occur during application of a uniaxial stress. Therefore, of all the average stress states to which the unit cell is likely to be exposed during HIPing, constrained uniaxial compression probably represents the one deviating the farthest from the purely hydrostatic. By calculating in addition, the response of the unit cell to a pure hydrostatic loading, we obtain upper and lower bounds on the average densification rates. In addition, by identifying the behavior for these two experimentally verifiable stress states, the model can later be reliably extended to allow treatment of general states of stress.

UNIT CELL CONSTITUTIVE BEHAVIOR

In developing a model for the deformation response of the composite unit cell undergoing constrained uniaxial compression, we will neglect elastic strains, which are small compared with the inelastic strains of around 30% needed to reach full density. The total uniaxial strain rate, $\dot{\epsilon}$, is taken as the sum of the strain rates from all deformation mechanisms occurring within each sub cell:

$$\dot{\epsilon} = \dot{\epsilon}_p^{pl} + \dot{\epsilon}_p^{c} + \dot{\epsilon}_p^{d} + \dot{\epsilon}_r^{pl} + \dot{\epsilon}_r^{c} + \dot{\epsilon}_r^{d} \tag{1}$$

Subscripts refer to the R– and P–elements of the composite unit cell and the superscripts refer to plastic, creep and diffusional deformation mechanisms. This scheme follows after the work of Ashby and co–workers for the modeling of powder compaction.[3-4] Here, we report on the development of constitutive models for plastic yielding and power law creep, and defer diffusional mechanism contributions.

P–element Plasticity

We seek a constitutive model to predict the densification resulting from plastic yielding of contacts between adjacent foils. The P–element is taken to be a rectangular cell containing a hemispherical asperity, which is itself fully dense and therefore incompressible. We wish to predict the applied stress at which the hemisphere plastically yields when being pressed between two rigid flat plates. We then relate the resulting change in density to this applied stress. The yield condition is taken to be of the form:

$$\sigma_c \left(=\frac{f_c}{a_c}\right) \geq \beta(D) \cdot \sigma_y \tag{2}$$

where σ_c is the stress at the contact, f_c the force at the contact, a_c the contact area, which is a function of density D, $\beta(D)$ a 'geometric hardening coefficient' and σ_y is the temperature dependent uniaxial yield stress of the matrix material. Eqn (2) implies that yielding occurs when the local stress at the contact exceeds a certain value which depends on the current shape of the deformed asperity. Previous analyses for the densification of powders have assumed a constant value for β of about 3, thus neglecting the influence of changing geometry on the yield condition.[4] However, as the P–element approaches full density, β must approach infinity if the matrix material is assumed to be plastically incompressible. One might calculate β using finite element analysis, however, here we have determined it empirically from uniaxial compression tests on hemispherical samples. We find that β increases approximately linearly from an initial value of about 0.5 to about 3 as the relative density approaches 0.9.

Eqn (2) becomes a constitutive relation if the contact area, a_c, can be expressed as a function of the uniaxial strain, (i.e., the relative density of the P–unit cell). We do this by analogy with a similar development for powder compaction modeling[5], by distributing the volume of material displaced from the contact as a uniform shell enclosing the deformed hemisphere. The final expression one derives by this means is lengthy, however a fairly accurate approximation is given by

$$a_c = \frac{3}{5} \cdot [6 + 9 \ln D] \tag{3}$$

Relating the externally applied load to the local force at the contact, f_c in (2) can be replaced by $4R^2\sigma$, where $4R^2$ represents the in–plane area of the unit cell and σ is the applied uniaxial stress. Combining this result with eqns (2) and (3) provides an expression which can be solved for the relative density as a function of the applied stress:

$$\left[\frac{\sigma}{\sigma_y}\right]R^2 - C_1 = C_2D + \frac{3}{2}(C_1+C_2D)\ln D \tag{4}$$

where the linear approximation, $\beta(D) = C_1+C_2D$, has been used. This model is strictly valid for densities less than about 90% of the theoretical density.

P–element Power Law Creep

The prediction of the initial (stage I) densification rate due to power law creep does not entail any new geometrical considerations beyond those introduced for plastic yielding. We need only the rate at which the height of the asperity, h, decreases as a function of the applied stress and current density. As a first approximation, we adapt an expression for the neck growth rate of two adjacent spheres undergoing power law creep[6]:

$$\dot{x} = -\frac{B \cdot x}{2\left[1-\left[\frac{x}{R}\right]^{\frac{3}{n}}\right]^{n}} \cdot \left[\frac{3}{2}\cdot\frac{A_f}{\pi x^2}\cdot\frac{\sigma}{n}\right]^{n} \tag{5}$$

where B is a material parameter, x is the neck radius, n is the stress exponent and $A_f = 4R^2$ is the area over which the external stress is applied. (Note the effective stress at the contact, σ_c, $= \frac{A_f}{a_c}\cdot\sigma$, where $a_c = \pi x^2$, the contact area.)

We can solve (3) for x and differentiate with respect to time. Combining this result with (5) gives the densification rate of the P–unit cell due to power law creep:

$$\dot{D} = -\frac{5\pi}{27}\cdot Bx^2D\cdot\left[\frac{6}{\pi x^2}\cdot\frac{|\sigma|}{n}\right]^{n} \tag{6}$$

where the approximation x << R has been made. Accordingly, this model becomes increasingly inaccurate beyond densities of about 0.75. Further work is needed here to extend the range of validity by introducing a more representative geometry or by refinements which explicitly account for the geometry–dependence of the constitutive behavior.

R–element Plasticity

As shown in Fig. 3, the R–element is a three–phase unit cell consisting of matrix, voids and fiber. The volume fraction of matrix containing voids has been found to be about 10%, at which density models developed for stage II powder compaction become applicable. We take the pores to be spherical and uniformly distributed throughout the matrix and apply an existing constitutive model developed for porous, isotropic continua.[7] The local influence of the fiber on void shrinkage is thereby neglected, however the dominant effect of the fiber is already accounted for by taking the strain along the fiber axis to be zero.

The components of the plastic strain tensor are taken as derivable from a plastic potential Φ^p according to:

$$d\epsilon_{ij} = \frac{\partial\Phi^p}{\partial\sigma_{ij}} \tag{7}$$

We consider a form for the plastic potential derived by Gurson[7]:

$$\Phi^{p} = \left[\frac{\sigma_e}{\sigma_y}\right]^2 - 2(1-D)\cosh\left[\frac{3\sigma_m}{2\sigma_y}\right] - [1+(1-D)^2] = 0 \quad (8)$$

where $\sigma_e, = \sqrt{\frac{3}{2}s_{ij}s_{ij}}$, is the equivalent stress and $\sigma_m, = \frac{\sigma_{kk}}{3}$, is the mean normal stress.

Since during constrained uniaxial compression, the plastic strain increment in the lateral direction $d\epsilon_1^p$ ($= d\epsilon_2^p$) is zero, (7) implies that the gradient of Φ in the 1–direction (or 2–direction) must be zero and therefore,

$$\frac{\partial\Phi}{\partial\sigma_1} = \frac{(\sigma_1-\sigma_3)}{\sigma_y} - (1-D)\sinh\left[\frac{(2\sigma_1+\sigma_3)}{2\sigma_y}\right] = 0 \quad (9)$$

where $\sigma_1 = \sigma_2$ has been introduced. For any given applied stress, σ_3, equations (8) and (9) are sufficient to solve for the lateral stress, σ_1, and the relative density, D.

R–element Power Law Creep

Limiting solutions for strain rate potential surfaces, or creep potentials, have been developed by solving appropriate minimization (maximization) problems. Based on the work of Castaneda[8], we consider the creep potential:

$$\Phi^c = \frac{\dot{\epsilon}_0\sigma_0}{n+1}\left[\frac{s}{\sigma_0}\right]^{n+1} \quad (10)$$

where $s^2 = a(D)^2\sigma_e^2 + b(D)^2\sigma_m^2$. The parameters a and b are known functions of the relative density only. Components of the strain rate are obtained from

$$\dot{\epsilon}_{ij} = \frac{\partial\Phi^c}{\partial\sigma_{ij}}, \quad (11)$$

so that the uniaxial strain rate due to power law creep during constrained compression is found to be:

$$\dot{\epsilon}_3 = \alpha(D)B\sigma_3^n \quad (12)$$

where

$$\alpha(D) = \left[a^2(1-\zeta)^2 + \frac{b^2}{9}(1+2\zeta)^2\right]^{\frac{n-1}{2}}\left[a^2(1-\zeta) + \frac{b^2}{9}(1+2\zeta)\right] \quad \text{with } \zeta(D) = \frac{a^2 - \frac{2}{9}b^2}{a^2 + \frac{4}{9}b^2}.$$

The factor ζ represents the ratio of the lateral stress to the axial stress and, as expected, goes to 1 as the density goes to 1, in agreement with the assumption of incompressibility of the fully dense material. If the factor α in eqn (12) is equal to 1, (12) reduces to the usual uniaxial power law creep relation for a fully dense material. However, since the strain rate must go to zero at full density for a material undergoing constrained compression, α must go to zero as the density approaches 1, which in fact it does. The densification rate may be determined from (12) by recalling that $\dot{D} = -D\dot{\epsilon}_{kk}$.

Composite Response

As an example of the approach we consider the case of an SCS–6 reinforced Ti–14Al–21Nb foil. The relative density of each sub–cell (P and R) is determined by first calculating the contribution due to plasticity, which is taken to occur instantaneously for any given applied stress and temperature. Further densification at a given pressure and temperature is determined by numerically integrating the equation for the densification rate due to power law creep. Now the density of the composite unit cell is obtained by applying the rule–of–mixtures to the sub–cell densities. Figure 4 illustrates, for a constant temperature of 800 C, the predicted densification as a function of pressure and time. (Time contours are shown at t = 0, 0.25, 0.5, 2 and 4 hours.) The initial densities of the P– and R–elements were taken as 0.5 and 0.9 respectively, with volume fractions of 0.55(P) and 0.45(R). Matrix material properties used are those shown in the table. The fiber volume fraction was taken to be 35%.

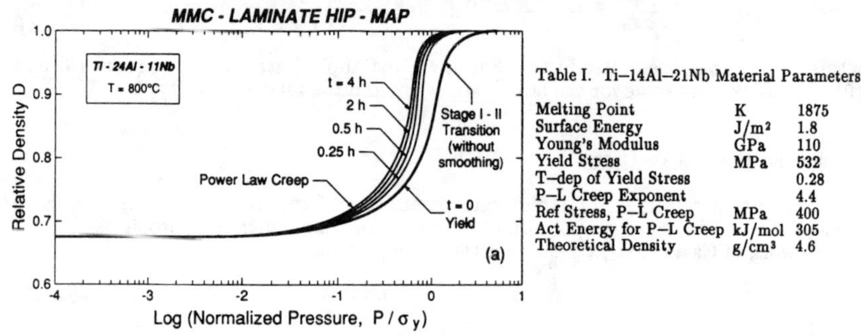

Table I. Ti–14Al–21Nb Material Parameters

Melting Point	K	1875
Surface Energy	J/m^2	1.8
Young's Modulus	GPa	110
Yield Stress	MPa	532
T–dep of Yield Stress		0.28
P–L Creep Exponent		4.4
Ref Stress, P–L Creep	MPa	400
Act Energy for P–L Creep	kJ/mol	305
Theoretical Density	g/cm^3	4.6

Fig. 4 Predicted density of a SCS–6/Ti–14wt%Al–21wt%Nb composite as a function of applied pressure.

SUMMARY

A preliminary model for predicting the HIP–consolidation behavior of intermetallic–matrix composite foils has been formulated. The problem is simplified by the identification of a representative volume element, whose behavior is approximated by summing deformations due to plastic yielding and power law creep. The analysis presented is for the case of a thin, flat plate geometry, for which the average stress state in the matrix corresponds approximately to that of constrained uniaxial compression. Continuing work is aimed at experimental validation of the model, extension of P–element models to arbitrary stress states and inclusion of diffusional deformation mechanisms.

Acknowledgements

The authors would like to acknowledge support for this work by the Defense Advanced Research Projects Agency (W.Barker, Program Manager), the National Aeronautics and Space Administration, and General Electric Aircraft Engines (Lynn).

References

1. R. Mehrabian, Mat. Res. Soc. Symp. Proc., 120 (3), (1988).
2. D.G. Backman, JOM, July 1990, p.17.
3. A.S. Helle, K.E. Easterling and M.F. Ashby, Acta Metall., 33, 2163, (1985).
4. E. Arzt, M.F. Ashby and K.E. Easterling, Met. Trans., 14A, 211, (1983).
5. H.F. Fischmeister and E. Arzt, Pow. Metall., 26 (2), 82, (1983).
6. D.S. Wilkinson, Acta Metall., 23, 1277, (1975).
7. A.L. Gurson, Trans. ASME, Jan. 1977, p.2.
8. P.P. Castaneda, J. Mech. Phys. Solids, (1990), in press.

PART III

Processing and Characterization of Cellular Materials

MICROCELLULAR FOAMS PREPARED FROM DEMIXED POLYMER SOLUTIONS

J. H. Aubert, Division 1813, Sandia National Laboratories, Albuquerque, New Mexico 87185

ABSTRACT

Low-density, microcellular polymer foams have numerous applications as structural supports in high-energy physics experiments, in catalysis, ion exchange, and filtration, and for a variety of biomedical uses. A versatile method to prepare such foams is by thermally-induced phase separation (TIPS) of polymer solutions. Demixed solutions can be transformed into a foam by freezing the demixed solution and removing the solvent by freeze-drying. The morphology of these foams is determined by the thermodynamics and kinetics of phase separation. A model of both the early and late stage structure development for demixed polymer solutions will be presented. For semi-crystalline polymers, gels can be prepared by crystallizing the polymer from solution, either a homogeneous solution or a demixed solution. Foams can be prepared from these gels by the super-critical extraction of the solvent. By understanding and utilizing the phase separation behavior of polymer solutions, engineered microcellular foams can be prepared. To design the foams for any application one must be able to characterize their morphology. Results will be presented on the morphological characterization of these foams and the relationship of the morphology to their processing history.

INTRODUCTION

687 million kilograms of polystyrene foams were manufactured in the U.S. during 1989 [1]. A great deal of this product was prepared by a physical process wherein a blowing agent (usually 6% pentane) was diffused into polystyrene beads and expanded at elevated temperature (above the glass transition temperature of polystyrene, $T_g = 100°C$). In this process each bubble nucleation site becomes the center of a foam cell which has roughly a spherical shape and a diameter in the range of 50 to 150 micrometer (μm) [2]. These foams enjoy great usage in packaging where their insulating and shock absorbing capabilities are exploited. About 14% of the polystyrene foam produced in 1989 was in the form of cups, such as coffee cups [1]. Figure 1 shows a scanning electron photomicrograph (SEM) of a

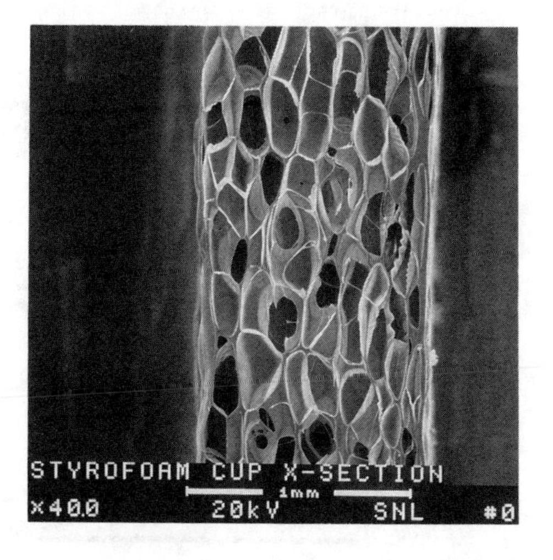

Fig. 1. Scanning electron photomicrograph of a polystyrene foam obtained from a coffee cup. The bar in the photograph is 1000 micrometers.

polystyrene bead foam obtained from a coffee cup. The cells have diameters of about 100 μm. What if one wanted to make a package with a thickness of only 200 μm?

This is not as uncommon of a request as one might think. A number of targets for high energy physics experiments require low-density foam coatings or small low-density foam supports with dimensions smaller than a millimeter (1000 μm). An example would be a target for inertial confinement fusion (ICF) containing deuterium-tritium fuel [3,4,5]. The foam coatings within such a target could be made with a conventional foam but since the entire thickness would contain as few as 2 or 3 cells it would look very nonuniform, i.e. the fluctuations of density within the foam (varying from about 1.0 g/cc in the solid phase to about 0.0 g/cc in the gas phase) would occur on a length scale comparable to the size of the part. This is undesirable for these physics applications which require structural supports which are very homogeneous. To address the need for small-celled foams, numerous authors [3,6,7] have studied other types of phase-separation processes which can create finer-scaled structures. These foams have been termed microcellular.

PROCESS DESCRIPTION

Foams are two phase systems which consist of a continuous polymer phase and either a continuous or discontinuous gaseous phase. Most foams begin as a single phase and undergo some type of phase separation to create the foams' structure. In the polystyrene foam (coffee cup) described above the polymer contains dissolved solvent (i.e. pentane), which phase separates by a nucleation process when the temperature is raised. To prepare a microcellular polymer foam, we take advantage of different types of phase separation processes which create finer structures. Two types of phase separation processes utilizing polymer solutions are liquid-liquid phase separation and polymer crystalization. Sometimes both can occur during the thermal processing of a polymer solution. A generic phase diagram (temperature-composition) of a monodisperse polymer solution showing these types of phase separations is shown in Fig. 2.

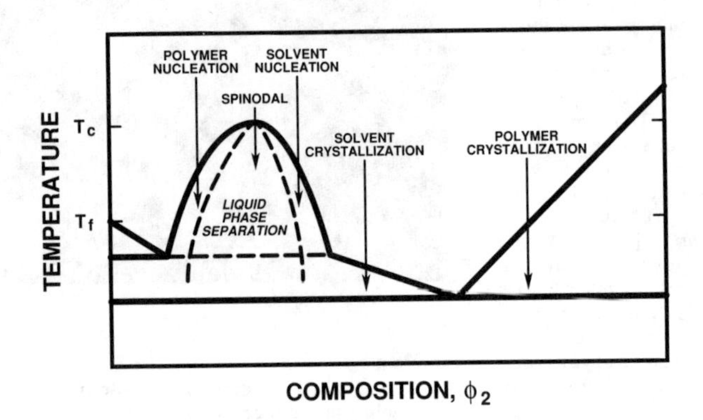

Fig. 2. Typical temperature-concentration phase diagram for a polymer and a solvent which can both independently crystallize and can also liquid phase separate.

At high temperature the polymer is in solution and only one phase exists on this phase diagram. As the temperature is reduced, polymer crystals grow from solution when the equilibrium crystallization curve is crossed. This temperature usually depends upon the concentration of the solution and can be described by polymer solution thermodynamic theories such as that due to Flory [8]. The solution may also undergo a liquid-liquid phase separation which results in the formation of two liquid phases in equilibrium; one phase is typically nearly pure solvent and the other phase is a semi-dilute or concentrated polymer solution. The assymetry in the liquid-liquid phase separation is a manifestation of the polymers high molecular weight compared to that of the solvent. Liquid-liquid phase separation of a polymer solution can also be described by polymer solution thermodynamic theories [8]. Both of these phase separation processes are termed **thermally-induced phase separations** (TIPS) because of the convienient quench used to induce phase separation.

The region of liquid-liquid phase separation can be further divided into two types which depend upon the kinetics of phase separation, the nucleated region which occurs between the binodal and spinodal curves on the phase diagram, and the region of spinodal decomposition which occurs below the spinodal curve. The nucleated region is characterized by phase separation occuring by the nucleation of a second phase. To the left of the critical point the nucleated phase would be a polymer-rich phase, while to the right of the critical point the nucleated phase would be a solvent-rich phase. In both cases only nuclei of a sufficient size would be stable enough to persist because of the energy associated with the interfacial area. Hence, nucleation is characterized by a lag time, prior to which few nuclei have grown to a sufficient size to be stable. The spinodal region is characterized by the critical nuclei size being exactly zero, and hence no lag time is associated with phase separation in this region. In addition, at high enough concentrations, phase separation in the spinodal region is characterized by the presence of bicontinuous phases, rather than discrete droplets usually associated with nucleation and growth.

Finally, for some polymer solutions the freezing point curve of the solvent can be crossed. This usually depends upon polymer concentration as there is a freezing point depression with any dissolved component in a solution (such as the polymer). Under the binodal, however, the freezing point is always constant because of the thermodynamic requirement that the solvent chemical potential be identical in the two equilibrated phases. This fact is often used to identify the binodal region; i.e. under the binodal the solvent freezing point remains constant.

If the TIPS process results in the formation of a continuous polymer-rich phase, then a microcellular foam can be prepared with two additional processing steps. First the morphology of the phase-separated solution must be preserved either through vitrification or crystallization of the polymer. This step preserves the small-scale morphology of the demixed solution. If the quench results in vitrification or crystallization of the polymer, without the solvent crystallizing, then a polymer gel results. If the solvent is crystallized it normally excludes polymer from the crystals. The polymer therefore becomes more concentrated and vitrifies, which also renders it immobile. After either of these possibilities, solvent freezing or solution gelation, the solvent must be removed.

In a freeze-drying process, the frozen solvent is removed by sublimation under vacuum. After solvent removal only polymer remains which has a morphology closely related to that of the phase-separated solution. Super-critical extraction is another technique in which the solvent can be removed while preserving the desired morphology. If the liquid in a gel is taken above its' critical point, then it can be removed in the single phase region without any liquid/vapor meniscus and the related interfacial tension. In the absence of interfacial tension, the solvent leaves the gel without collapsing the gel morphology. In actual practice, the original solvent is often exchanged for one which is amenable to freeze-drying or to super-critical removal. In the latter case, super-critical carbon dioxide has been invaluable [7,9].

MODEL FOAMS

Polystyrene microcellular foams have been prepared using the TIPS process with a wide range of controlled morphologies. The morphologies can be directly correlated to the solution phase diagrams and processing conditions. By controlling how the phase separation occurs, foam morphology, cell size, and density can be engineered for a specific application [6,9,10].

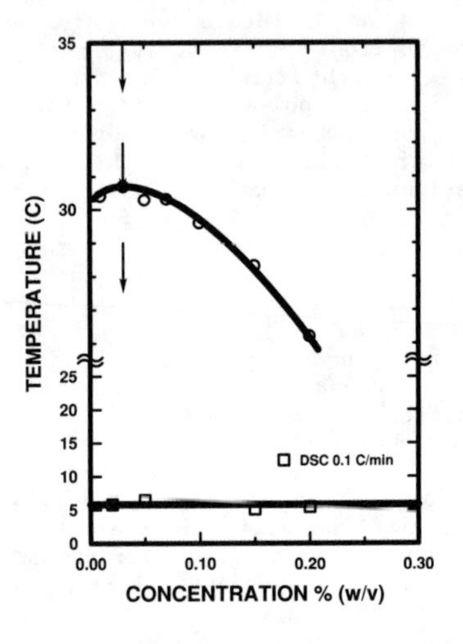

Fig. 3. Cloud point curve of a solution of polystyrene (molecular weight of 2.23×10^6) in cyclohexane solvent.

Figure 3 shows a phase diagram (cloud point curve) determined for polystyrene ($M_w = 2.23 \times 10^6$, $M_w/M_n = 1.12$) in cyclohexane. This polymer is not crystalline but does have a region of liquid-liquid phase separation which is indicated by the cloud point curve shown in the figure. The freezing point curve of cyclohexane in the solution is also shown. If a solution with a concentration near critical is quenched through the cloud point curve, and then frozen and freeze-dried, a foam with a morphology consistent with a spinodal structure is obtained (bicontinuous phases). The cell sizes obtained for fast quenches are consistent with predictions of spinodal wavelengths for demixed polymer solutions [6]. The cell size can be varied by varying the quench rate through the cloud point curve. Fig. 4 shows the cell sizes which can be obtained from four different quench rates. This variation is possible because when the solution enters the two-phase region a demixed structure forms via a preferred kinetic route. The equilibrium state, however, correspondes to two separate solutions, each of which has the tie-line concentrations. The structure which initially forms in the two-phase region will evolve to this equilibrated structure. By freezing the solution at different times (i.e. by varying the quench rate) different structures can be preserved during this evolution. A model for the structural coarsening of demixed polymer solutions has been presented which is based upon a diffusive coarsening mechanism with the structural size growing with time to the 1/3 power [10].

Using different polystyrene concentrations, foam morphologies which resemble those expected for a nucleated process can be obtained. Below the critical concentration, phase separation proceeds by the nucleation of droplets of the polymer-rich phase. A foam prepared in this region has the

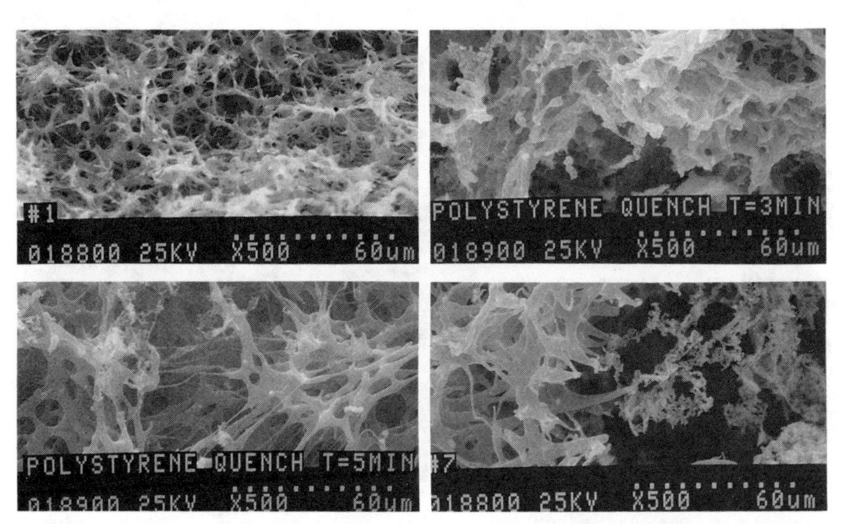

Fig. 4. Scanning electron photomicrographs of polystyrene foams obtained from 3%w/v solutions of polystyrene (M_w = 2.2x10^6) in cyclohexane quenched at different rates. The quench rates were varied by holding each solution in the two-phase regions for 1, 3, 5, or 7 minutes prior to freezing the solution.

morphology shown in Fig. 5. As is seen, the morphology consists of polymer balls weakly connected together, which is consistent with a nucleation process. At concentrations higher than the critical concentration, one expects that the phase separation will proceed by the nucleation of the solvent-rich phase. In the dried material then, one expects to see roughly spherical voids which correspond to the solvent-rich regions of the demixed solution. Fig. 6 shows a foam prepared in this region and one does indeed see that the morphology has a more traditional foam-like morphology with roughly spherical voids.

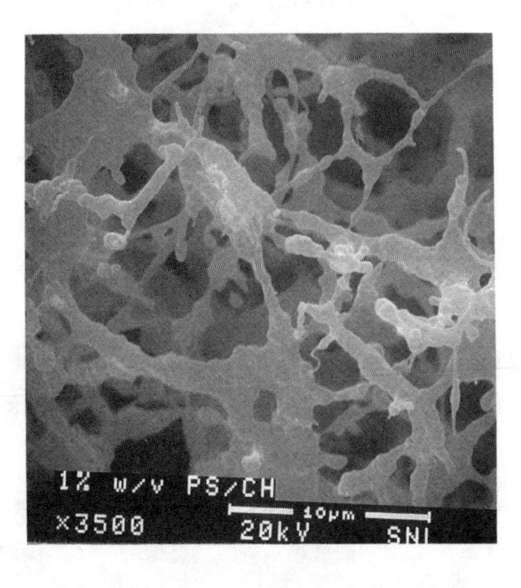

Fig. 5. Polystyrene foam prepared from a 1%w/v solution in cyclohexane (lower than critical). The phase separation was expected to occur by polymer droplet nucleation.

In Fig. 7 schematic phase diagrams for polystyrene in different ratios of the cosolvent mixture 1,4-dioxane and isopropanol are shown. Below the phase diagrams are SEM photomicrographs obtained from foams made from 5%w/v solutions in the various cosolvent mixtures. By changing the solvent ratios, one can vary the foam morphology from one that is anisotropic to one that is isotropic. One can vary the cell sizes of both the anisotropic and the isotropic morphologies. In a solvent composed of 80% 1,4-dioxane and 20% isopropanol, the phase diagram consists of just the freezing point curve of 1,4-dioxane from solution. There is no region of liquid-liquid phase separation. As 1,4-dioxane freezes, dendritic crystal growth excludes polymer which is displaced to the grain boundaries. This results in the anisotropic morphology shown. As more non-solvent (e.g. isopropanol) is added to a solution, the solvent quality worsens and eventually a region of liquid-liquid phase separation occurs.

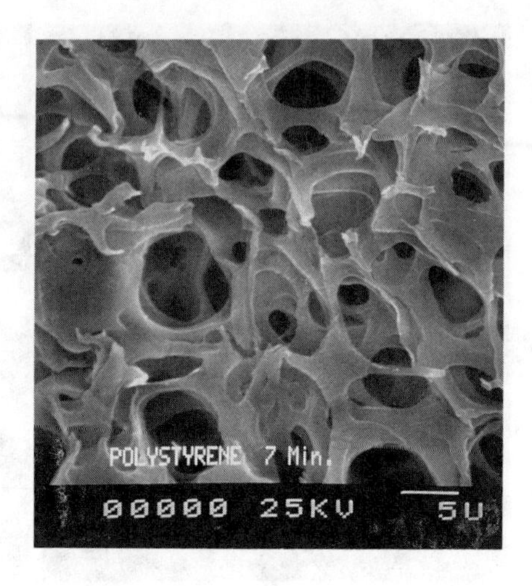

Fig. 6. Polystyrene foam prepared from a 10%w/v solution in cyclohexane (higher than critical). The phase separation was expected to occur by solvent droplet nucleation.

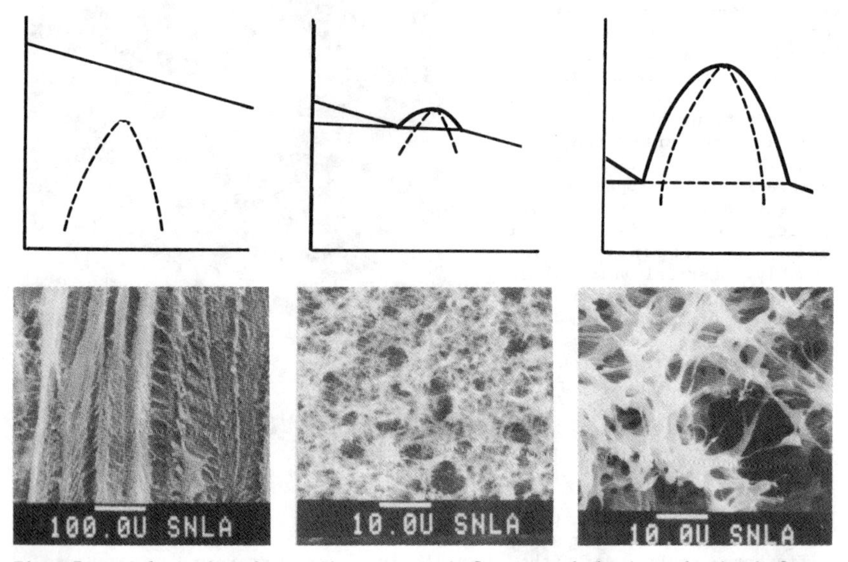

Fig. 7. Schematic phase diagrams and foam morphologies obtained from solutions of atactic polystyrene ($M_w = 2.2x10^6$) in a cosolvent of 1,4-dioxane and isopropanol in different ratios: left 80% 1,4-dioxane and 20% isopropanol, center 60% 1,4-dioxane and 40% isopropanol, right 55% 1,4-dioxane and 45% isopropanol.

For low concentrations of isopropanol only a small region exists as shown in the schematic phase diagram for 60% 1,4-dioxane and 40% isopropanol. A quench of a solution through this region is therfore necessarily quick and a small-celled foam results. If additional nonsolvent is added to the solution the region of liquid-liquid phase separation is larger as depicted for the cosolvent of 55% 1,4-dioxane and 45% isopropanol. If a solution is quenched through this region the quench will be relatively slower since the distance to traverse through the region of liquid-liquid phase separation is greater. Hence, the foam cell size obtained is larger as is shown in the photomicrograph. By changing the solvent, the polymer concentration, or the quench a great many thermodynamic histories can be obtained for a given polymer solution, each of which may result in a different foam morphology.

Isotactic (crystallizable) polystyrene (iPS) has been utilized to prepare microcellular foams with even smaller cell sizes and higher surface areas [9] than obtained with atactic (noncrystallizable) polystyrene. Figure 8 shows a SEM photomicrograph of an iPS foam at very high magnification which demonstrates the ability of this system to produce extremely small-celled and high surface area foams. The phase diagrams have been established for some solutions. One example is that of an iPS of molecular weight 160000 in nitrobenzene solvent, which is shown in Fig. 9. This phase diagram has some of the features depicted in Fig. 2 [9]. At quite high temperatures a crystallization curve is found below which polymer crystals grow from solution The rate of crystallization is very slow for this polymer. However, if sufficient time is allowed for polymer crystals to grow, they will physically interact and the solution will gel. The solvent in the gel can be extracted by replacing it with super-critical CO_2, which is then removed as described above. The dry-gel (foam) has a morphology as shown in Fig. 10. This is a common morphology obtained for polymer crystallization from solution.

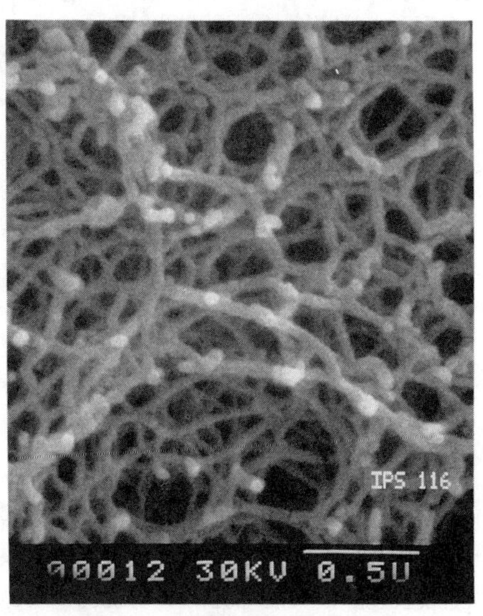

Fig. 8. Dried gel of isotactic polystyrene formed from a solution in nitrobenzene and quenched into the spinodal region. The gel was dried by exchanging the solvent with supercritical CO_2.

Because of the slow crystallization kinetics of this polymer, large supercoolings of the solutions are possible. At sufficient supercoolings, a different phase boundary can be crossed which has been interpreted as a buried spinodal [9]. Below the spinodal liquid-liquid phase separation occurs quickly and the structure formed is immediatly gelled because of polymer crystal growth. Dried gels formed in this region have a morphology as shown in Fig. 8. This morphology is consistent with a spinodal process and similar to that obtained with atactic polystyrene quenched near the critical point.

At even larger supercoolings, the solvent (in this case nitrobenzene) can be frozen. As in the case of atactic polystyrene, foams formed by quenches directly into this region are anisotropic due to dendritic crystal growth as shown in Fig. 11. Foams formed by quenches through the spinodal region and ending in solvent freezing, can have either morphology depending upon the strength of the gel formed in the spinodal region.

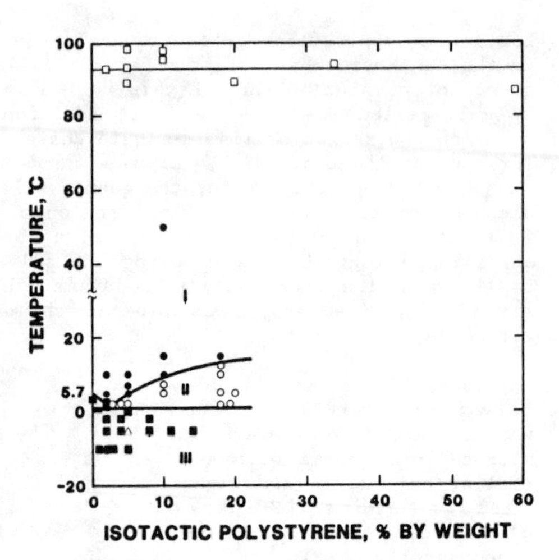

Fig. 9. Phase diagram of isotactic polystyrene (M_w=1.6x10^5) in nitrobenzene: spinodal region, 0; polymer crystallization, ●; atactic polystyrene cloud point (M_w=2.2x10^6), △ nitrobenzene freezing, ■; gel melting points by DSC, □.

Polystyrene micro-cellular foams prepared using the TIPS process have been studied in some detail and excellent correlation between the solution phase diagrams, the kinetics of phase separation, and the observed morphologies was obtained. Numerous other microcellular polymer foams have also been prepared with this method. Usually this has been accomplished without detailed knowledge of the phase diagram and kinetics. In some cases the techniques developed for polystyrene foams have been successfully applied to other systems. A number of other microcellular foam systems have been developed for specific applications which take advantage of the unique mechanical or chemical properties of the polymer chosen. To effectively use these materials in any of the applications it is necessary to characterize the cell size distribution. This is not an easy task since often the cells are very random or the walls are nonuniform or strut-like.

CHARACTERIZATION

An exact stereological relationship exists between the surface area per unit volume, S_V, of a material and the number of intersections of test line of unit length with the surfaces, P_L [11,12,13].

$$S_V = 2 P_L \quad (1)$$

We define N_L as the number of intersections of the solid phase of the foam per unit length of test line. For a low-density foam, a test line will almost always intersect both sides of the solid phase (once entering and

Fig. 10. Dried gel of isotactic polystyrene (M_w=1.6x10^5) formed in nitrobenzene in the region of polymer crystallization. The solvent was exchanged with supercritical CO_2.

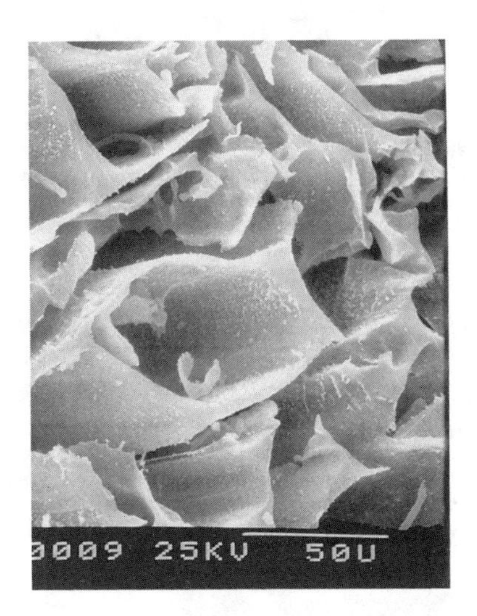

Fig. 11. Dried gel of isotactic polystyrene (M_w=1.6x10^5) formed in nitrobenzene in the region of solvent freezing. The solvent was exchanged with supercritical CO_2.

once leaving) and therefore, N_L, is equal to one half of P_L. Typical applications of these stereological relationships involve placing test lines on a photograph of a foam, counting the number of surface intersections, P_L, or solid phase intersections, N_L, and then calculating the surface area per unit volume or some other property related to these quantities. Our technique is to physically measure the surface area, with a nitrogen adsorption technique or with mercury intrusion, and with this measurement to calculate P_L or N_L. The motivation behind this is that $1/N_L$ is equal to the average distance between the solid phases of the foam, d:

$$d = 1/N_L = 4/S_V \tag{2}$$

This is an exact morphological feature which can be calculated from a measured value of the surface area per unit volume, S_V. The average distance between the solid phases of a foam can be thought of as a generalized cell size. This is a particularly useful property to characterize a foam which has a morphology that does not consist of easily-defined cells.

For a single surface area measurement obtained from BET nitrogen adsorption a single average cell size is obtained. This is called the surface-area-average cell size [12], and is defined as,

$$< d >_S = [\Sigma d_i S_i] / \Sigma S_i , \quad (3)$$

where S_i is the surface area of the cells having average spacing between surfaces of d_i. For rigid foams, such as carbonized polyacrylonitrile (PAN) [14], a cell size distribution exists which can be determined through mercury intrusion experiments. Various average cell sizes can be determined from the intrusion profile including the surface-area-average and the volume-average cell size, which is defined as [13],

$$< d >_V = [\Sigma d_i V_i] / \Sigma V_i , \quad (4)$$

where V_i is the volume of cells having average spacing between surfaces of d_i. Figure 12 shows a typical intrusion curve for a bimodal cell size distribution carbon microcellular foam prepared from PAN. Such techniques to characterize the cell size distribution of microcellular foams are invaluable to the development of microcellular foams, to comparisons of the cell size distribution of a foam to models of foam formation, or to relate the performance of a foam to its morphological structure.

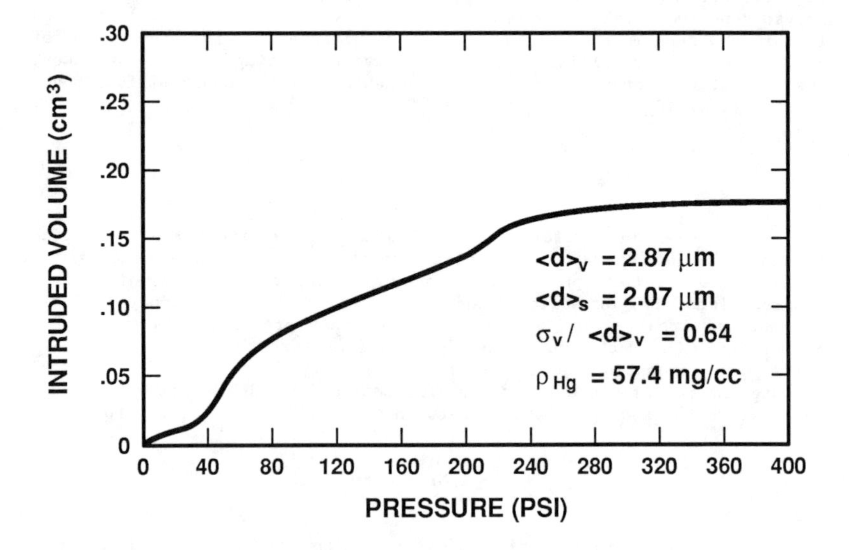

Fig. 12. Mercury intrusion curve of a carbon microcellular foam of density $0.057 g/cm^3$. The bimodal cell size distribution is evident from the intrusion curve and also from photomicrographs of the foam.

CONCLUSIONS

A large variety of polymers can be made microporous using the TIPS process. The morphology of these foams is determined by the thermodynamics and kinetics of phase separation. Hence, by controlling the phase separation process, one can engineer the morphology of the foams for a given application. Foam morphology can be varied from an anisotropic-tubular morphology to a very isotropic strut-like morphology. For a given morphology, the cell size can also be varied. With the TIPS process, microcellular foams have been prepared with a cell size below 0.1 μm, and a corresponding surface above 300 m^2/g. These foams have a large number of potential applications which span the range from somewhat esoteric uses such as high energy physics targets to biomedical uses.

ACKNOWLEDGEMENTS

The author wishes to thank E. Russick for technical assistance. L. Maestas, D. Husskinson, and B. McKenzie provided the SEM assistance. This work was performed at Sandia National Laboratories supported by the U. S. Department of Energy under contract number DE-AC04-76DP00789.

REFERENCES

1. Data from Modern Plastics 67, 99 (1990).

2. C. J. Benning, Plastic Foams, (Wiley-Interscience, NY 1969).

3. A. T. Young, D. K. Moreno, and R. G. Marsters, J. Vac. Sci. Technol. 20, 1094 (1982).

4. A. Coudeville, P. Eyharts, J. P. Perrine, L. Rey, and R. Rouillard, J. Vac. Sci. Technol. 18 (3), 1227 (1981).

5. A. T. Young, J. Cellular Plastics 23, 55 (1987).

6. J. H. Aubert and R. L. Clough, Polymer 26, 2047 (1985).

7. J. M. Williams and J. E. Moore, Polymer 28, 1950 (1987).

8. P. J. Flory, Principles of Polymer Chemistry (Cornell University Press, Ithaca N.Y., 1953).

9. J. H. Aubert, Macromolecules 21, 3468, (1988).

10. J. H. Aubert, Macromolecules 23, 1446, (1990).

11. E. E. Underwood, Quantitative Stereology, (Addison-Wesley, Reading, MA 1970).

12. J. H. Aubert, J. Cellular Plastics 24, 132 (1988).

13. J. H. Aubert and A. P. Sylwester, submitted to J. Mat. Sci. (1990).

14. A. P. Sylwester, J. H. Aubert, P. B. Rand, C. Arnold, Jr., and R. L. Clough, ACS Polym. Mat. Sci. and Eng. 57, 113 (1987).

SYNTHESIS AND CHARACTERIZATION OF CELLULAR SiO_2 MATERIALS BY FOAMING SOL GELS

Josephine Covino and Allen P. Gehris, Jr.
Research Department, Naval Weapons, Center China Lake, Ca. 93555-6001

ABSTRACT

A variety of cellular SiO_2 materials have been synthesized using a foaming sol-gel process and their properties have been characterized. The process uses the rapid viscosity change during gelation to stabilize the structure of a foamed silica sol. It was found that the properties of these cellular materials are determined by method used. For example, the porosity and strength of these porous oxides depend on method of agitation and addition of Freon during the foaming process.

Density measurements, viscosity measurements as a function of pH, optical characterization, x-ray crystallography, ultimate compressive strength, dielectric constant measurements and thermal diffusivity were used to characterize these porous SiO_2 materials. This paper will discuss the synthetic processes used to develop the porous silicas and properties of these materials.

INTRODUCTION

Today lightweight ceramics are finding many varied applications due to their unique properties, including high specific stiffness, high damping capacity, excellent dimensional stability, high thermal shock resistance, high surface area, low thermal conductivity, and low dielectric permittivity. This mix of properties makes these materials useful in either structural or functional applications such as high temperature insulation, catalyst supports, chemical and moisture sensors, and as possible candidates for high speed computer device packaging. Additionally, these porous materials offer a potential alternative for fabrication of composites by their infiltration with polymers, metals, fibers, or ceramics. Due to ease of processing and varied properties these highly porous materials may lead to a new generation of lightweight ceramics and composites.

Processing methods for cellular inorganic materials include: foaming of molten glasses [1] and cements [2], sintering of hollow glass spheres [3] and replication of polymer foams [4]. Unfortunately, these processing methods are not reproducible, and materials with small, uniform cell size and with a tailored cell structure are very difficult to make.

Recently, Fujiu, et al. reported the use of a modified sol-gel process for making porous inorganic materials with tailored cell structure [5]. Their method applies the polymer foaming concept to sol-gel systems. Polymer foaming is performed by decomposition or evaporation of a foaming (Freon) agent followed by a rapid polymerization reaction to stabilize the foam structure. In this paper we will use the synthetic process of Fujiu, et al. to synthesize SiO_2 ceramics having tailored cell structures for use as lightweight insulation materials. Characterization of these materials includes: microstructure as a function of foaming process, density measurements, viscosity measurements as a function of time at different temperatures and varying pH, x-ray crystallography, dielectric constant measurements and thermal diffusivity measurements.

EXPERIMENTAL

Synthesis

Samples were synthesized using the Fujiu, et al. [5] foaming process. A schematic of the integral steps in this process is illustrated in Figure 1. In the first step, a colloidal silica (LUDOX HS-40, DuPont De Nemours and Co.) containing 40 wt% SiO_2 and stabilized with Na_2SO_4 is pH adjusted from 9.7 to a pH of 5.2-6.0 with H_2SO_4. By lowering the pH of the silica hydrosol with H_2SO_4, gelation rapidly takes place within 15 to 20 minutes. Step 2 consists of addition of surfactant, co-surfactant and foaming agent. The surfactant, sodium dodecyl sulfate {(SDS) $[CH_3(CH_2)_{11}OSO_3Na]$} is added in step 2 at a concentration of 0.1 wt%. SDS acts to disperse Freon and stabilize the gas bubbles during foaming. Fujiu, et al. [5] found that foams are unstable if less than 0.1 wt% of SDS was added. Methanol (0.1 cc/g sol) and Freon (Freon-11 (CCl_3F) 0-0.01 cc/g sol) are also added to the hydrosol in Step 2. Methanol acts as a co-surfactant, increasing surface viscosity and foam life [6]. Freon is the foaming agent. Freon-11 was chosen for its low solubility in water, low boiling point (23.8°C), and availability.

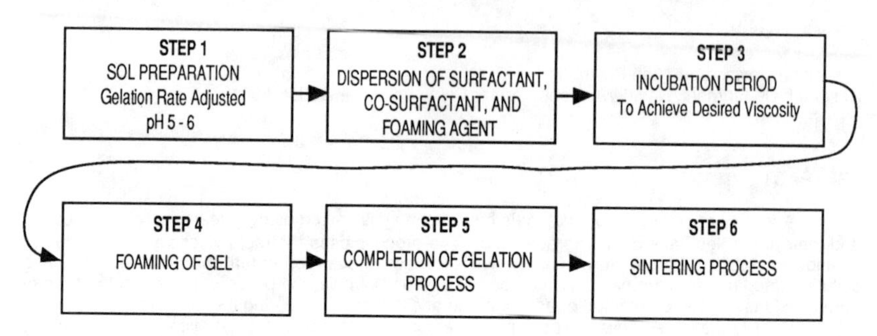

Figure 1. Schematic of the sol-gel foaming process as developed by Fujiu, et al. [5]

Step 2 takes place within 15 minutes of adjusting the pH, at a sol temperature of 20°C± 2°C. Temperature control below 23.8°C is necessary to prevent premature volatilization of Freon. A magnetic stirrer is used to mix surfactant, methanol and Freon additions. This sol is then allowed to go through an incubation period (Step 3) by placing it in a temperature controlled bath at 30°C for 20-40 minutes (depending on pH). Once the desired foaming viscosity is reached the gel is agitated using either a magnetic stirrer or a laboratory stirrer to promote foaming (Step 4). Maximum foaming occurs usually after one minute, at which point the beaker containing the foamed gel is sealed and aged for 12 hours in a water bath at a temperature of 25°C - 40°C (Step 5). The foamed gels are then carefully dried by sealing and storing the samples at room temperature for 5 days and one additional day unsealed. This procedure helps to insure uniform drying and prevents cracking. In Step 6, the foamed gels are sintered. They are sealed and stored in a drying oven at 70°C for seven days. Then the foam is heated to 400°C for forty-eight hours to remove residual organics and sintered in air at 1000°C - 1100°C for 30 minutes.

Characterization

The foam materials were characterized using the following techniques: viscosity measurements as a function of pH and temperature, density measurements, optical microscopy, strength measurements, x-ray diffraction, dielectric constant and thermal emissivity.

Viscosity measurements were made using a Brookfield viscometer model #DV-II with a helipath TB spindle, which is specifically designed for gels and other high viscosity materials. Viscosity measurements were made as a function of pH and temperature for the SiO_2 sols.

The density and porosity of sintered samples were measured using the ASTM C 737 procedure. [7] This method permits direct measurement of bulk density, water absorption, open porosity and an assessment of closed porosity by weighing the dry sample, the wet sample in air, and the wet sample suspended in water.

Strength measurements were performed using ASTM C 773-88 procedures [8]. A compression technique was used on cylinders of heat treated foam with a length-to-diameter ratio of 2:1. They were crushed on an Instron using two platens and the load was recorded.

X-ray data were obtained on powdered samples with an automated Scintag PAD V powder diffractometer system using nickel filtered CuKα (λ =1.5405 Å) radiation, 45 kV, and 31 mA. The Scintag PAD V system is equipped with a liquid nitrogen cooled solid state germanium detector and a 2θ: θ goniometer radius set at 220 mm. Each sample was mounted as a thin layer on a very thin smear of petroleum jelly on a "zero-background" off-axis-cut quartz single crystal plate. The data were background corrected and α_2 stripped before peak finding with Scintag supplied software.

Dielectric constants were derived from the ratio of capacitances between electrodes separated by the sample and electrodes separated by air. Flat aluminum rods (3/8" diameter) which could be adjusted to accommodate samples of different thickness were used as electrodes. The capacitance of the samples was measured using a Hewlett-Packard 4192 LF Impedance Analyzer set at 10^5Hz. All dielectric measurements were made at 25°C.

Thermal diffusivity measurements for the heat treated foams were performed as a function of Freon concentration using the "flash method". [9,10] The thermal diffusivity, α, of a material is a measure of the rate that a thermal wave passes through such a material. In the "flash-method," a very rapid pulse of heat is applied to the front face of a thin piece of material (approximately 500 μm thick samples were used) and the temperature rise of the back face is monitored as a function of time. The time it takes the back face to reach half of its maximum temperature rise ($t_{1/2}$) after the heat pulse is applied is directly related to the thermal diffusivity [9,10].

RESULTS AND DISCUSSION

Fujiu, et al. stated in their work that the gelation kinetics of the sols are a function of both pH and temperature [5]. Specific data for pH dependence and temperature dependence of our silica sol are reported in Figures 2 and 3, respectively. Figure 2 shows viscosity vs. time as a function of pH (5.0-6.0). The gelation rate varies from 17-33 minutes depending on the pH of the sol. Within this pH range the system rapidly gels after an incubation period. If bubble generation from the dispersed liquid Freon is performed at this stage, quick gelation fixes the foamed structure before it collapses due to the foam decay. Similarly, Figure 3, shows the gelation time as a function of temperature at a fixed pH of 5.6-5.7. The data shows that the gelation time for the silica sol increases with decreasing temperature. At 60°C, the gelation time is 3 minutes, while at 30°C the gelation time is about 20-22 minutes. As a result of these gelation kinetic results, we chose a temperature of about 25°C and a pH of 5.6-5.7 for the remaining experiments.

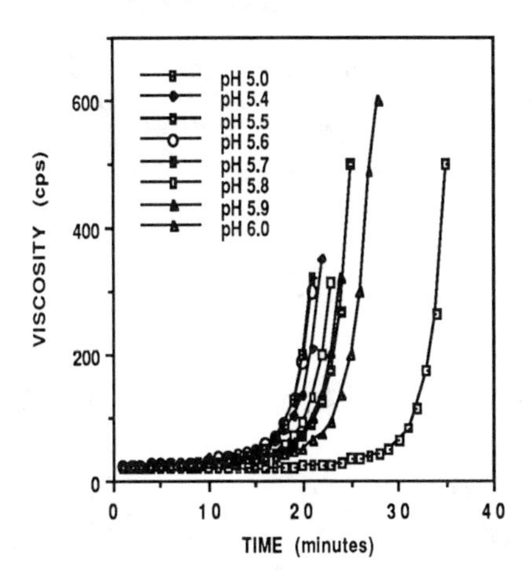

Figure 2. Viscosity vs. time as a function of pH for SiO_2 sols at 25°C.

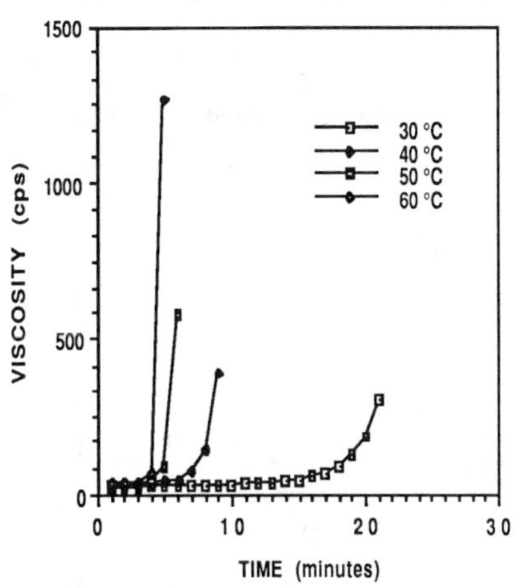

Figure 3. Viscosity vs. time at different temperatures for SiO_2 sols produced at a pH of 5.6-5.7.

In order to understand what effects the stirring method and Freon concentration would have on the foam properties, silica foams were prepared using both a magnetic stirrer and a laboratory stirrer and varying Freon concentration (0-0.01 cc/g). Figures 4 and 5 are high magnification photographs of incubated samples foamed with the two different stirring methods. In general, we found that the stirring method influenced the microstructure of the foam. Both stirring methods produced microstructures largely composed of open porosity. The magnetic stirring induces spherical pore formation with a uniform distribution of large and small pores. However, the use of a laboratory stirrer causes oval shaped pores which are much smaller, more uniform and larger in quantity. The difference between the two microstructures can be attributed to the larger shear forces introduced by the laboratory stirrer. High shear tends to increase the number of nucleation sites for void formation and also tends to distort the shapes of these voids. Regarding the role of Freon concentration, it was found that the number of pores increased with increasing Freon, but the size and shape of the pores did not change. Figure 6 and 7 represent density and porosity data as a function of Freon concentration for the two stirring methods used. In general, the density falls with increasing Freon concentration. Conversely, the porosity increases with higher Freon concentration. The data also shows that the foams produced by the magnetic stirring method are more dense and have higher porosity than the foams made by the laboratory stirring method. The density for the foams produced by the magnetic stirring process range from 0.88 to 0.67 g/cc with increasing Freon concentration. While the densities of the foams produced by the laboratory stirring process vary from 0.74 to 0.54 g/cc. Similarly, the porosity of the foams produced by the magnetic stirring process varies from 60-70 % with increasing Freon concentration and the porosity of the foams produced by the laboratory stirring process varies from 38-58 % with increasing Freon concentration.

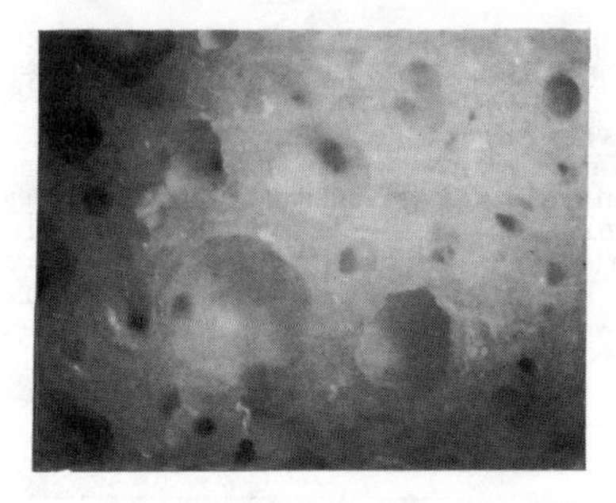

Figure 4. High magnification (50X) photographs of cellular silica produced by the magnetic stirring process. Sample was foamed without Freon.

Figure 5. High magnification (50X) photographs of cellular silica produced by the laboratory stirring process. Sample was foamed without Freon.

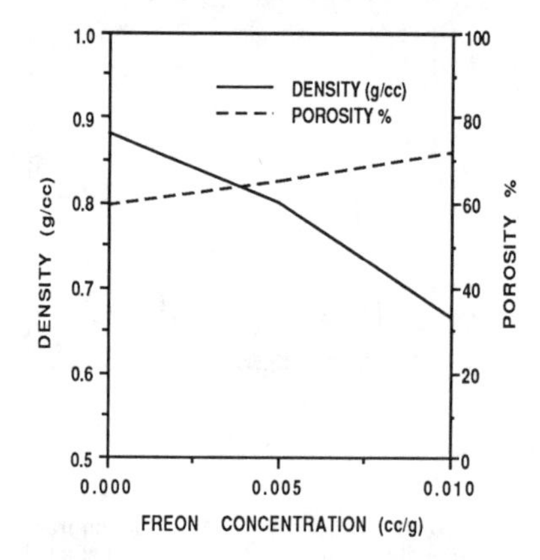

Figure 6. Density and porosity data as a function of Freon concentration for cellular SiO_2 produced using the magnetic stirring process.

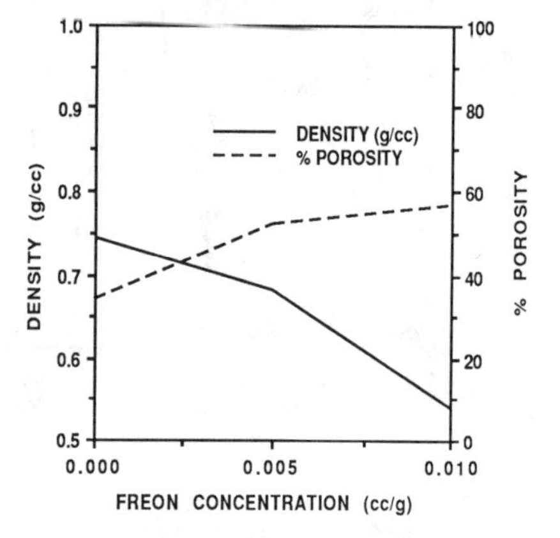

Figure 7. Density and porosity data as a function of Freon concentration for cellular SiO_2 produced using the laboratory stirring process.

X-ray experiments were performed on samples made at a pH of 5.6-5.7, using a laboratory stirrer and annealed between 1000-1100°C for 40 minutes. Figure 8 illustrates these data as a function of Freon concentration. In all cases, the samples are a crystalline form of SiO_2. The samples made using 0.005 and 0.01 cc/g of Freon have the cristobalite crystal structure, while the sample which was made using no Freon contains both the cristobalite [11] and tridymite [12] crystal structures. Cristobalite is a high temperature form of SiO_2 (stable from 1470°C to 1728°C) at atmospheric pressure. On the other hand, tridymite is a lower temperature form of SiO_2 which is stable from 870°C to 1470°C. Having produced crystalline samples using this low temperature process is quite unusual. Wu, et al. [13] stated that their cellular materials were non crystalline when made by the mechanical stirrer technique. However, the laboratory stirrer technique gives very crystalline materials. This confirms that the processing parameters can tailor the material structure and properties. In some of our earlier work, we were able to crystallize Al_2O_3 at low temperature using a sol-gel approach and a sonication technique.[14]

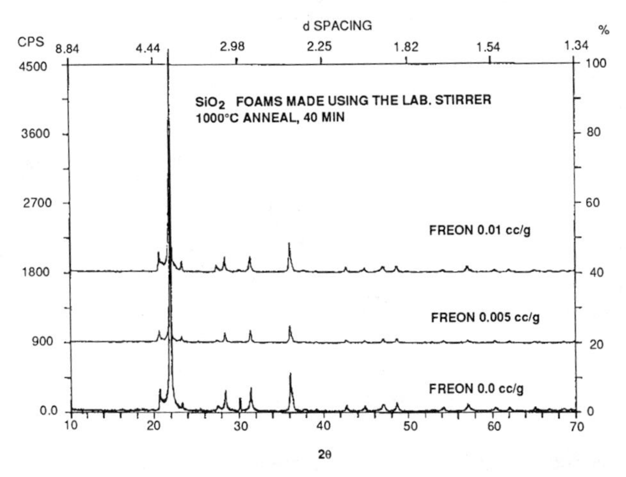

Figure 8. X-ray data for cellular SiO_2 as a function of Freon concentration.

Dielectric constant measurements were also made on both sintered foams and incubated forms and are summarized in Table 1. The data shows that the less dense samples have a lower dielectric constant when compared to the more dense samples. The dielectric constant is defined as the ratio of the permittivity of a material to that of free space. It is a measure of a material's polarizability (e.g., ionic, electronic polorizability).

Compression strength measurements on sintered cellular SiO_2 produced by the laboratory stirring process are tabulated in Table 2. In all cases, except for the 0.005 cc/g of Freon samples, 10-20 samples were measured. The more dense samples are stronger (have a larger ultimate compressive strength). In the case of the 0.005 cc/g of Freon samples there were only two measurements made and the data is not too accurate.

Thermal diffusivity data for sintered cellular SiO_2 is plotted in figure 9. The scatter can be attributed to the irregular porosity of the samples. The data shows a decrease in diffusivity with decreasing density. The Cellular SiO_2 samples have a slightly higher diffusivity when compared to the value of fused silica ($\alpha = 8.48 \times 10^{-3}$ cm^2/s at 25°C [9-10]). However, the thermal diffusivity for these samples is smaller than the value for air (α for air at 25°C is 0.222 cm^2/s [9-10]). A simplified equation for thermal diffusivity is

$$\alpha = \frac{k}{C_p \rho} \quad (1)$$

with $k = 3.3 \times 10^{-3}$ cal-cm/cm^2-s-°C
$C_p = 0.177$ cal/g-°C
and ρ = density in g/cc

Using equation (1) to calculate thermal diffusivity with ρ adjusted to match the amount of silica in the sample gives values close to those measured. This simplistic calculation does not take into account that C_p changes for these samples.

Table 1. Dielectric Constants for Cellular SiO_2 taken at 10^5Hz

Type of SiO_2 Foams	Freon Concentration (cc/g)	ϵ
Sintered Foams	0	1.72
	0.005	1.69
	0.01	1.52
Incubated Foams	0	1.75
	0.05	1.63
	0.01	1.58

*SiO_2 dielectric constant is 4.5

Table 2. Ultimate Compressive Strength for Cellular SiO_2

Freon Conc. (cc/g)	Ultimate Compressive Strength (psi)
0.0	487 ± 192
0.005	133 (only two samples tested)
0.01	251 ± 91

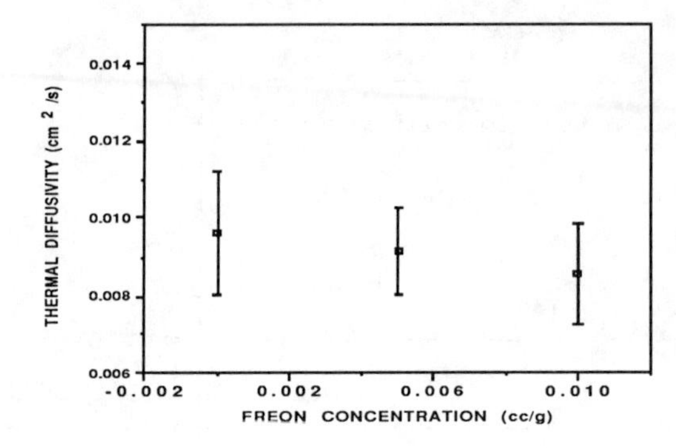

Figure 9. Thermal diffusivity as a function of Freon concentration. Data taken at 25°C.

CONCLUSIONS

A variety of cellular SiO_2 with tailored properties have been synthesized using the sol-gel process. The method uses viscosity control during sol-to-gel transition to stabilize the foam structure introduced by Freon volatilization. By careful control of the process, cellular SiO_2 having densities ranging from 0.9-0.5 g/cc, ultimate compressive strengths from 130-490 psi, dielectric constants from 1.5-1.7 and varied microstructure can be obtained. This process also offers the capability of making both crystalline or amorphous SiO_2 by careful control of method of foam agitation. Such control of the synthetic process offers many opportunities for the synthesis of new materials with advanced and desired properties.

REFERENCES

1. R.R. Hengst and R.E. Tressler, Cement Concrete Res. 13 , pg.127 (1983).

2. J.S. Morgan, J.L. Wood and R.C. Bradt, Mater. Sci. Eng. 47, pg. 37 (1981).

3. H. Verwell, G. Dewith and D. Veeneman, J. Mater. Sci. Eng. 20 , pg. 1069 (1985).

4. F.F. Lange and K. T. Miller, Adv. Ceram. Mater. 2, pg. 827 (1987).

5. T. Fujiu, G. L. Messing and W. Huebner, J. Am. Ceram. Soc. 73 (1), pg. 85 (1990).

6. A.G. Brown, W.C. Thuman, and J.W. McBain. J. Colloid Sci. 8, pg. 491 (1953).

7. ASTM C373-72 (Reapproved 1982). "Standard Test Method for Water Absorption, Bulk Density, Apparent Porosity, and Apparent Specific Gravity of Fired Whiteware Products."

8. ASTM C 773-8, 1988. "Standard Test Method for Compressive (Crushing) Strength of Fired Whiteware Materials."

9. Y.S. Touloukian, R.W. Powell, C.Y. Ho and M.C. Niolaou, "Thermal Diffusivity" in Thermophysical Properties of Matter, 10, pp. 15a-37a. (IFI/Plenum, New York 1973).

10. K. Kobayaski, "Simultaneous Measurement of Thermal Diffusivity and Specific Heat at High Temperatures by a Single Rectangular Pulse Heating Method," International Journal of Thermophysics 7, 181-195 (1986).

11. W. Wong-Ng, H. McMurdie, B. Paretzkin, C. Hubbard, A. Dragoo, National Bureau of Standards (USA), JCPDS Grant-in-Aid Report, (1988); JCPDS #39-1425.

12. Sato, Mineral J. (Japan) 4, pg. 215; Dana's System of Mineralology, 7th Ed., 273, (1962); JCPDS #18-1170.

13 M. Wu, T. Fujiu and G.L. Messing, J. of Non-Crystalline Solids 1-3, pg. 407, (1990).

14. J. Covino and R.A. Nissan, "Better Ceramic Through Chemistry" Mat. Res. Soc. Symp. Proc. 73, pg. 565, (1986).

Improvement of a Porous Material Mechanical Property by Hot Isostatic Process

Atsushi TAKATA, Kozo ISHIZAKI and Shojiro OKADA*

Department of Materials Science and Engineering, School of Mechanical Engineering, Nagaoka University of Technology, Nagaoka 940-21, Japan
* Japan Grain Institute Co., Ltd., 1047-9, Sue, Ryonan-cho, Ayauta-gun, Kagawa 767-01,Japan

ABSTRACT

Open porous materials of higher strength and higher porosity were produced by a hot isostatic process (HIP). A grinding wheel of fused alumina grain with powdered frit for bridge is discussed as an example. The samples were directly heated to a normal sintering temperature under HIPping conditions. The HIPped products were characterized by higher bending strength in spite of having more open porosity than the normally sintered ones.

INTRODUCTION

Hot isostatic process (HIP) is usually used to densify materials, and considered not to be suitable for porous materials production [1,2]. However, porous materials, mainly with open pores, can be obtained by means of HIP. The products are characterized by higher strength and more open porosity than those produced by conventional methods [3,4].

The configuration of pores, flaws and defects play sensitive roles in the mechanical properties of porous materials. The porosity and strength are generally supposed to be contradictory properties. Several researchers have investigated the relationship between the strength and the porosity of porous materials [5,6]. The pores of their materials were produced as residue vacant holes after burning out organic additives. According to these methods, however, the programmed porosity, and consequently the expected strength can hardly obtained within reasonabl error because of various uncontrollable pore producing factors.

In this study, a porous material with various porosities was prepared by controlling the HIPping pressure without adding any pore formation additives.

EXPERIMENTAL

Powder Preparation and Pretreatment for Sintering

For this study the authors chose a vitrify-bonded grinding wheel as example. The raw materials used were white fused alumina #60 (Nippon Kenma Co., Type WA) and powdered frit as the abrasive grains and bonding agent respectively. Table 1 shows the composition of the frit. The raw materials were mixed in liquid emulsion by a mullite ballmixer with mullite balls for 1 h.

Table 1. Chemical composition of the frit powder used

Composition	wt. %
Ignition loss	3.90
SiO_2	63.60
Al_2O_3	16.63
B_2O_3	2.02
Na_2O	1.71
Fe_2O_3	0.39
CaO	3.07
K_2O	8.57
MgO	0.11

Table 2 gives the composition of the mixture, which was then dried at 393 K for about 12 h. The green body was pressed uniaxially in a stainless steel mold under 10 MPa for 60 s, and then the green body was heated by an electric furnace of normal atmosphere at 1073 K for 1 h to burn out the temporary organic binder. Three samples with porosities between 43 and 44 % were chosen for each HIP sintering temperature.

Table 2. Composition of the mixture

Composition	wt. %
Abrasive grain	84
Frit powder	12
Dextrin	2
Water	2

Sintering and Assessment

The HIP sintering was carried out using an O_2-HIP apparatus (Kobe Steel, O_2-Professor-HIP) without encapsulation in an argon (80 %) and oxygen (20 %) atmosphere. Table 3 gives the sintering conditions. In order to compare the effect of HIP process, several samples were also sintered in normal atmosphere (pressure:0.1 MPa).

The samples were pressurized to 30 MPa at room temperature, and heated at a rate of 400 K/h. The pressurizing rate was fixed to reach the final pressure and temperature simultaneously. The samples were cooled in the furnace, from 1673 K to room temperature in about 1 h.

Several properties of the sintered specimens were analyzed. The density and porosity were measured by the Archimedean method. The bending strength of sintered specimens were measured by using a mechanical testing machine (Auto graph, Shimazu Seisakusyo Co).

Table 3. Sintering conditions

Temperature, K	1273	1473	1573	1673
Pressure, MPa				0.1
				20
	100	100	100	100
Holding Time, h	1	1	1	1

RESULTS AND DISCUSSION

The strength of porous materials

Figure 1 shows the effects of various sintering temperatures on the open porosities of normal and HIP sintered samples. The open mark ○ indicates open porosity of the pre-sintered body after removal of the temporary binder. The open porosities of these samples were between 43 and 44 %. The solid marks ● and ▲ correspond to samples HIPped under 100 and 20 MPa respectively. The open mark □ indicates the normally sintered specimen.

By increasing the sintering temperature from 1073 to 1673 K, the open porosity decreased about 1 to 5 %. Comparing the normal and HIP treated samples at 1673 K for 1 h, the HIPped ones show 3 to 4 % higher porosity.

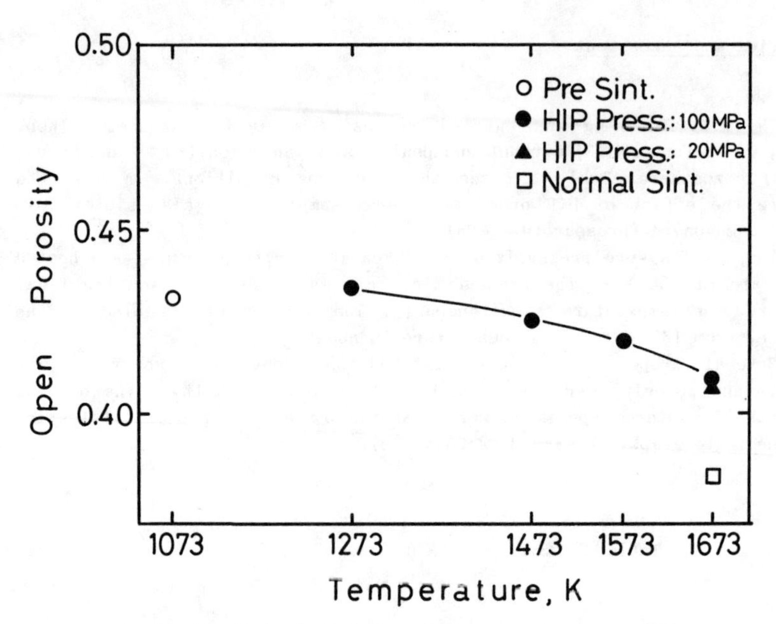

Figure 1. Sintering temperature vs. open porosity.

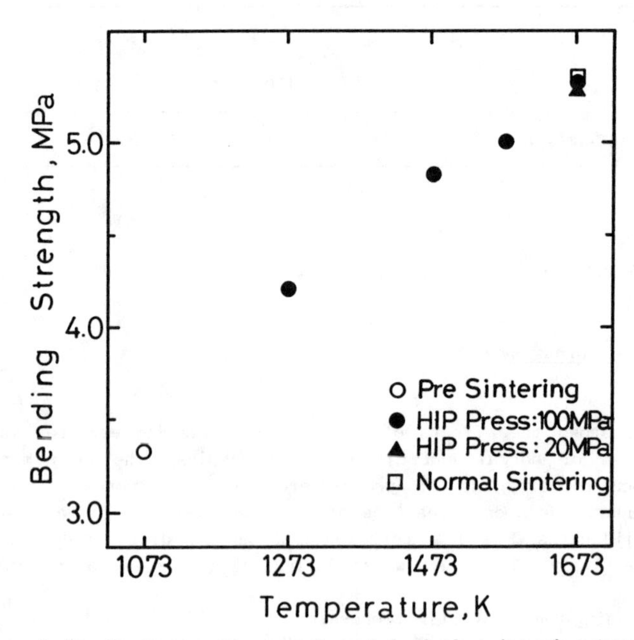

Figure 2. Bending strength as a function of the sintering temperature.

Figure 2 shows the relationship between sintering temperature and bending strength of the samples. The bending strength increases by increasing the sintering temperature. There is only a small difference in bending strength between the normal and HIP treated samples at 1673 K. This implies that the HIPped samples have higher strength in spite of higher open porosity than the normal ones.

Figure 3 indicates the bending strength as a function of the porosity. Normally, the porosity was considered to be a contradicting property of the strength, but this figure indicates that for a given porosity, the strength can be enhanced by HIPping.

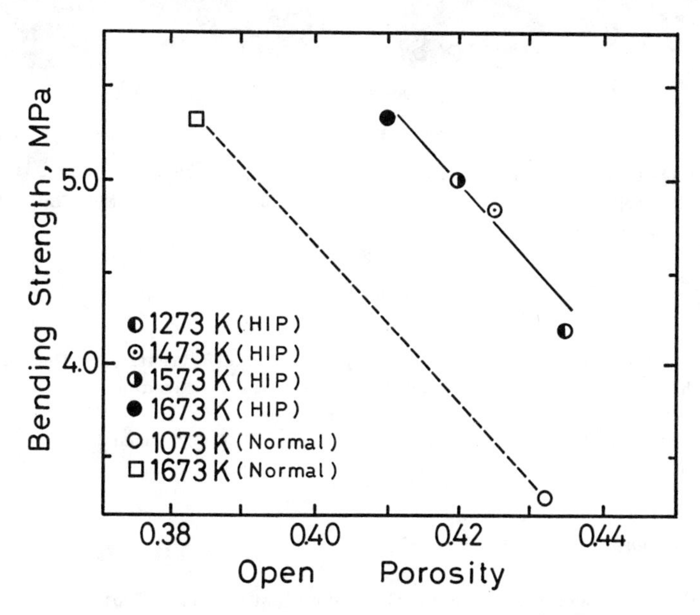

Figure 3. Bending strength as a function of open porosity . The solid line indicates HIPped samples and the dashed line indicates normally sintered samples. For a given porosity, the bending strength was enhanced by HIPping.

To explain the cause of this enhanced mechanical property by HIPping, SEM micrographs were taken as seen in figure 4. Samples were sintered normally at 1473K to consolidate enough for SEM observation. Fractured surfaces were observed by SEM. A typical observation is shown in Fig. 4(a). The same position is shown in Fig. 4(b) after HIPping at 1273 K under 100 MPa for 1 h. Many cracks were observed in the sample, Fig. 4(a), which were removed by the HIP treatment, Fig. 4(b). By the HIP process the interparticle bridges of the samples were remarkably consolidated and liberated from flaws and cracks which were not cured by normal sintering even at a temperature of 1473 K. This bridge enhancement is believed to be the cause of the increased strength of the HIPped porous materials.

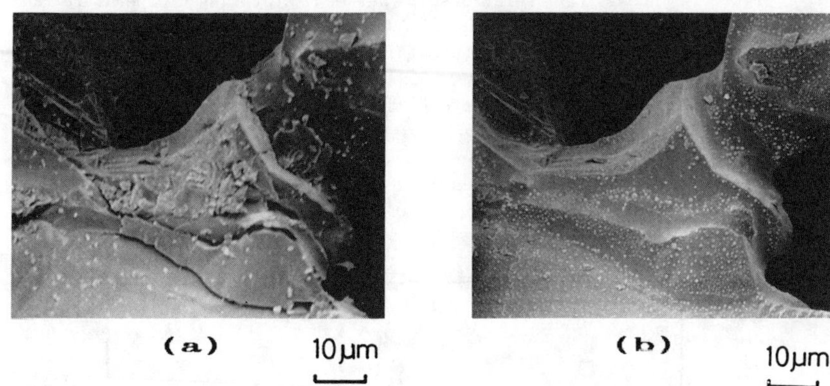

Figure 4.
(a) SEM photograph of normally sintered sample at 1473 K.
(b) SEM photograph of (a) after HIPped at 1273 K under 100 MPa for 1 h.The bar indicates 10 mm.

CONCLUSIONS

Porous materials were improved by using a HIP treatment. The bending strength of normal and HIP sintered specimens was compared.

1) The bending strength was enhanced by HIPping.
2) The internal flaws and defects were substantially reduced by the HIPping treatment.

ACKNOWLEDGMENT

This study was supported by the "Cooperative Program" of Monbusyo (the Ministry of Education & Culture) and Kobelco (Kobe steel).

REFERENCES

1) P. E. Price and S. P. Kohler, Hot Isostatic Pressing of Metal Powders, in Metal Handbook 9th ed. vol 7 Powder Metallurgy, (ASM, Metals Park, Ohio 1984) pp.418-443
2) K. Ishizaki and K. Tanaka, Sintering of Advanced Materials - Applications of Hot Isostatic Pressing -, (Uchida Rokakuho, Tokyo, 1987)
3) K. Ishizaki, A. Takata and S. Okada, J. Japan Ceram. Soc.,98, [6], 533-540 (1990)
4) K. Ishizaki, S. Okada, T. Fujikawa and A. Takata, Patent pending. (Jpn. 205421, 1989, U.S., Germany)
5) E. Ryshkewitch, J. Am. Ceram. Soc., 36, [2], 65-68
6) J. G. J. Peelen, B. V. Rejda and J. P. W. Vermeiden, Philips Tech. Rev., 37, [9/10], 234-236 (1977)

CHARACTERIZATION OF POROUS CELLULAR MATERIALS FABRICATED BY CHEMICAL VAPOR DEPOSITION

ANDREW J. SHERMAN, BRIAN E. WILLIAMS, MARK J. DELAROSA, AND RAFFAELE LAFERLA
Ultramet, 12173 Montague Street, Pacoima, CA 91331

ABSTRACT

The flexibility of chemical vapor deposition (CVD) permits the fabrication of a large number of materials in various geometric forms, one of which is the porous cellular structure. CVD fabrication of such a structure begins with the pyrolysis of a resin-impregnated thermosetting foam to obtain a reticulated carbon foam skeleton. The foam ligaments can then be coated with a variety of materials (metals, oxides, nitrides, carbides, borides, silicides, etc.), either singly or as hybrid, layered, alloyed, or graded structures. During this process, 10 to 1000 microns of the desired material(s) are deposited onto the foam ligaments by a variation of CVD known as chemical vapor infiltration (CVI). The thermomechanical properties of the resultant structure are dominated by the properties of the deposit, becoming independent of the carbon properties at very small material loadings. With precise control over the variables available, it is possible to obtain the simultaneous optimization of stiffness, strength, thermal conductivity, overall weight, and environmental resistance. This paper discusses the fabrication and properties of various CVD foam materials investigated to date.

REFRACTORY FOAMS

Ultramet has developed a new class of structural composite materials, refractory foams, fabricated by infiltrating a porous vitreous carbon foam with a refractory material or combination of materials (metal, ceramic, or both) by CVI [1,2,3]. The density of the coated foam structure remains less than theoretical in order to enhance its specific strength and thermal insulating properties for applications requiring very lightweight, strong structural components.

Techniques now exist for fabricating foams not only of polymers, but of metals, ceramics, and glasses as well. These newer foams are increasingly being used structurally, for insulation, and in systems for absorbing the kinetic energy of impacts. Their uses exploit the unique combination of properties offered by cellular solids, properties that ultimately derive from the cellular structure of the material.

Several techniques are used to produce engineering (structural) foams. With the exception of syntactic foams, these materials are produced using a foamed polymer as the starting material. From these economical precursors, three processing routes have been established for the production of ceramic and metallic foams. Ceramic foams can be produced by dipping the polymer foam in a slurry containing an appropriate binder and ceramic phases, followed by pressureless sintering at elevated temperatures. A second process used to make metallic foams utilizes an electroless process for the deposition of a metal onto the polymer foam precursor via electrolytic deposition.

A third process, used exclusively at Ultramet, begins with the pyrolysis of the thermosetting polymer foam to obtain a carbonaceous foam skeleton [4]. These carbonaceous foam materials are themselves attractive for many aerospace and industrial applications, including thermal insulation, impact absorption, catalyst support, and metal filtration. They are thermally stable, low in weight and density, and are chemically pure; they have low thermal expansion, resist thermal stress and shock, and are relatively inexpensive [5]. These Ultrafoam™ materials are becoming readily available, can be furnished in various sizes and configurations, and are easy to maintain and repair.

Ultramet takes this process several steps farther by infiltrating the vitreous carbon Ultrafoam skeleton with refractory material(s) using CVD/CVI. In this process, 10-1000 microns of the desired refractory metal or ceramic are deposited onto the interior surfaces of the reticulated carbon foam. The structural integrity of the resultant refractory foam composite material (Ultrametal™) is greatly enhanced by the deposit, and the properties of the composite are dominated by the structure and properties of the deposit. The mechanical properties for a given material and density are often one to two orders of magnitude higher compared to slurry-cast materials because the CVD deposit is typically 100% dense with grain sizes less than 1-5 microns (although larger or smaller grains can be deposited) and <0.05% impurities.

The Ultrametal process utilizes the high deposition rates of CVD (100-400 μm/hr) while depositing material within the foam structure via a combination of CVD and CVI techniques. The open-pore carbon foam precursor is heated to the temperature suitable for the desired deposition reaction, while the reactant gases are pulled through it. The gaseous precursor compound is reduced or decomposed at the foam surfaces, forming a uniform deposit throughout the internal structure of the foam. The carbon lattice functions only as a substrate for the material being deposited. Figure 1 shows an artist's conception of the ligamental structure of an infiltrated foam, while Figure 2 shows an individual coated foam ligament.

It is important to note that the structural integrity of the fabricated foam composite is provided by the deposited thin films, rather than the carbon substrate. These films have much higher elastic moduli than the thin sections of vitreous carbon in the foam. Their high stiffness relative to the carbon results in their supporting the mechanical load for the entire body, ensuring that failure does not occur in the carbon. Because of the superior properties of the deposited films, the individual ligaments act as microcellular materials, with 70-100% of the strength being contributed by the deposit.

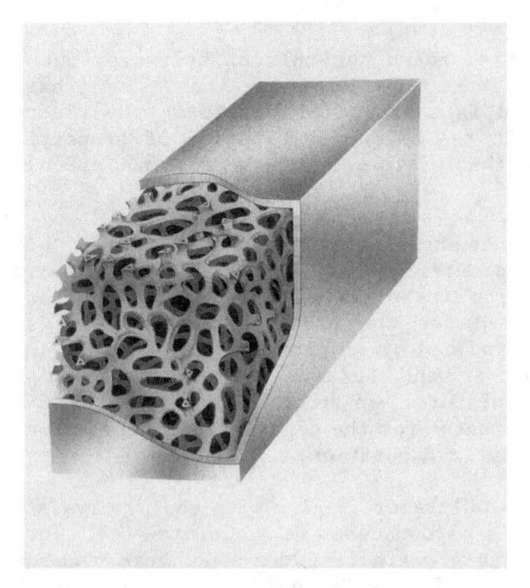

Figure 1. Artist's conception of infiltrated carbon foam

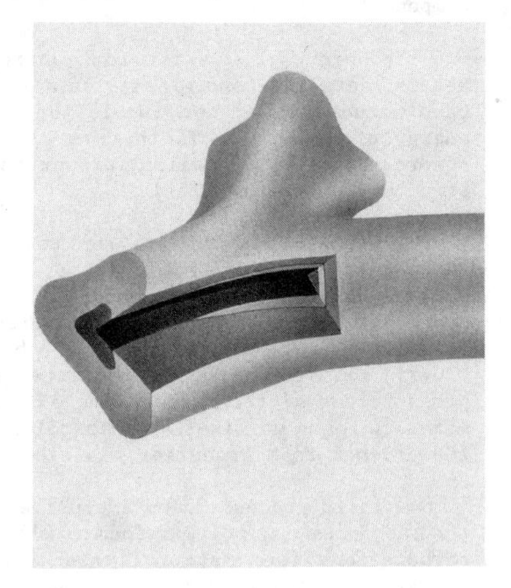

Figure 2. Artist's conception of coated foam ligament

In addition to providing increased strength and durability, the Ultrametal process is capable of producing foams from over 150 different materials, and of providing hybrid, layered, alloyed, or graded materials. Ultrametal foam composites can be fabricated from any material or materials combination (either simultaneously deposited or layered) that can be deposited by CVD/CVI. Densities up to 50% of the theoretical density of the deposited material can readily be achieved. It is possible to obtain the simultaneous optimization of stiffness, strength, weight, and thermal conductivity for a given application.

Structural applications often benefit from the low density, low thermal conductivity, and excellent thermal shock resistance of the foam composite, but require greater mechanical properties (flexural, tensile) than the basic Ultrametal can provide. Flexural and tensile properties can be greatly enhanced if an adherent, continuous sheet is applied to the surfaces of the foam. Face sheets have been applied directly via CVD by changing process conditions and gas flow patterns to promote surface deposition, as well as by diffusion bonding, brazing, mechanical fasteners, and adhesives. The specific attachment method is determined by use conditions (e.g. temperature, stress, strain, environment, weight requirements) and can be optimized for strength, weight, adhesion, environmental resistance, etc. The applied face sheet(s) can be the same material as the foam, a totally different material, or a combination of both, as conditions require.

MECHANICAL PROPERTIES

The elastic properties at small deformation in foams can be calculated from the linear elastic bending of a beam of length l loaded at its midpoint for a regular square area of cells [6,7]. In this manner, the relative density can be related to the Young's modulus, shear modulus, and Poisson's ratio respectively:

$$E^{*} = \frac{\sigma}{\epsilon} = \frac{C_1 E_s I}{l^4} \approx E_s \left(\frac{\rho^{*}}{\rho_s}\right)^2 \tag{1}$$

$$G^{*} = \frac{\tau}{\gamma} = \frac{C_2 E_s F}{l^4} \approx \frac{3}{8} E_s \left(\frac{\rho^{*}}{\rho_s}\right)^2 \tag{2}$$

$$\nu = \frac{C_1}{2C_2} \approx \frac{1}{3} \tag{3}$$

This analysis necessarily includes a number of approximations; for example, the way in which density is calculated double counts as the cell vertices, and the axial and shear displacements of the cell walls have been neglected.

A similar model results in equations for the elastic collapse strength, plastic collapse strength, and brittle fracture strength of foam materials. These relationships can be expressed as follows.

For elastic collapse:

$$\sigma^{*}_{el} = .05 E_s \left(\frac{\rho^{*}}{\rho_s}\right)^2 \tag{4}$$

For plastic collapse in early stages:

$$\sigma^*_{pl} = .30\sigma_{ys}\left(\frac{\rho^*}{\rho_s}\right)^{3/2} \tag{5}$$

For plastic collapse, maximum tension:

$$\sigma^*_{pl_{max}} = \sigma_{ys}\left(\frac{\rho^*}{\rho_s}\right) \tag{6}$$

For brittle crush strength:

$$\sigma^*_{cr} = .65\sigma_{fs}\left(\frac{\rho^*}{\rho_s}\right)^{3/2} \tag{7}$$

For fracture toughness (brittle fracture):

$$\frac{K_{1c}}{\sigma_{fs}\sqrt{\pi l}} = .65\left(\frac{\rho^*}{\rho_s}\right)^{3/2} \tag{8}$$

The results of mechanical testing conducted on CVD foams agree extremely well with values predicted from the relationships above. Compressive strength data for a refractory metal foam is presented in Figure 3 [8]. The deposit morphology clearly affects the strength of the deposit. The small disagreement with the actual theoretical values for silver and gray rhenium is due to inaccuracies in the measured strengths of the deposited material. Black rhenium, however, has an extremely dendritic structure in which only a small portion of the deposit actually contributes to the overall material properties. Figure 4 shows SEM micrographs of these three rhenium foam microstructures. The coated foam ligaments act in a manner similar to a microcomposite, in which the deposit acts independently from the carbon substrate, as seen in the tungsten foam shown in Figure 5.

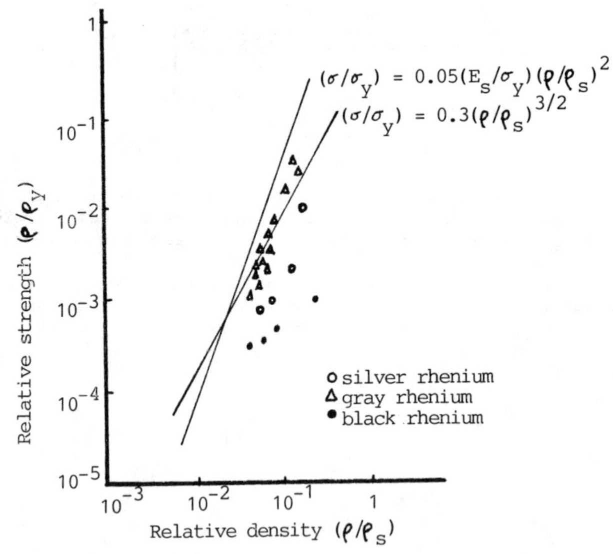

Figure 3. Compressive strength data for rhenium foam [8]

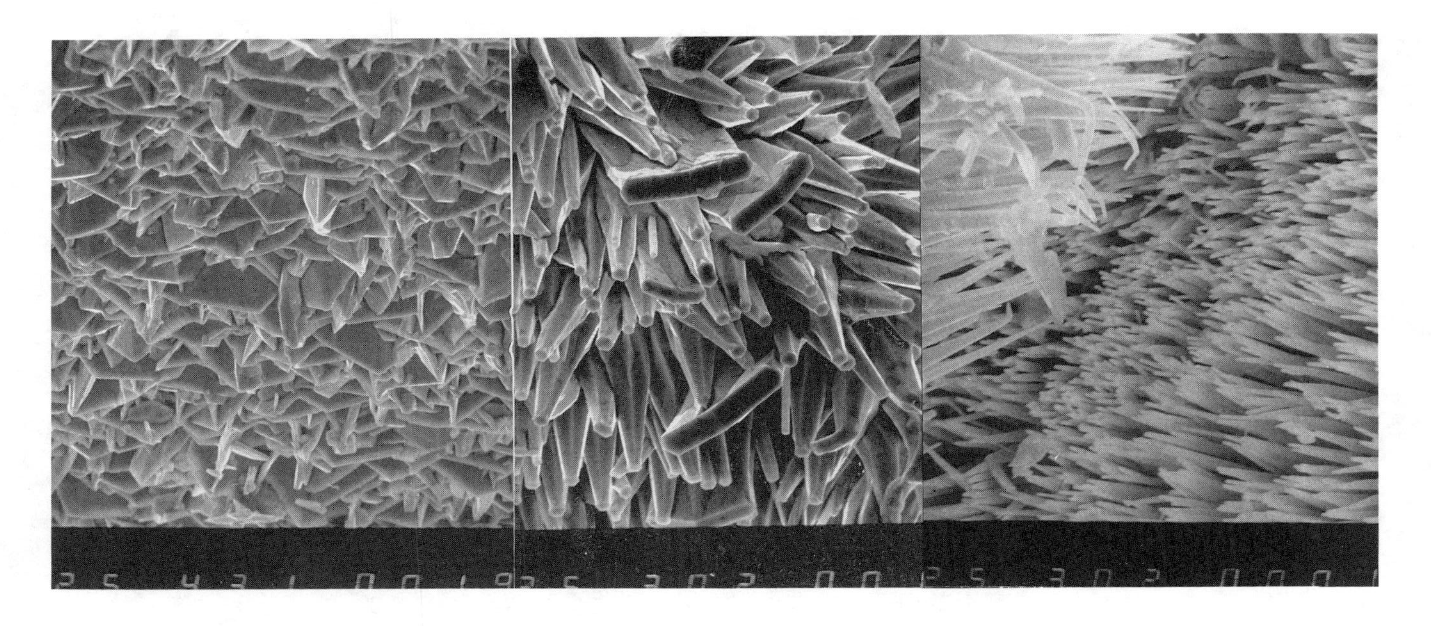

Figure 4. SEM micrographs of silver (430x), gray (3000x), and black (3000x) rhenium microstructures

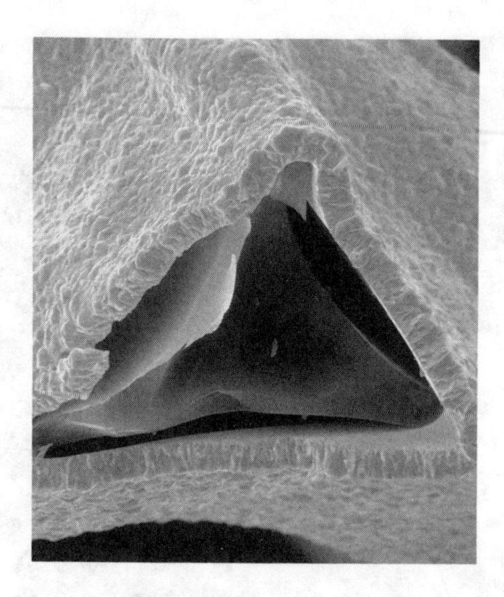

Figure 5. SEM micrograph (1200x) of tungsten foam ligament

Flexure modulus measurements corresponded to the theoretically predicted modulus within the accuracy of the test. Flexure testing of sheathed rhenium foams revealed core shear failure under both three- and four-point loading. Doubling the density of the foam core resulted in an approximately fourfold increase in core strength. 0.2% proof testing values for a 2.1 g/cm^3 rhenium foam were measured at 25 MPa, while a 4.1 g/cm^3 rhenium foam was proof tested in flexure at 96 MPa.

Ceramic foams are much more sensitive to the testing method and sample configuration than are metal foams. Two sample geometries and testing methods were studied for ceramic foams. Two sample geometries, 0.5" x 0.5" x 1.0" with epoxy-mounted ends and 2" x 2" x 0.5", were selected based on reported testing methods. The 2" x 2" sample was selected to minimize shear and bending moments on the sample, while the 0.5" x 0.5" sample was selected to conform more closely to ASTM standards for compression testing. Drastically different stress-strain profiles were observed for each sample geometry, as illustrated in Figure 6. Unmounted samples exhibited the classic foam crushing behavior characterized by linear elastic, plateau, and densification regions. Epoxy-mounted samples failed within the gauge length at 45° to the load axis, and exhibited stress-strain behavior analogous to a monolithic ceramic material, 100% linear elastic to failure. Crush test values for each method were independent of whether or not a face sheet had been applied to one surface.

Modulus and strength values obtained from each method and compared to theoretically predicted values (using deposit strengths representative of CVD SiC) are shown in Figures 7 and 8 respectively [9]. Strength values followed the predicted relationship, with epoxy-mounted samples measured at values three to five times higher than unmounted samples. Recorded modulus values did not follow the predicted trend with increasing relative density. Comparison with measured values for bare carbon foam, shown in Table I, revealed that strengths measured using the first method were in some instances lower than the strength of the starting materials. While this is not a physical impossibility, it is highly unlikely (qualitative tests indicate that it is not true), such that the second method has been adopted as the standard test method at Ultramet.

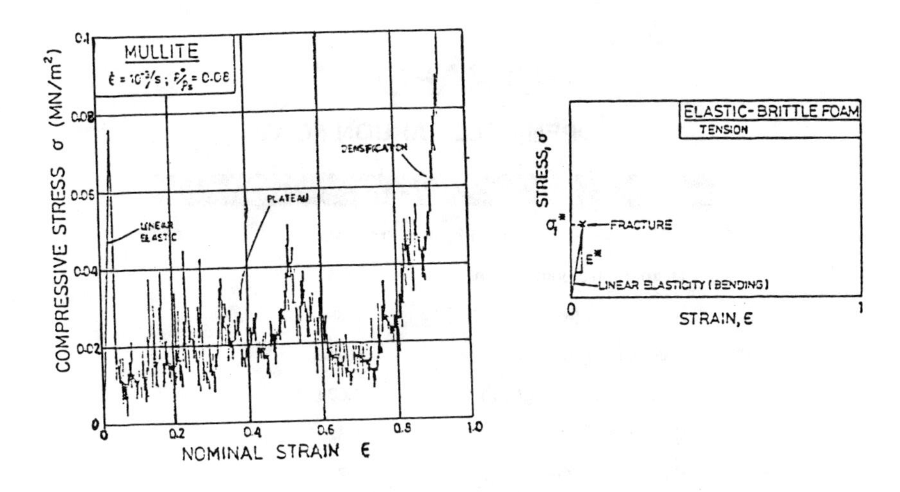

Figure 6. Stress-strain profiles obtained for two sample geometries: 2" x 2" x 0.5" (left) and 0.5" x 0.5" x 1.0" (right) [6]

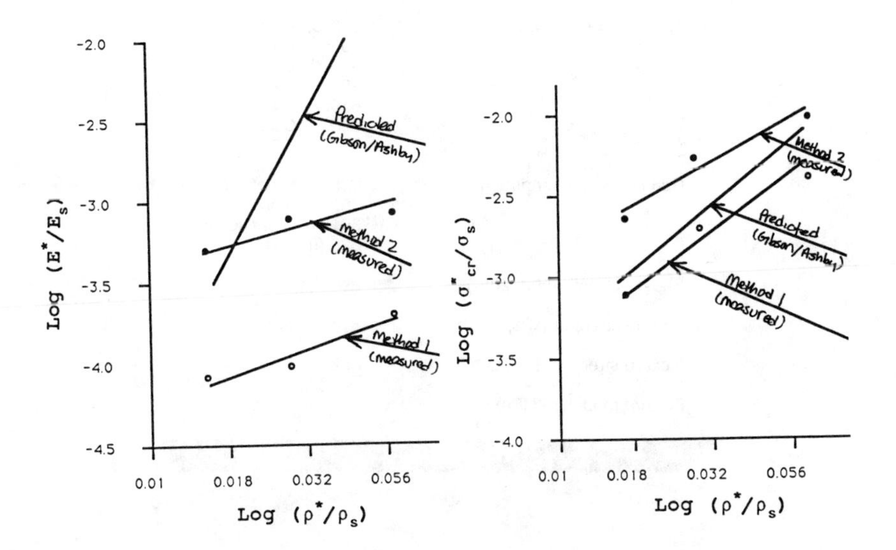

Figure 7. Modulus of SiC foams

Figure 8. Strength of SiC foams

Table I. Properties of Carbon Foam

$ULTRAFOAM_{60}$

OPEN-CELL CARBON FOAM

Typical Physical Properties

Micrographic porosity (ppi)	58.8
Ash content (wt%, 1000°C)	0.39
Bulk density (g/cm^3)	0.042
Ligament density (g/cm^3)	1.538
Surface area (m^2/g)	1.623
Resistivity (ohm·cm)	0.75
Specific heat (cal/g/°C)	0.30
Maximum use temperature (°C)	in air: 350 inert: 3500
Thermal expansion (ppm/°C)	0-200°C: 1.15 0-500°C: 1.65 0-1000°C: 1.65
Thermal conductivity (W/m·K)	200°C: 0.085 300: 0.125 400: 0.180 500: 0.252 650: 0.407 800: 0.625 950: 0.882
Compressive strength (kPa, 20°C)	625 (10% deflection) 763 (ultimate)
(kPa, 1000°C)	391 (10% deflection) 628 (ultimate)
Shear strength (kPa, 20°C)	290
Tensile strength (kPa, 20°C)	810
Flexure strength (kPa, 20°C)	862
Flexure modulus (MPa)	58.6

REFERENCES

1. R.H. Tuffias and R.B. Kaplan, Research & Development 31, 116 (1989).

2. A.J. Sherman, R.H. Tuffias, and R.B. Kaplan, presented at the 1990 American Ceramic Society Annual Meeting, Dallas, TX, 1990 (unpublished).

3. A.J. Sherman and R.H. Tuffias, to be published in Amer. Cer. Soc. Bulletin 70 (1991).

4. H.L. Harped et al., in Oak Ridge Y-12 Plant Report Y-DA-2654, 1969.

5. A.E. Sands and M.E. Scribner, in Oak Ridge Y-12 Plant Report Y-DA-2654, 1969.

6. M.F. Ashby, Metall. Trans. A 14, 347 (1983).

7. L.J. Gibson and M.F. Ashby, Cellular Solids: Structure and Properties (Pergamon Press, New York, 1986).

8. A.J. Sherman and R.B. Kaplan, Ultramet report no. ULT/TR-88-6598, 1988.

9. A.J. Sherman, Q. Jang, and R. LaFerla, Ultramet report no. ULT/TR-90-7422, 1990.

STRESS-DENSITY VARIATIONS IN ALUMINA SEDIMENTS: EFFECTS OF POLYMER CHEMISTRY

C. H. SCHILLING,* J. J. LANNUTTI,# W.-H. SHIH,# and I. A. AKSAY#

*Pacific Northwest Laboratory,+ Richland, WA 99352
#Department of Materials Science and Engineering; and
Advanced Materials Technology Center, Washington Technology Center,
University of Washington, Seattle, WA 98195

INTRODUCTION

Achieving spatially uniform and hierarchically structured microstructures during the shape-forming of colloidal ceramics depends largely on (*i*) the magnitude of the effective stresses (i.e., stresses that are supported by the particulate network) and (*ii*) plastic properties, which in turn are significantly altered by processing parameters affecting interparticle friction and adhesion. To quantify the effects of processing parameters on consolidation, we present a novel approach for analyzing sediments by gamma-ray densitometry[1] and a fluid mechanics model. This method enables us to correlate processing parameters with spatial variations of the packing density and the local effective stress. These correlations are difficult to achieve by traditional techniques (e.g., rheometry, sedimentation kinetics modeling, soil mechanics tests), especially for the low stresses (< 1000 Pa) that are typically encountered in sediments. Aside from being destructive to samples, these techniques also tend to measure volume-averaged properties, and as a result they usually fall short of describing localized consolidation phenomena.

Significant variations in plasticity can arise through modification of interparticle forces by changing the chemistry of surface-adsorbed polymers or ions. In the present study, we address the question of how to design surface-adsorbed polysiloxanes to enhance particle rearrangement into densely packed structures. Polysiloxanes are excellent lubricants (e.g., silicon oil) due to the low rotational energy of the Si–O–Si bonds comprising the polymer backbone. These inorganic polymers are also available with a broad range of chemical structures that can pyrolyze to various inorganic silicon-based compounds under appropriate conditions.[2] This research is part of a larger program examining the use of polymeric additives that may form useful inorganic phases within the pores of a ceramic compact upon pyrolysis.[3]

We analyzed stress-density correlations of alumina sediments as a function of two major variations in additive chemistry: (*i*) the attachment of polar (carbonyl) pendants to the siloxane backbone, and (*ii*) changes in the molecular weight of nonpolar $[MeHSiO]_x$ (hydrosiloxane). We suspect that the carbonyls may enhance particle rearrangement since these moieties are the primary constituents of organic polymers (e.g., polymethacrylic acid) which are used to increase packing density.[4] A sample containing fluorinated polyester, an organic polymer which has previously been shown to promote the formation of high green densities in alumina cakes,[5] was also analyzed for comparison with the inorganic polymers. In addition, we analyzed a sample prepared with an aluminosiloxane polymer that pyrolizes to mullite ($3Al_2O_3 \cdot 2SiO_2$).[2]

+ Pacific Northwest Laboratory is operated for the U.S. Department of Energy by Battelle Memorial Institute under Contract DE-AC06-76RLO 1830.

EXPERIMENTAL PROCEDURE

Five starting solutions were prepared using dry chloroform and one of the following polymers: fluorinated polyester (FPE), polyacryloxypropylsiloxane (PAS), a mullite-forming aluminosiloxane (MF), and hydrosiloxanes of two different molecular weights.++ Figure 1 shows the chemical structures of the polymers. Each of the hydrosiloxane solutions contained polymers with either four (HS4) or approximately twenty-two (HS22) monomers per chain. Each solution contained approximately 4 x 10^{18} polymer molecules per ml. Four vol% α-Al_2O_3 particles+++ was added to each solution, which was followed by 10 minutes of ultrasonication. Of each suspension, 180 ml were subsequently poured into Pyrex sedimentation tubes (3 cm inside diameter). Each tube was pre-treated with $(CH_3)_3SiCl$ solution to reduce shear stresses at the inner wall.[2,3]

During sedimentation, local measurements of packing density were obtained by gamma-ray densitometry.[1] Each sediment was irradiated by a 3.2-mm-diameter collimated beam of photons (661 keV) emitted from cesium-137. Photon transmission was measured at each elevation, z, and related to the volume fraction of solids, ϕ, (accuracy $\pm$ 0.015) using the Beer-Lambert law. We assume that ϕ is uniform within the irradiated volume at each elevation.

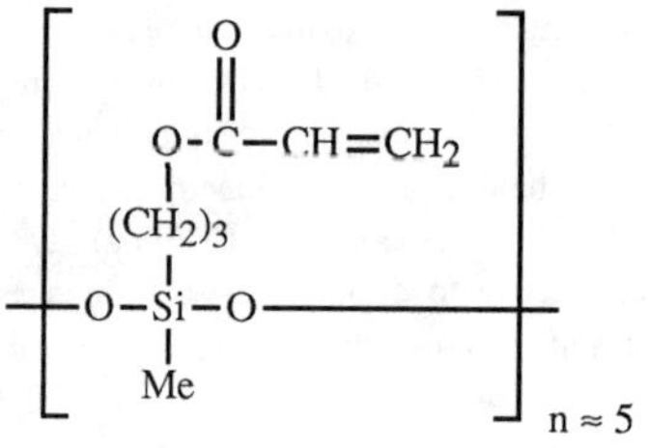

$$\left[-O-\underset{\substack{| \\ Me}}{\overset{\substack{O-\overset{\displaystyle O}{\overset{\|}{C}}-CH=CH_2 \\ | \\ (CH_2)_3 \\ |}}{Si}}-O- \right]_{n \approx 5}$$

Figure 1a. Polyacryloxypropylsiloxane (PAS).

$$Me-\underset{\substack{| \\ Me}}{\overset{\substack{Me \\ |}}{Si}}-O-\left[\underset{\substack{| \\ H}}{\overset{\substack{Me \\ |}}{Si}}-O\right]_n-\underset{\substack{| \\ Me}}{\overset{\substack{Me \\ |}}{Si}}-Me \qquad n = 4 \text{ or } 22$$

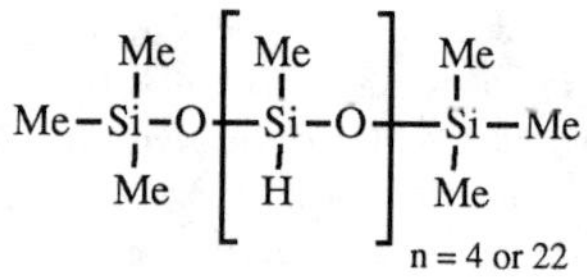

Figure 1b. Hydrosiloxane.

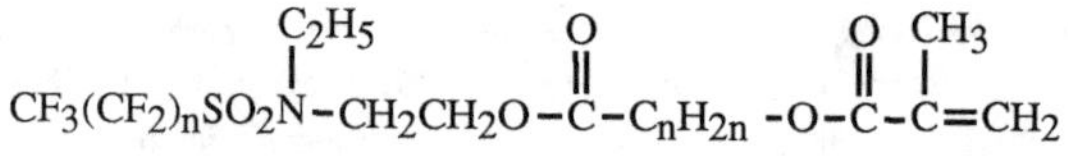

$$CF_3(CF_2)_nSO_2\overset{\substack{C_2H_5 \\ |}}{N}-CH_2CH_2O-\overset{\substack{O \\ \|}}{C}-C_nH_{2n}-O-\overset{\substack{O \\ \|}}{C}-\overset{\substack{CH_3 \\ |}}{C}=CH_2$$

Figure 1c. Fluorinated polyester (FPE); molecular weight ≈ 800.

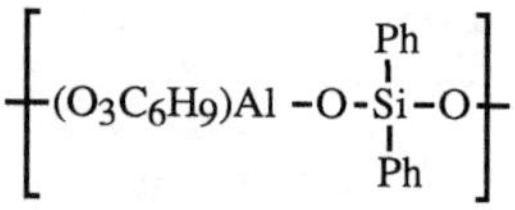

$$\left[-(O_3C_6H_9)Al-O-\underset{\substack{| \\ Ph}}{\overset{\substack{Ph \\ |}}{Si}}-O- \right]$$

Figure 1d. "Mullite-former" (MF); molecular weight ≈ 800.

RESULTS AND DISCUSSION

All suspensions appeared to be flocculated, based on the rapid initial settling rates (≈ 1 cm/min) and the visibly transparent supernatants that remained above each sediment throughout the experiment. At the conclusion of visible sedimentation, the density profiles remain time-invariant, as shown in Figure 2. The hydrosiloxane samples exhibited nonuniform packing and relatively low solids fractions ($0.16 < \phi < 0.38$) that increased with decreasing elevation, although HS4 had slightly greater densities.

++ FPE was obtained from 3M Corporation, Minneapolis, Minn. The remaining polymers were obtained from Huls Petrarch Systems, Bristol, Penn.

+++ 0.4 μm average diameter, Type AKP-30, Sumitomo Chemical America, New York, N.Y.

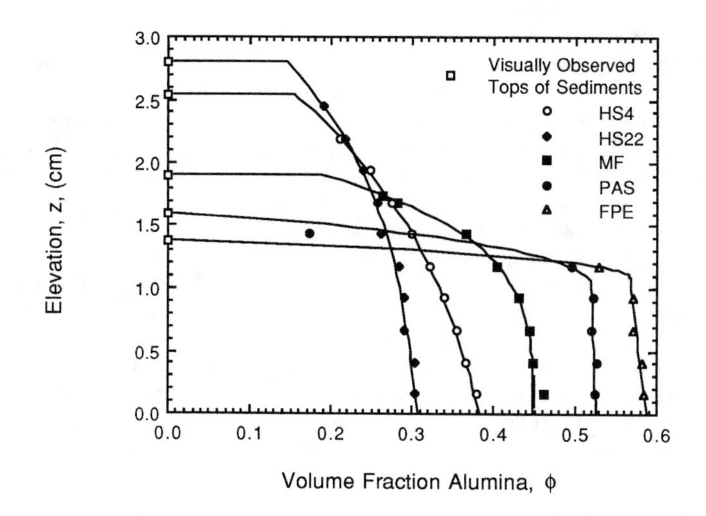

Figure 2. Density profiles for the equilibrium sediments as a function of polymer chemistry.

The MF sample appeared to consist of a densely packed lower layer ($\phi \approx 0.45$) of approximately 1 cm in thickness and an upper layer having lower densities and approximately the same thickness. The FPE and PAS samples exhibited greater maximum densities of 0.58 and 0.53, respectively.

As shown in Figure 2, we curve fitted the density functions by setting the integrated area under each curve approximately equal to the known true volume of powder (7.2 cm^3) in each sample:

HS4: $$\phi = -6.23544(10^{-3})z^3 - 4.08715(10^{-3})z^2 - 4.000812(10^{-2})z + 0.3847906; \ (0 \le z \le 2.54 \text{ cm}); \tag{1}$$

HS22: $$\phi = -6.55993(10^{-3})z^3 + 5.09235(10^{-3})z^2 - 2.147357(10^{-2})z + 0.3092194; \ (0 \le z \le 2.79 \text{ cm}); \tag{2}$$

PAS: $$\phi = -5.590529(10^{-3})z + 0.5265175; \ (0 \le z^* \le 1.0 \text{ cm}); \tag{3}$$

$$\phi = -5.5200292(10^{-1})z^4 + 1.18777395z^3 - 7.8881246(10^{-1})z^2 + 1.5629941(10^{-1})z + 0.5226283; \ (1.0 < z \le 1.6 \text{ cm}); \tag{4}$$

FPE: $$\phi = -1.854329(10^{-2})z + 5.878838(10^{-1}); \ (0 \le z \le 1.105 \text{ cm}) \tag{5}$$

$$\phi = -3.31288603z^6 + 10.2322178z^5 - 11.65428686z^4 + 6.02684442z^3 - 1.41215531z^2 + 1.12755(10^{-1})z + 0.5846; \ (1.105 < z \le 1.37 \text{ cm}); \tag{6}$$

MF: $$\phi = -2.4279(10^{-2})z^4 + 2.1792(10^{-2})z^3 - 3.0148(10^{-2})z^2 + 8.1956(10^{-3})z + 0.44995; \ (0 \le z \le 1.905 \text{ cm}). \tag{7}$$

It was not possible to resolve the exact shapes of the density profiles within the uppermost regions of each sample because of the relatively large diameter of the photon beam. However, based on the experimental curve fits above, we can infer that the PAS and FPE samples consisted of two layers: a thick, lower layer of uniform high density, and a thin, upper layer of lower densities. As further evidence, we also analyzed a set of curve fits that assumed the absence of a two-layer structure in the PAS and FPE samples; in each case ϕ was expressed as a linear function of z at all elevations, and the integrated areas significantly exceeded the known mass of powder in each sample. It should be mentioned that more accurate density profiles are possible by collimating the beam to a smaller size.

Equations (1) through (7) were used to calculate spatial variations of the effective stress using a fluid statics model.[6-8] The model assumes that, during settling, flocculated particle networks form an elastic continuum that transmits the weight, buoyancy, and viscous drag forces from one elevation to the next. Shear stresses acting from the walls of the sedimentation tubes are also neglected. The net force transmitted by the network at a particular elevation is simply equal to the dynamic balance of the above forces summed over the entire network above that elevation. For stable cakes, the viscous drag component is zero:

$$P(z^*) = \int_{z^*}^{z_{max}} \Delta\rho \; g \; \phi \; dz \qquad (8)$$

where $P(z^*)$ is the effective stress at an elevation of z^*, z_{max} is the sediment height, $\Delta\rho = 2.47$ g/cm^3 is the density difference between the solid and liquid, and g is the gravity acceleration constant. After substituting equations (1) through (7) into (8), we plotted $P(z^*)$ as a function of the local density at z^* (Figure 3). Results indicate that ϕ plots as a series of approximately linear functions of $\ln P(z^*)$, a trend that was previously reported for alumina cakes by measuring volume-averaged densities during drained uniaxial compression.[9,10] Similar behavior was also reported for flocculated sediments of thoria[7] (ThO_2) and microbarite[8] ($BaSO_4$).

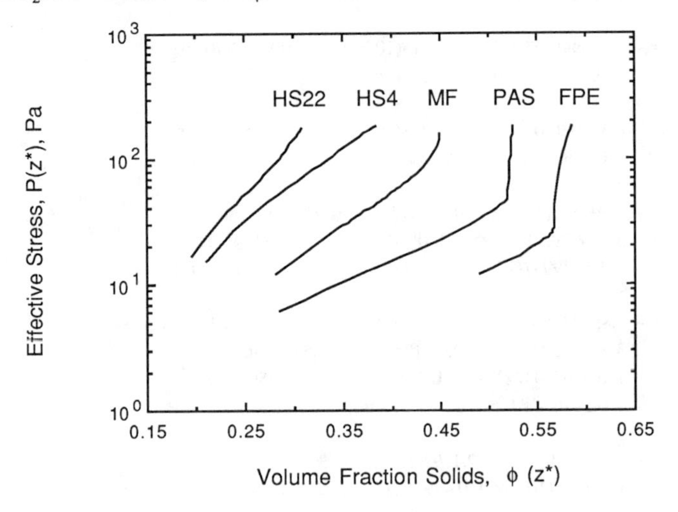

Figure 3. Effective stress as a function of packing density and polymer chemistry.

The slope and position of a given ϕ – ln P(z*) plot reveal important information regarding the effects of suspension chemistry on particle rearrangement. The slope, d[lnP(z*)]/dϕ, can be interpreted as an activation barrier for rearrangement, α, based on a model reported by Shih et al.[10,11] This model proposes that a change in density, dϕ, due to an increment in effective stress, dP(z*), is proportional to the probability of overcoming an energetic barrier whose height is proportional to the local density. Thus, $d\phi/dP(z^*) = \exp[-\alpha\phi]$. Integrating and rearranging, one obtains $P \approx \exp(\alpha\phi)$, as indicated in Figure 3. The activation barrier can be affected by changes in processing parameters that influence interparticle attraction.

A striking feature of Figure 3 is that α for the MF, PAS, and FPE samples abruptly increases at a critical point that is uniquely characteristic of the polymer chemistry. These abrupt increases in α correspond to the approximate elevations of the interfaces between the low- and high-density layers in these samples. Also, when successively comparing MF to PAS to FPE, we observe that (*i*) the critical density increases, (*ii*) the thickness of the low-density layer decreases, and (*iii*) below the critical densities, α decreases from approximately 20 to 10. We believe that these trends may be attributed to a reduction in interparticle attraction, although additional experiments with well-defined attractive interparticle potentials are needed to confirm this hypothesis.

The presence of a two-layer structure is not suggested by the density profiles of the hydrosiloxane samples; as a result, these samples exhibit α values ($\approx$ 20) that were approximately independent of z, ϕ, and P(z*). A decrease in the hydrosiloxane molecular weight, however, decreases the α value and hence increases the local density at a particular effective stress. It is possible that the hydrosiloxane samples may exhibit an abrupt increase in α by increasing the effective stress, e.g., by increasing the alumina concentration of the starting suspension or by using taller sedimentation columns.

We learned that slight differences in curve fitting the density functions will not influence the general shapes of the plots in Figure 3, but they may change the values of α and the exact locations of the critical points. For example, the above calculations were repeated by varying the curve fit within the low-density layer of the PAS sample, and we observed small effects on both the critical density and α values for densities below the critical density. The critical effective stress, however, varied by as much as 10 Pa. We also learned that slight variations in the curve fit for densities above the critical density may produce large uncertainties in α above the critical density. As a result, we cannot specify reliable α values for densities above the critical density because of experimental measurement uncertainty.

Additional studies are needed to further investigate the presence of layered microstructures in sediments and to establish correlations between α, interparticle attraction-repulsion forces, and microscopic mechanisms of particle network restructuring in these layers. Direct imaging, e.g., by high-resolution computerized tomography or electron microscopy, would be beneficial to quantify spatial variations of structural features, especially those that may be responsible for the existence of layered structures in sediments. In addition, useful information can be obtained by the use of theoretical models that relate macroscopic properties, such as stress and packing density, to interparticle attraction/repulsion energies and microstructural deformation processes.[11-15]

CONCLUSIONS

A novel analytical technique is presented, which uses densitometry and fluid mechanics modeling to correlate spatial variations of the packing density and local effective stress in sediments. Using this technique, we show that the consolidation behavior of chloroform-alumina sediments is significantly affected by the chemistry of polysiloxane additives. Nonpolar hydrosiloxanes resulted in relatively poor packing behavior with a smooth gradient in density from the bottom to the top of each sample. The activation barriers in these samples remained at high values that were independent of elevation and density for solids fractions of up to 0.38. Slight improvements in packing were obtained by decreasing the hydrosiloxane molecular weight. In contrast, samples containing polyacryloxypropylsiloxane, fluorinated polyester, and a mullite-forming aluminosiloxane exhibited "two-layer" microstructures consisting of (*i*) a densely packed bottom layer and (*ii*) a top layer having lower densities. At the interfaces between these two layers, the activation barriers exhibited an abrupt transition from low values, occurring within the top layers, to high values occurring in the bottom layers. Variations in the thickness of each layer and the critical densities associated with the activation barrier transitions were uniquely characteristic of the polymer chemistry and may be attributed to changes in interparticle attraction.

ACKNOWLEDGMENTS

This work was supported by the Office of Basic Energy Sciences, U.S. Department of Energy, through a subcontract by Pacific Northwest Laboratory under Contract No. 063961-A-F1. The authors wish to thank Dr. Wan Shih for valuable discussions.

REFERENCES

1. C. H. Schilling, G. L. Graff, W. D. Samuels, and I. A. Aksay, in ***Atomic and Molecular Processing of Electronic and Ceramic Materials: Preparation, Characterization, and Properties, MRS Conf. Proc.***, edited by I. A. Aksay, G. L. McVay, T. G. Stoebe, and J. F. Wager (Materials Research Society, Pittsburgh, Pennsylvania, 1988), p. 239.
2. J. J. Lannutti, Ph.D. Thesis, University of Washington, 1990.
3. J. J. Lannutti, C. H. Schilling, and I. A. Aksay, in ***Processing Science of Advanced Ceramics, MRS Symp. Proc.,*** Vol. 155, edited by I. A. Aksay, G. L. McVay, and D. R. Ulrich (Materials Research Society, Pittsburgh, Pennsylvania, 1989), p. 155.
4. J. Cesarano III, I. A. Aksay, and A. Bleir, *J. Am. Ceram. Soc.*, **71**, 240 (1988).
5. D. Gallagher, Ph.D. Thesis, University of Washington, 1988.
6. R. Buscall, *Colloids and Surfaces*, **5**, 269 (1982).
7. H. A. Kearsey and L. E. Gill, *Trans. Inst. Chem. Engrs.*, **41**, 296 (1963).
8. F. M. Tiller and Z. Khatib, *J. Colloid Interface Sci.*, **100**, 55 (1984).
9. F. F. Lange and K. T. Miller, *Bull. Am. Ceram. Soc.*, **66**, 1498 (1987).
10. W.-H. Shih, S. I. Kim, W. Y. Shih, C. H. Schilling, and I. A. Aksay, in ***Better Ceramics Through Chemistry IV, MRS Symp. Proc.,*** Vol. 180, edited by C. H. Brinker, D. E. Clark, D. R. Ulrich, and B. J. J. Zelinski (Materials Research Society, Pittsburgh, Pennsylvania, 1990), p. 167.
11. W. Y. Shih, W.-H. Shih, and I. A. Aksay, in ***Physical Phenomena in Granular Materials***, Vol. 195, edited by G. D. Cody, T. H. Geballe, and P. Sheng (Materials Research Society, Pittsburgh, Pennsylvania, 1990), p. 477.
12. W. Y. Shih, I. A. Aksay, and R. Kikuchi, *Phys. Rev. A*, **36**, 5015 (1987).
13. W.-H. Shih, W. Y. Shih, S. I. Kim, J. Liu, and I. A. Aksay, *Phys. Rev. A*, **42**, 4772 (1990).
14. L. T. Kuhn, R. M. McMeeking, and F. F. Lange, in ***Powders and Grains***, edited by J. Biarez and R. Gourves (Balkema, Rotterdam, 1989), p. 331.
15. P. A. Cundall, J. T. Jenkins, and I. Ishibashi, in ***Powders and Grains***, edited by J. Biarez and R. Gourves (Balkema, Rotterdam, 1989), p. 319.

ENERGY ABSORPTION BY EXPANDED BEAD POLYSTYRENE FOAM: DEPENDENCE ON FRACTURE TOUGHNESS AND BEAD FUSION

P.R. STUPAK AND J.A. DONOVAN
University of Massachusetts, Department of Mechanical Engineering, Amherst, MA 01003

INTRODUCTION

Increased energy absorption and load result when the contact area between an object and a foam cushion is less than the foam area because of an increased foam deformation volume (i.e. "Load Spreading"). The deformed volume is trapezoidal (i.e. not prismatic) and is a function of the foam dimensions and the object geometry. The load spreading effect is greatest when the ratio of the object area to the foam thickness is smallest; a common configuration in commercial packages (Figure 1).

Polymeric foam energy absorbers frequently fracture during impact due to tensile stresses normal to the trapezoid boundary. Does the fracture process increase the energy absorption capacity of a protective component by dissipating energy, or decrease it by limiting the volume of foam deformed during impact?

Gibson and Ashby [1,2] recently described brittle fracture of homogeneous foams using linear elastic fracture mechanics. The fracture toughness was determined as a function of foam density, maximum bending stress in the cell wall, and cell diameter, and was applied to data from independent studies of rigid polyurethane [3,4], glass [5,6] and polymethacrylamide [7] foams. However, the relationship of foam fracture toughness to energy absorption has not been fully addressed [8].

A commercially important polymer foam, expanded polystyrene bead foam (EPS), is the material of choice for most packaging and insulating applications with over 200Gg used annually [9]. EPS has a bi-level microstructure of large scale fused beads, each containing small scale cells. Compression induced fracture and poor bead fusion are frequently observed in EPS packaging. The effect of fracture and bead fusion on energy absorption was determined as a function of time, steam pressure, and foam density, with a specimen representative of current EPS package design.

FRACTURE

Procedure

Polystyrene beads (0.5 mm Dia.) containing ten percent pentane were initially expanded in steam for 30-60 seconds. The resulting cellular, spherical (0.5-3mm Dia.) "pre-expanded" beads, with a closed cell microstructure, were divided into three foam densities (23,26,29 Kg/m^3) by sieving. The larger and smaller bead diameters corresponded to lower and higher foam densites, respectively. Three point bend foam fracture specimens were prepared using a 0.023 x 0.04 x 0.227m^3 rectangular aluminum mold.

The molding parameter matrix consisted of six molding times and steam pressures and three foam densities. The large number of unique samples required by a classical testing approach suggested the use of a factorial testing plan [10]. The parameters were divided into two 3x3 Latin squares, allowing 18 samples to represent 54 possible molding conditions (Figure 2). Two samples were made for each molding condition.

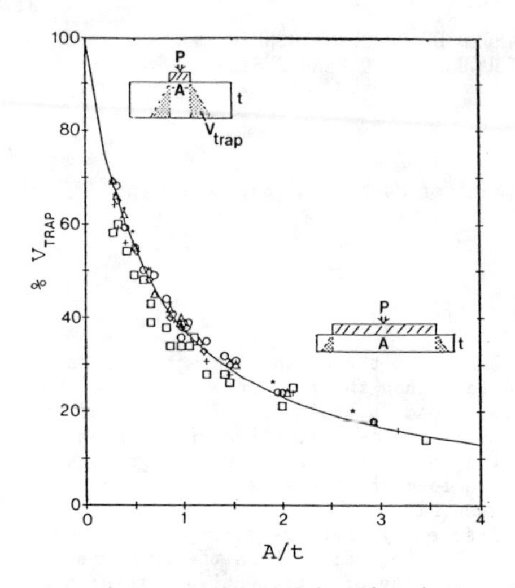

Figure 1 Percentage of deformed region due to "load spreading" (Vtrap) as a function of the indenter width to foam thickness ratio.

PRESSURE

Time	P1	P2	P3	P4	P5	P6
t6	ρ3	ρ2	ρ1			
t5	ρ1	ρ3	ρ2			
t4	ρ2	ρ1	ρ3			
t3	DENSITY			ρ3	ρ2	ρ1
t2				ρ1	ρ3	ρ2
t1				ρ2	ρ1	ρ3

Figure 2 Experimental design factorial testing scheme for molding foam fracture mechanics specimens.

The molded samples were pre-cracked to the specimen midpoint with a fresh razor blade. All fracture tests were three point bend tests using a uniaxial testing machine instrumented with a 50Kg load cell and at a crosshead rate of 0.5m/min.

Linear elastic fracture mechanics could not be used to analyze the fracture data because the load-deflection curves were non-linear. Rather, the elastic-plastic fracture toughness parameter, the J-integral, was employed according to,

$$J_m = 2U/(Bb) \tag{1}$$

where U is the energy or area beneath the load-deflection curve up to the point of crack initiation, minus the energy expended by indentation of the cylindrical supports; b is the remaining ligament length, and B is the specimen thickness. Since crack initiation could not be clearly identified, the fracture energy was calculated at the point of maximum load, yielding, J_m.

Results and Discussion

The fracture toughness increased with increasing molding time (T), pressure (P) and decreasing foam density (ρ). An empirical relation between the fracture toughness and the molding parameters was determined from the experimental design analysis, which assumes the dependent variable, fracture toughness, was the product of the individual functions of the independent variables, T,P, and ρ (Figure 3).

$$J_m = 770\ T^{1.25}\ P^{6.7}\ (\rho/(-2.5+0.13\rho)) \tag{2}$$

While useful for making "engineering" predictions, Equation (2) does not give insight into the processes responsible for the time, temperature (pressure), and density dependence of the fracture toughness. Past work on crack healing and polymer welding in solid polymers [11-13] suggest that the establishment of interfacial contact area and the increased molecular mobility at temperatures above the polymer's glass transition temperature results in polymer diffusion across contacting bead-bead interfaces, giving bead fusion.

A fractographic analysis of the fracture surfaces revealed two types of bead morphology: (1) smooth beads, exhibiting flat facets due to bead impingement, but little or no bead fusion, termed brittle interbead (BI), and (2) cellular fracture (CF), resulting from partial bead fusion, where material was transfered from one contacting bead face to another by local fracture of the near surface cellular microstructure (Figure 4).

The establishment of bead-bead interfacial contact area is time dependent and the existence of the brittle interbead (BI) fracture mode demonstrates that interfacial contact alone does not imply fusion. The smooth BI mode indicates it is the least fused of the two modes and dissipates the least energy when the interface separates. Interfacial contact over sufficient time to allow polymer diffusion gives fusion, identified by the cellular fracture (CF) mode.

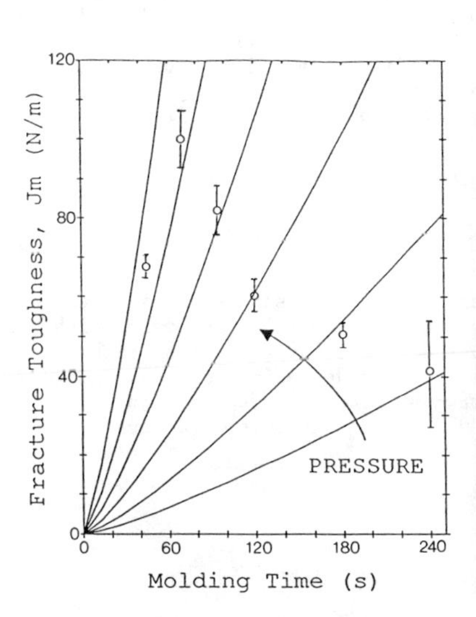

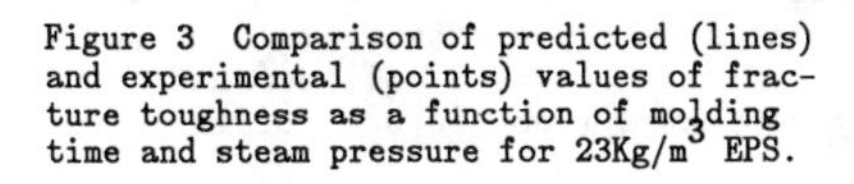
Figure 3 Comparison of predicted (lines) and experimental (points) values of fracture toughness as a function of molding time and steam pressure for 23Kg/m^3 EPS.

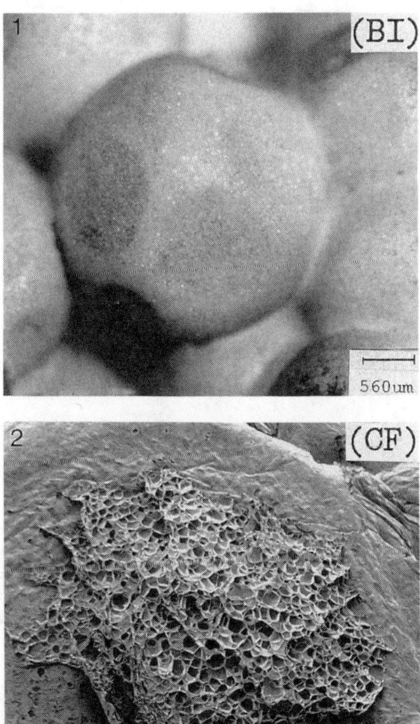

Figure 4 The two modes of bead failure: 1) brittle interbead (BI), 2) cellular fracture (CF).

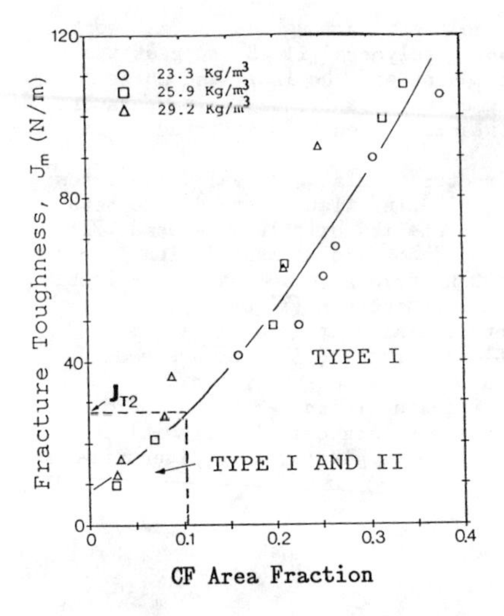

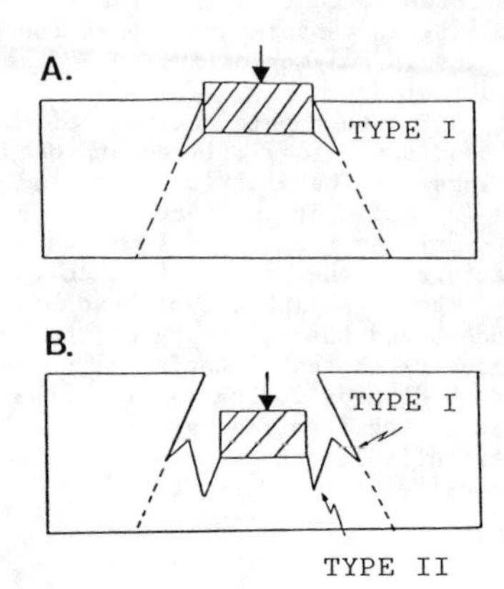

Figure 5 Fracture toughness as a function of cellular fracture (CF) area fraction for three densities of EPS.

Figure 6 Compressively induced fracture modes: A) Type I, B) Type I and II.

The fracture toughness (J_m) increased with increasing CF mode area fraction measured from the fracture surfaces (Figure 5). The initially low observed fracture toughness values are attributable to an initially dominant BI population. Increasing the molding time and pressure (i.e. temperature) at a given density increased the toughness through conversion of BI failure to the more fully fused, energy dissipative CF fracture mode.

ENERGY ABSORPTION

Procedure

The post-fractured three-point bend spans were used as the indentation specimens. All indentation tests were conducted to 75% compression with a 0.025 x 0.05m rectangular indenter mounted on a uniaxial testing machine and at a cross-head velocity of 0.5m/min. Load-deflection curves, energy absorption and indentation crack length data were recorded for two specimens of each molding condition and foam density (i.e. fracture toughness).

Results and Discussion

Two fracture modes were identified: (1) Type I and (2) Type II (Figure 6).

Type I fracture was observed for all specimens after 10-20% deflection. Crack initiation occurred several millimeters from the edge of the indenter and propagated slowly downward at a nearly constant angle of 35 degrees from the vertical (Figure 6a). The result was a well defined trapezoidal deformed region, partially separated from the bulk of the specimen.

Type II fracture only occurred below a critical value of fracture toughness (J_{T2}), independent of foam density. During compression, Type I fracture was established initially, becoming Type II between 35-70% deflection. The lowest toughness samples initiated Type II cracks earliest and propagated farthest. They formed closer to the indenter edge than Type I and rapidly propagated vertically (Figure 6b).

The occurrence of both Type I and II or only Type I fracture is ultimately determined by the degree of bead fusion in the sample. Plotting J_{T2} on Figure 5 shows that the origin of Type II fracture is insufficient bead fusion below a critical value.

The energy absorbed at 75% compression increased with increasing toughness and foam density (Figure 7). Below J_{T2}, the rapid increase in absorbed energy with increasing toughness was attributable, in part, to the energy lost (area under the load-deflection curve, Figure 8) through Type II fracture. As Jm increased toward J_{T2}, Type II fracture released less of the load spreading region and at larger indenter deflections. Above J_{T2}, shorter Type I cracks correlated with increasing fracture toughness and energy absorption through increased load spreading from the trapezoidal region to the bulk specimen (Figure 9). Longer Type I cracks in lower toughness samples physically separates the trapezoidal deformed region from the bulk of the specimen more than cracks in higher toughness samples, thereby limiting load spreading and energy absorption.

Whereas, the dissipated fracture energy (the product of J_m and the fracture area) was at most an order of magnitude too low to account for the increase in absorbed energy above J_{T2}.

In brief, increased bead fusion increases fracture toughness that causes increased energy absorption through decreased compressive fracture.

CONCLUSIONS

The energy absorbed, during compression of low density polystyrene bead foam by a rectangular indenter, increased with increasing fracture toughness because compression induced fracture was reduced. Shorter cracks correlated with increased fracture toughness and energy absorption through increased load spreading from the trapezoidal region to the bulk specimen.

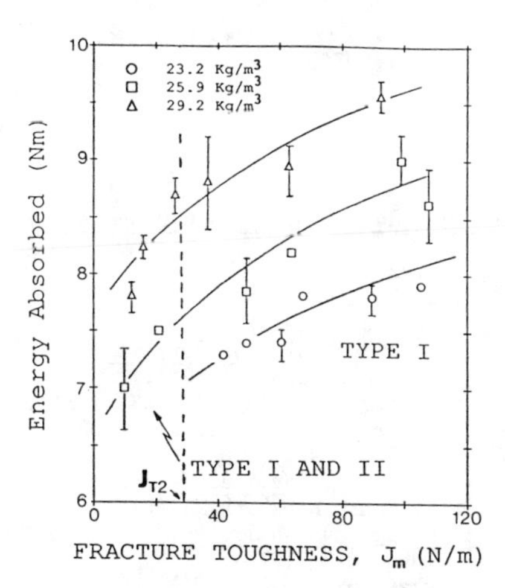

Figure 7 Energy absorbed at 75% compression as a function of fracture toughness and foam density.

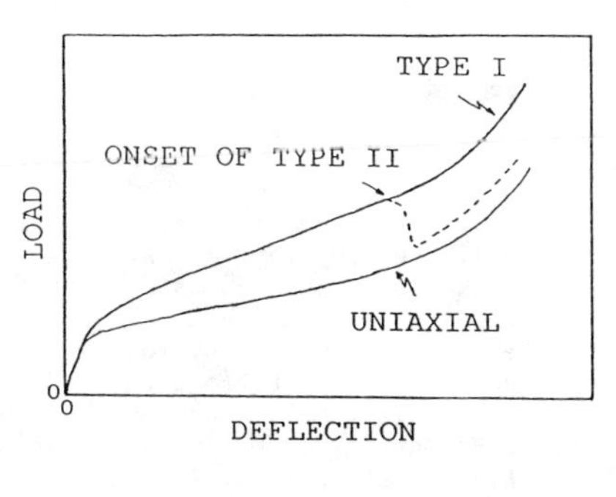

Figure 8 Load-deflection curves for typical specimens exhibiting Type I, Type I and II fracture. A uniaxial curve is included for comparison.

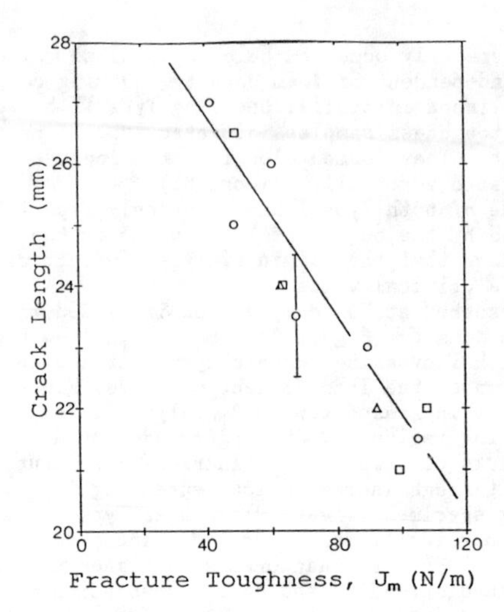

Figure 9 Type I crack length as a function of fracture toughness for three densities of EPS.

The fracture toughness increased with increasing molding time, steam pressure (i.e. temperature), and decreasing foam density because bead fusion increased. Bead fusion was evaluated from the fracture surface as the "cellular fracture" area fraction which increased with increasing toughness.

ACKNOWLEDGEMENT

The authors acknowledge the generous support of this work by the Center for UMass-Industry Research on Polymers (CUMIRP). We also thank John G. Klepic (ARCO Chemical Co.) for supplying the polystyrene beads.

REFERENCES

1 L.J. Gibson and M.F. Ashby, Cellular Solids: Structure and Properties, (Pergamon, Oxford, 1988).
2 L.J. Gibson, Mater. Sci. Eng. A110, 1 (1989).
3 C.W. Fowlkes, Int. J. Fract. 10, 99 (1974).
4 A. McIntyre and G.E. Anderton, Polymer 20, 247 (1979).
5 J.G. Zwissler and M.A. Adams, in Fracture Mechanics of Ceramics, edited by R.C. Bradt, Vol. 6 (Plenum, New York, 1983) p. 211.
6 J.S. Morgan, J.L. Wood and R.C. Bradt, Mater. Sci. Eng. 47, 37 (1981).
7 S.K. Maiti, M.F. Ashby and L.J. Gibson, Scrip. Metal. 18, 213 (1984).
8 R.F. Harris, J. Cell. Plast. Sept.-Oct., 221 (1970).
9 A.E. Platt and T.C. Wallace, in Encyclopedia of Chemical Technology, edited by H. Mark, Vol. 21 (John Wiley & Sons, New York, 1983) p. 836.
10 C. Lipson and N.J. Sheth, Statistical Design and Analysis of Engineering Experiments, (McGraw Hill, New York, 1973).
11 R.P. Wool and K.M. O'Conner, J. Appl. Phys. 52, 5953 (1981).
12 R.P. Wool, B.L. Yuan and O.J. McGarel, Poly. Eng. Sci. 29, 1340 (1989).
13 K. Jud, J.G. Williams and H.H. Kausch, J. Mat. Sci. 16, 204 (1981).

TESTING OF IMPACT LIMITERS FOR TRANSPORTATION CASK DESIGN

A. K. Maji *, S. Donald ** and K. Cone **
* Assistant Professor of Civil Engineering, University of New Mexico.
** Graduate Research Assistants, Departments of Civil and Mechanical Engineering, University of New Mexico, Albuquerque, NM 87131.

ABSTRACT

'Soft Impact Limiters' such as polyurethane foams and aluminum honeycombs are being studied to assist the Sandia National Laboratory's Transportation Base Technology Program. The aim of this research is to study the mechanical behavior of these materials, which are being used as impact absorbers in nuclear waste transportation containers.

A series of tests were performed along various loading paths using an Instron, servo-controlled, multi-axial loading machine and Soiltest triaxial testing apparatus. Static tests included uniaxial tension, uniaxial compression, triaxial compression, hydrostatic compression and fracture toughness testing. The purpose of using different loading paths was to generate an extensive test data which is being used to develop constitutive models for these materials, under a separate research program.

Dynamic tests were conducted at strain rates of 100 strains/sec., to generate experimental data relevant to accident situations. These tests were conducted on an instrumented Charpy impact testing apparatus. Results of these tests were subsequently used to conduct scale-model tests of transportation casks of different industrial designers.

Different densities of Polyurethane foams, aluminum honeycombs, and corrugated aluminum honeycombs were tested in different orientations. The paper discusses the experimental program, instrumentation and test results for the nuclear waste transportation industry and other potential applications.

INTRODUCTION

Protecting the nuclear waste transportation casks from damage in case of accidents is one of the prime issues underlying the safe and successful disposal of nuclear waste. The casks that are used to transport nuclear waste are padded with cellular materials like Aluminum honeycombs and Polyurethane foams to absorb impact energy that may result from the accidents during transportation and handling. For many years the mechanical behavior of these materials was studied and many theories were put forth. Tests have been conducted on diverse types of loadings and have involved different types of materials. Although the theoretical formulations developed and experiments performed by Gent and Thomas[1] and Lederman [2], in the earlier years, were useful to some extent, they were disregarded as the formulations assumed that the cell walls bear only the axial load. Gibson et al [3] studied the mechanical properties of honeycombs in a more general manner by studying the properties in terms of bending, elastic buckling and plastic collapse of cell walls. Gibson and Ashby [4] treated the mechanical properties of foams as functions of cell wall properties and cell geometry. they used an electron microscope to validate their theories with physical observations of cell wall bending of polyurethane foams subjected to uniaxial compression.

These theories were used to predict the behavior of cellular materials subjected to uniaxial stresses only. However, in many practical situations, most of the cellular materials are subjected to multiaxial stress. Therefore, it is also necessary to consider results from multiaxial tests. An experimental study was done by Shaw and Sata [5], Triantafillou et al [6], Triantafillou and Gibson [7] to predict the properties of the materials subjected to multiaxial forces. Shaw and Sata studied the failure of polystyrene foams subjected to compressive, tensile and hydrostatic forces. They observed Luder like bands and predicted that the foams yield according to a maximum principal stress criterion. Gibson et al [8] have indicated that fully dense plastic solids yield according to the Von Mises criterion which involves all three principal stresses. Hence to compare the results with theoretical modeling, Triantafillou et al tested two flexible foams; an open cell polyurethane and a closed-cell polyethylene subjected to uniaxial compression, biaxial compression, axisymmetric compression and hydrostatic compression. Recently, Miller and Eichinger [9] have calibrated Near-infrared (NIR) diffuse reflectance spectroscopy to

Table I. Static Test Schedule (number of tests performed in each category)

MATERIALS TESTED	UNIAXIAL COMP	UNIAXIAL TENSION	HYDRO-STATIC COMP	TRIAXIAL COMP 50	100	150
3 Pcf foam	6	6	4	4	-	-
5 Pcf foam	6	6	4	4	4	-
10 Pcf faom	6	6	4	-	4	4
20 Pcf foam	6	6	4	-	-	4
3.1 Pcf Honeycomb	9	9	-		-	
5.1 Pcf Honeycomb	9	9	-		-	
12.1 Pcf Honeycomb	9	6	-		-	
22.1 Pcf Honeycomb	9	3	-		-	

nondestructively evaluate the compressive strength of polyurethane foams. As cellular materials find more diverse applications, it has become necessary to obtain experimental data pertaining to a number of different tests so as to aid in a more general description of the cellular materials, and constitutive modeling.

The purpose of this paper is to present the experimental techniques employed and to provide results of the static and dynamic tests that were carried out on impact limiter materials. The experiments that were conducted along various loading paths on the test specimen included uniaxial compressive and tensile tests of honeycombs and foams, and hydrostatic and triaxial compressive tests of foams.

EXPERIMENTAL PROGRAM

Static Testing

An extensive experimental database containing the results from all the tests mentioned above was obtained (Table I). The mechanical behavior of these materials was closely studied by observing the physical phenomena of deformation, and through the Load-deflection curves. Two types of materials were tested. Aluminum honeycombs made from 5052 H39 alloy were of four densities of 3.1,8.1,12.0 and 22.1pcf (0.05,0.13,0.19 and 0.35 gm./cc). The former two were made by the expansion process, while the latter two were made by the corrugated process. The second material was rigid, closed cell, thermoset polyurethane foam made from polymeric isocyanate polyether polyol. Four densities of 3,5,10 and 20pcf (0.048,0.08,0.16 and 0.32 gm/cc.) were tested. While the 3 and 5pcf foams were made by a gas blown process by using freon, the 10 and 20pcf were made by the water blown process. The foams were manufactured by General Plastics Corp. and the Honeycombs were manufactured by Hexcel Corp.

A series of tests along various loading paths were performed on both the Aluminum honeycombs and Polyurethane foams. While the honeycombs were tested for compressive and tensile loads only, the foams were also tested for hydrostatic and triaxial loads. All the four densities of honeycombs were tested in triplicate along three directions as shown in figure 1. The four densities of foams were tested in two directions; one perpendicular to the cell growth, and the other parallel to the cell growth. Uniaxial compression tests (ASTM C 365) and flatwise tensile tests (ASTM C 297) were performed on a multi-axial servo controlled Instron loading machine, the rate of deflection being 0.095 cm/s. Special loading fixtures as recommended in ASTM C 297, which recommends using aluminum blocks connected to grips through a 1.1 cm. rod were used for testing the specimens in tension. A scotch weld epoxy adhesive (1838) from 3M company was used for bonding the samples to the aluminum blocks.

A Soil-test triaxial chamber capable of withstanding a pressure of 2.76MPa and a load of 44.5 KN was used for pressurizing the specimen for hydrostatic and triaxial compression tests. Pressure was applied through water and volume change of the sample was measured through a burette. Since the volume change in foams is far greater than those typically expected in soils, the burette had to be modified to measure the volumes accurately. The foam specimen were kept dry by enclosing them in a waterproof membrane. Cubical samples were used, and measured 10.2cm and 7.6 cm respectively for the compressive and tensile tests. Cylindrical samples having both diameter and height of 7.1cm were used for the hydrostatic and triaxial tests. Triaxial tests involved first applying hydrostatic pressure, followed by application of deviatoric pressure.

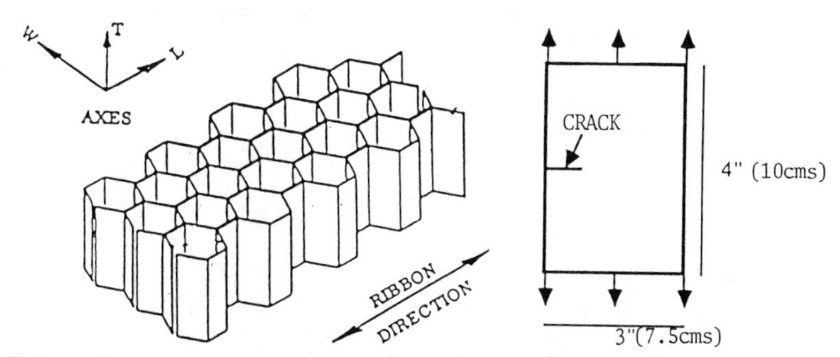

Figure 1. Honeycomb Anisotropy Directions

Figure 2. Fracture Toughness Test Specimen

Dynamic Testing

A conventional Charpy impact apparatus was instrumented to perform dynamic tests. The strain rate hence achieved was approximately 100 strains /sec. The samples were 2.5cm in diameter and 2.5cm long. A dynamic load cell (Kistler model 9041A) was used, along with a RVDT (Rotational Variable Displacement Transformer), to measure the displacements. An accelerometer and a LVDT were also used to obtain displacement readings. The LVDT suffered from inertial effects. Especially, once the materials densify and start to lock up and deceleration is high, the LVDT does not respond fast enough and the resulting deflection readings are different from those actually seen on the specimens after the test. The integration of the accelerometer data to obtain displacements also proved to cause errors.

The RVDT provided a more accurate data than the other two methods. It measures the rotational swing of the Charpy pendulum and is continuously is contact with the moving parts. The angular rotation time history is subsequently converted to displacements. Since the specimen size was much smaller than the hammer arm (about 0.9m), the resulting displacement during the impact was linear. A comparison between the plastic plateau stress levels obtained from static and dynamic loading have been shown in Table II.

Fracture Toughness Testing

Notched specimens were tested under uniaxial tension to obtain fracture toughness values. Specimens with a cross-section of 7.6 x 7.6cm were used. Notches 1.27 and 2.54 cm deep were cut into the samples by a fine saw to ensure a sharp notch tip (Figure 2).

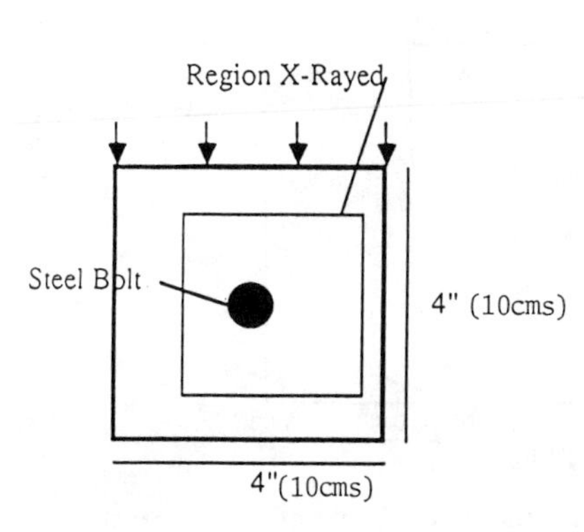

Figure 3a Specimen Tested with X-Rays

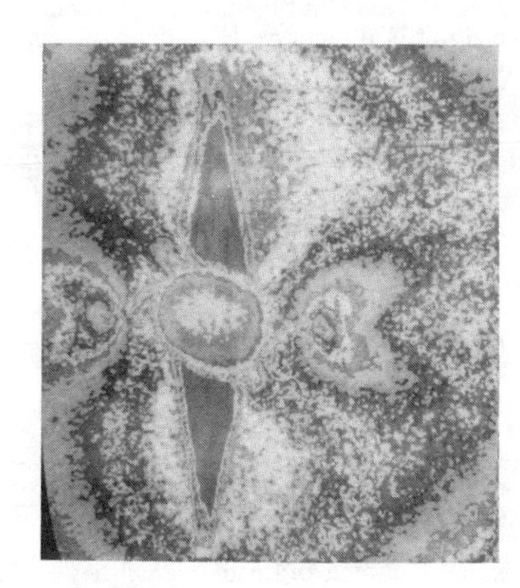

Figure 3b. Internal Densification Image

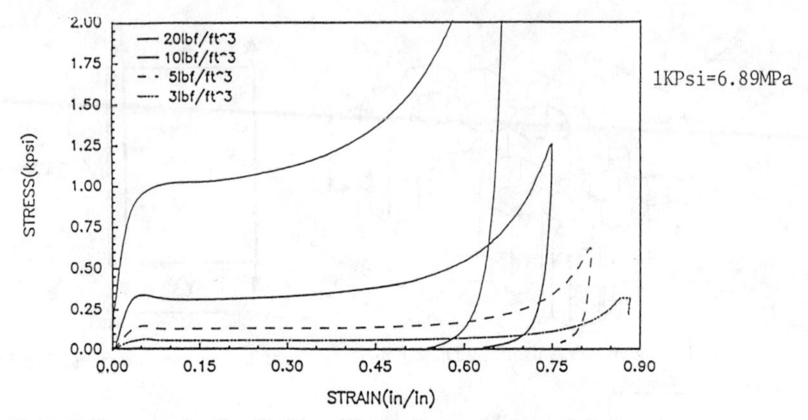

Figure 4. Compression Test Results of Foams Perpendicular to Cell Growth

Internal Densification Study with X-Rays

External deformations were monitored with a video camera and measured with conventional instrumentation. In order to measure internal deformations, an IRT IXRS 160/32000 industrial x-ray system was used The x-ray system has a maximum output capacity of 3200 watts, and can be use to penetrate high density foams. The x-ray system was used while a specimen (Figure 3a) was loaded in the Instron machine. This allowed for real time observation of internal densification of the foam. The densification can then be quantified by image analysis of the amount of x-ray radiation passing through the material. The densification patterns shown in the figure seems to indicate the formation of plastic yield zones (Figure 3b).

EXPERIMENTAL RESULTS

Typical uniaxial compression test results of different densities of foams have been shown in Figure 4. The variation in behavior of the honeycombs with the orientation of loading is shown in Figure 5. The honeycombs demonstrate a more steady plateau stress than the foams. In the T direction, the crushing occurs progressively from one end of the honeycombs until densification. The densification of the foams is more uniform, although some bulging and localization is evident. Bands of localized deformation are prominent in the honeycombs loaded in the L and W directions. High density foams (20 pcf or 0.32 gm./cc.) exhibit tensile splitting during densification.

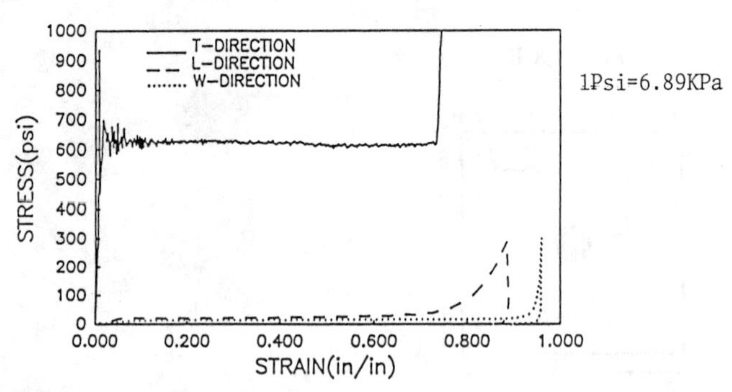

Figure 5. Effects of Load Orientations on Compression Test Results of 8 pcf Honeycombs

The foam materials are elastic and brittle in tension. Figure 6 show typical tensile behavior of foams and the effect of the anisotropy. The nature of failure in the honeycombs in tension vary widely depending upon the orientation and the material density. Failure could occur by progressive tensile fracture of cell walls, longitudinal tensile debonding of the honeycomb's adhesives, or lateral splitting by adhesive failure. The resulting stress-strain behavior also varies widely as a consequence. Results of Hydrostatic compression are shown in Figure 7. Note that the 20pcf foam does not densify because of the limited pressurizing capacity of the loading cell. Typical triaxial test results (initial confining pressure of 172kPa), have been included in Figure 8.

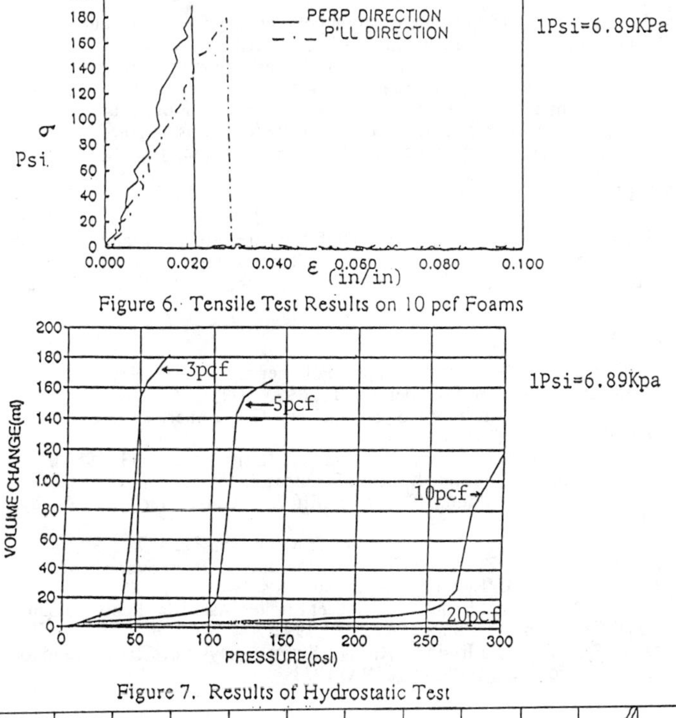

Figure 6. Tensile Test Results on 10 pcf Foams

Figure 7. Results of Hydrostatic Test

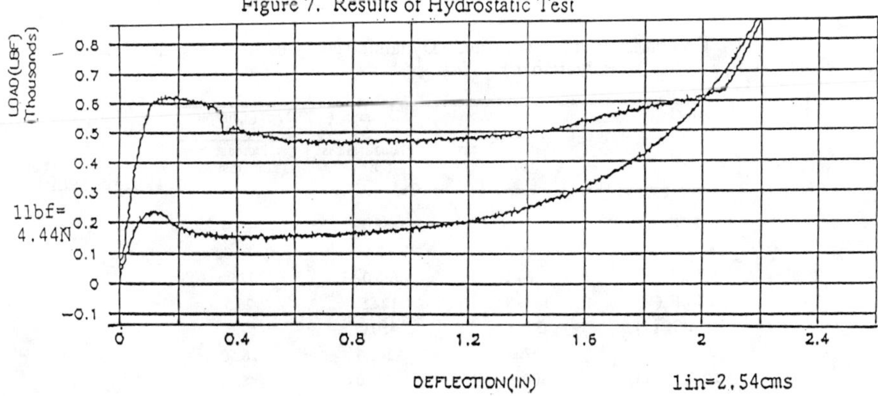

Figure 8. Results of Triaxial Tests on 5 pcf Foams Perpendicular to growth Parallel to growth

This paper provides only a general description and a sample of the test data due to the space limitations. All data pertaining to the test program described on table I are available in details in reference [10].

Fracture toughness values were calculated from the tensile strengths of the precracked specimens using closed form solutions [11]. The fracture toughness values (K_{IC}) parallel to the direction of cell growth (weaker direction) for the 0.08gm/cc foam was about 1.0 mPa$\sqrt{cm}$. Some comparison between static and dynamic testing data pertaining to foams is shown in table II.

CONCLUSIONS

1. A wide variety of tests were carried out on polyurethane foams and aluminum honeycombs. The test program and instrumentation has been described and some typical data has been presented. More extensive results of the test data are reported elsewhere.
2. Dynamic test results indicate the rate effect on the materials. The honeycombs were practically insensitive, whereas the foams were somewhat stronger under impact conditions.
3. Typical fracture toughness value for a low density foam was obtained.
4. Internal densification could be monitored using x-ray and image analysis systems.

ACKNOWLEDGEMENT

This research was funded by the DOE's Waste Education and Research Consortium (WERC), and the Sandia National Laboratory's Transportation Systems Technology Division (6322, Bob Glass, coordinator). We thank Dr. M.L. Wang for the use of the X-ray system.

REFERENCES

1. A. N. Gent and A. G. Thomas, J. appl. Polymer Sci. 1,107 (1959).
2. J. M. Lederman, J. appl. Polymer Sci. 15, 693 (1971).
3. L. J. Gibson, M. F. Ashby, G. S. Schager and C. I. Robertson, Proc. R. Soc. (London) A382, 25 (1982).
4. L. J. Gibson, and M. F. Ashby, Proc. R. Soc. (Lond.) A382, 43 (1982).
5. M. C. Shaw and T. Sata, Int. J. Mech. Sciences 8, 469 (1966).
6. T. C. Triantafillou, J. Zhang, T. L. Shercliff, L. J. Gibson and M. F. Ashby, Int. J. Mech. Sci 31, 665 (1989) .
7. T. C. Triantafillou and L. J. Gibson, Int. J. Mech. Sci. 32, 479 (1990).
8. L. J. Gibson, Ashby, Zhang and Triantafillou, Int. J. Mech. Sci., 31, 635 (1989).
9. C. E. Miller and B. E. Eichinger, Appl. Spectrosc. 44, 887 (1990).
10. Maji A. K. "Mechanical Behavior of Impact Limiter Materials" Report to Sandia National Laboratories, Albuquerque, NM, 1991.
11. Tada H., Paris P.C. and Irwin G. R. "The Stress Analysis of Cracks Handbook", Del Research Corporation, Hellertown, PA, 1973.

Table II. Comparison of Static and Dynamic compression test results of Polyurethane foams

Sample type	Static	Dynamic (100.0 / s) (Charpy)	
	Plateau Stress (psi)	Plateau stress (psi)	% increase over static results
3 PCF (I)	54.0	68.0	25.9
3 PCF (II)	40.5	62.5	54.32
5 PCF (I)	124.1	160.0	28.90
5 PCF (II)	96.8	122.5	26.5
10 PCF (I)	304.6	480.0	57.6
10 PCF (II)	270.0	341.6	26.50
20 PCF (I)	1009.4	1500.0	48.60
20 PCF (II)	1126.7	1350.0	19.80

1Psi=6.89KPa

MECHANICAL ANALYSES OF WIPP DISPOSAL ROOMS BACKFILLED WITH EITHER CRUSHED SALT OR CRUSHED SALT-BENTONITE

RALPH A. WAGNER*, G. D. CALLAHAN*, AND B. M. BUTCHER**
*RE/SPEC Inc., P.O. Box 725, Rapid City, SD 57709
**Sandia National Laboratories, Div. 6345, P.O. Box 5800, Albuquerque, NM 87185

ABSTRACT

Numerical calculations of disposal room configurations at the Waste Isolation Pilot Plant (WIPP) near Carlsbad, NM are presented. Specifically, the behavior of either crushed salt or a crushed salt-bentonite mixture, when used as a backfill material in disposal rooms, is modeled in conjunction with the creep behavior of the surrounding intact salt. The backfill consolidation model developed at Sandia National Laboratories was implemented into the SPECTROM-32 finite element program. This model includes nonlinear elastic as well as deviatoric and volumetric creep components. Parameters for the models were determined from laboratory tests with deviatoric and hydrostatic loadings. The performance of the intact salt creep model previously implemented into SPECTROM-32 is well documented.

Results from the SPECTROM-32 analyses were compared to a similar study conducted by Sandia National Laboratories using the SANCHO finite element program. The calculated deformations and stresses from the SPECTROM-32 and SANCHO analyses agree reasonably well despite differences in constitutive models and modeling methodology. These results provide estimates of the backfill consolidation through time. The trends in the backfill consolidation can then be used to estimate the permeability of the backfill and subsequent radionuclide transport.

1.0 INTRODUCTION

The U. S. Department of Energy is planning to dispose of transuranic wastes (TRU) at the Waste Isolation Pilot Plant (WIPP) near Carlsbad, New Mexico. The current mission of the WIPP is to provide a research and development facility to demonstrate the safe management, storage, and disposal of TRU wastes generated by U. S. government defense programs. Sandia National Laboratories is conducting procedural and technical activities to assess WIPP compliance with regulatory requirements. Performance of seal, barrier, and backfill materials is being studied as part of these activities. A key candidate component of the room backfill system is crushed salt. Crushed salt is an attractive backfill material because it is readily available from the mining operations, it is compatible with the host rock, and it is expected to reconsolidate into a low permeability mass comparable to the intact salt as a result of the creep closure of the surrounding rock mass. Therefore, an understanding of the mechanical behavior of backfill is important at the WIPP. Optimization of the backfill emplacement is necessary to promote room stability, to enable sufficient backfill consolidation to reduce brine flow and retard the transport of soluble radionuclides, and to maintain sufficient gas permeability to avoid gas pressurization [1].

The study presented herein was conducted to investigate the consolidation of crushed salt and a crushed salt-bentonite mixture in disposal room configurations at the WIPP. Numerical simulations of the backfill and creeping host rock system to 200 years were performed. The finite element program, SPECTROM-32 [2], was modified [3] to incorporate a crushed salt consolidation model [4]. The model selected for implementation combines nonlinear elastic behavior and volumetric creep consolidation. The creep consolidation model was modified to include a deviatoric component. Results from these analyses are compared to results from similar analyses obtained with the finite element program SANCHO [5]. A notable difference between the two codes is that the SANCHO finite element program is based on finite strain theory and the SPECTROM-32 finite element program is based on infinitesimal or small strain theory.

2.0 PROBLEM DESCRIPTION

2.1 Problem Parameters

The basic problem parameters consist of the room and pillar geometry, depth of the room, mesh refinement of the modeled region, temperature, and boundary conditions (see Figure 2-1). Except for a variation in the mesh refinement, the basic problem parameters in the SPECTROM-32 analyses are identical to the SANCHO analyses. The total strain rates in the crushed salt and the intact salt were assumed to be the sum of elastic and inelastic components as discussed in the next sections. The initial state of stress in the SPECTROM-32 analyses was established by excavating the room instantaneously into an initially lithostatic stress field. Subsequently, the room was assumed to be completely backfilled with either crushed salt or crushed salt-bentonite.

2.2 Backfill Consolidation Model

Development of the creep consolidation constitutive equation used in SPECTROM-32 was guided by general considerations with specific functional forms taken from empirical relations matched to available laboratory data. From the application of thermodynamic concepts, the three-dimensional generalization for creep strain rates is given by [6]. Following this approach, two continuum internal variables were assumed, the average inelastic volumetric strain, $\epsilon^c_{eq_1}$, and the average equivalent inelastic shear strain, $\epsilon^c_{eq_2}$.

$$\dot{\epsilon}^c_{ij} = \dot{\epsilon}^c_{eq_1} \frac{\partial \sigma^f_{eq_1}}{\partial \sigma_{ij}} + \dot{\epsilon}^c_{eq_2} \frac{\partial \sigma^f_{eq_2}}{\partial \sigma_{ij}} \tag{1}$$

For the volumetric portion of Equation 2-1, the invariant strain-rate measure is

$$\dot{\epsilon}^c_{eq_1} = \dot{\epsilon}_v \left(\sigma_m\right) \tag{2}$$

The sign convention adopted assumes that tensile stresses and elongation (dilation) are positive. The volumetric strain rate $\dot{\epsilon}^c_v$ is described empirically [4] based on laboratory test data on hydrostatic consolidation of crushed salt as

$$\dot{\epsilon}^c_v = \frac{\left(1+\epsilon_v\right)^2}{\rho_0} B_0 \left[1 - e^{-B_1 \sigma_m}\right] e^{\frac{A \rho_0}{1+\epsilon_v}} \tag{3}$$

where

$$\begin{aligned}
\epsilon_v &= \epsilon_{kk}, \text{ total volumetric strain} \\
\epsilon_v^c &= \epsilon_{kk}^c, \text{ volumetric creep strain} \\
\sigma_m &= \tfrac{\sigma_{kk}}{3}, \text{ mean stress} \\
\rho_0 &= \text{initial density} \\
B_0,\, B_1,\, A &= \text{material constants.}
\end{aligned}$$

The invariant stress measure is given by

$$\sigma_{eq_1}^f = \sigma_m \tag{4}$$

For the deviatoric portion of Equation 2-1, the invariant strain-rate measure is taken to be

$$\dot{\epsilon}_{eq_2}^c = \beta \dot{\epsilon}_{eq_1}^c = \beta \dot{\epsilon}_v \left(\sigma_m\right) \tag{5}$$

and the invariant stress is assumed to be a scalar multiple of the octahedral shear stress

$$\sigma_{eq_2}^f = \sigma_e = \sqrt{3J_2} \tag{6}$$

where J_2 is the second invariant of the stress deviator $\left(J_2 = \frac{1}{2} S_{ij} S_{ij}\right)$. Substituting into Equation 2-1 and performing the required differentiation gives

$$\dot{\epsilon}_{ij}^c = \dot{\epsilon}_v^c \frac{\delta_{ij}}{3} + \beta \dot{\epsilon}_v^c \frac{3S_{ij}}{2\sigma_e} \tag{7}$$

β is selected such that in a uniaxial test the lateral components of $\dot{\epsilon}_{ij}^c$ equal zero. This requires that $\beta = -\frac{2}{3}$. After substituting for $\dot{\epsilon}_v^c$ in Equation 2-7, the strain rate components are given by

$$\dot{\epsilon}_{ij}^c = \frac{\left(1 + \epsilon_v\right)^2}{\rho_0} B_0 \left[1 - e^{-B_1 \sigma_m}\right] e^{\frac{A\rho_0}{1+\epsilon_v}} \left\{\frac{\delta_{ij}}{3} - \frac{S_{ij}}{\sigma_e}\right\} \tag{8}$$

Obviously, this creep consolidation equation will allow unlimited consolidation. Therefore, a cap is introduced that eliminates further consolidation when the intact material density ρ_∞ is reached. As an option, the crushed salt material may behave either as a nonlinear elastic material or creeping intact salt following complete consolidation.

2.3 Intact Salt Constitutive Model

A recently proposed WIPP reference constitutive relation for intact salt creep [7] and the previous WIPP reference law [8] were used in the calculations. The WIPP reference elastic-secondary creep model [8], along with a Mises-type flow potential, represented the mechanical behavior of the intact salt in the SANCHO analyses. The SPECTROM-32 analyses used the Munson-Dawson constitutive relation [7] for intact salt creep along with a Tresca flow potential. This combination was recommended [7] because of favorable comparisons with WIPP field tests in a previous study [9].

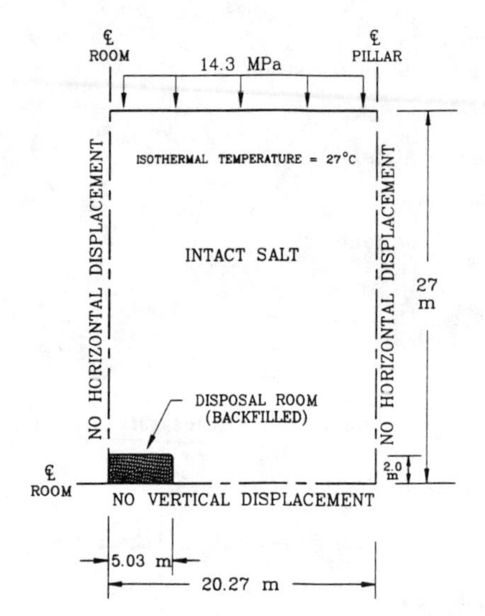

Figure 2-1. Description of Basic Problem Parameters.

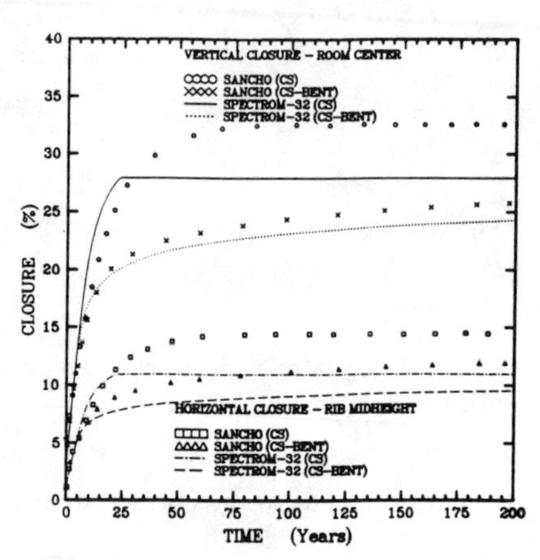

Figure 3-1. Vertical and Horizontal Room Closures for a Backfilled Disposal Room.

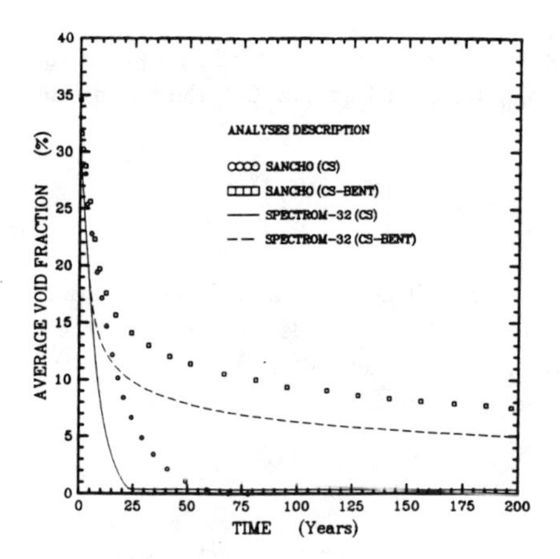

Figure 3-2. Average Void Fraction for a Backfilled Disposal Room.

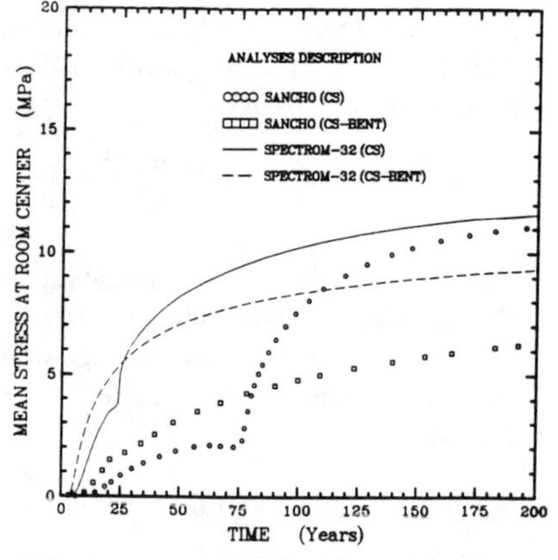

Figure 3-3. Mean Stress (Pressure) at the Center of a Backfilled Disposal Room.

2.4 Elastic Model

The elastic models used in the SPECTROM-32 and SANCHO analyses were identical. The crushed salt was described using a nonlinear elastic model [3, 4] and the intact salt was assumed to be linear elastic. However, corresponding elastic material constants for crushed salt and intact salt used in the SANCHO analyses were reduced by a factor of 12.5 [5]. This reduction in elastic properties stems from a recommendation [10] that is based on the usage of the early version of the reference constitutive relation for salt creep [7] that produced good agreement between calculated and measured deformations of WIPP field experiments. Since test data do not exist for crushed salt-bentonite mixtures, its elastic properties were assumed to be identical to those of crushed salt. Specific values for the elastic constants may be found in references 3, 4, and 11.

3.0 RESULTS

Results from the SPECTROM-32 analyses of the disposal room backfilled with either crushed salt or crushed salt-bentonite are compared to previously reported results using the SANCHO finite element program [5]. Results from the SANCHO analyses were plotted by digitizing the graphical results.

3.1 Room Closures

Figure 3-1 compares the SPECTROM-32 and SANCHO vertical and horizontal closures along the room periphery corresponding to the roof/floor centerline and the rib midheight, respectively. During the initial 15 years, the closures predicted by SPECTROM-32 are greater and, thereafter, are less than the closures predicted by SANCHO. The closure curves from both analyses for crushed salt backfill are nearly flat after the density of the crushed salt reaches the density of the intact salt. This behavior is not exhibited for the crushed salt-bentonite analyses since full compaction is not reached. Generally, agreement in the calculated deformations from the SPECTROM-32 and SANCHO analyses is within 10 percent (except when the crushed salt becomes fully consolidated) despite distinct differences in the constitutive relations, mesh refinement, material constants, and theoretical basis (finite versus small strain).

3.2 Average Void Fractions

Figure 3-2 provides the time history of the average void fraction remaining in the backfill. The average void fraction [5] is essentially a measure of porosity in the backfill. The trend and magnitude of the average void fractions from both codes in the two analyses agree closely during the initial 5 years. Thereafter, the rate of consolidation is considerably slower in both SANCHO analyses. The crushed salt backfill consolidates completely in 25 years in the SPECTROM-32 analyses; whereas, the corresponding consolidation takes 65 years in the SANCHO analyses. Full consolidation of the crushed salt-bentonite backfill is not reached in either the SPECTROM-32 or SANCHO analyses. The average void fraction remaining in the crushed salt-bentonite backfill after 200 years of simulation is 5.0 and 7.5 percent for SPECTROM-32 and SANCHO analyses, respectively.

3.3 Mean Stresses

Comparison of mean stress (pressure) is of interest since mean stress is the driving force in the backfill consolidation process. Figure 3-3 shows the mean stress history corresponding to the center of the disposal room. During the initial 5 years, pressure rise in the crushed salt is negligible throughout the disposal room. Over the next 50 to 75 years, the magnitude of the mean stress increases more rapidly in the SPECTROM-32 analyses than the SANCHO analyses. This response is indicative of the deformational behavior plotted in Figures 3-1 and 3-2 which show that closure (consolidation) occurs more rapidly in the SPECTROM-32 analyses. The cusps appearing in Figure 3-3 occur after full consolidation of the crushed salt. At this point, the consolidation process ceases. Cusps do not appear in the crushed salt/bentonite curves because full consolidation is not reached within the 200 year simulation period as shown by Figure 3-2. Mean stresses determined in the SPECTROM-32 analyses are 2 to 4 time greater than the mean stresses determined in the SANCHO analyses during the initial 80 years and only 1 to 1.5 greater in the final 100 years. This difference can be attributed to more than an order of magnitude difference in the elastic constants used in the two codes with the SPECTROM-32 constants being the greater of the two. Despite some disparity in magnitude, the trends of the mean stress determined from the two analyses compare favorably. Both analyses of crushed salt backfill show a significant increase in mean stress once full compaction is reached.

4.0 CONCLUSIONS

Based on the comparison of results from the SPECTROM-32 and SANCHO analyses of a disposal room backfilled with crushed salt and crushed salt-bentonite, void fractions in the backfill material are shown to decay significantly such that the backfill becomes an integral part of the sealing system. The creep consolidation model provides a method to estimate the long-term behavior of backfill materials in a disposal room. The calculated deformations and stresses from the two analyses agree reasonably well despite differences in methodology such as the consolidation model, intact salt creep model, strain theory, and material properties. The volumetric behavior of the backfill is based on hydrostatic laboratory tests, but the deviatoric response included is hypothetical. Deviatoric testing of crushed salt specimens is presently underway. Data obtained from these tests will be used to refine the deviatoric response included in the crushed salt constitutive model.

5.0 REFERENCES

1. **B. M. Butcher, 1990.** Preliminary Evaluation of Potential Engineering Modifications for the Waste Isolation Pilot Plant, SAND89-3095, prepared by Sandia National Laboratories, Albuquerque, NM.

2. **G. D. Callahan, A. F. Fossum, and D. K. Svalstad, 1990.** Documentation of SPECTROM-32: A Finite Element Thermomechanical Stress Analysis Program, prepared by RE/SPEC Inc., Rapid City, SD, RSI-0269, for Office of Nuclear Waste Isolation, Battelle Memorial Institute, Columbus, OH, February.

3. **G. D. Callahan, 1990.** Crushed Salt Consolidation Model Adopted for SPECTROM-32, prepared by RE/SPEC Inc., Rapid City, SD, RSI–0358, for Sandia National Laboratories, Albuquerque, NM, September.

4. **G. D. Sjaardema and R. D. Krieg, 1987.** A Constitutive Model for the Consolidation of Crushed Salt and Its Use in Analyses of Backfilled Shaft and Drift Configurations, SAND87-1977, prepared by Sandia National Laboratories, Albuquerque, NM.

5. **J. R. Weatherby and W. T. Brown, 1990.** Closure of a Disposal Room Backfilled With a Salt/Bentonite Mix, Sandia National Laboratories Internal Memorandum to B. M. Butcher, Division 6332, Albuquerque, NM.

6. **A. F. Fossum, G. D. Callahan, L. L. Van Sambeek, and P. E. Senseny, 1988.** How Should One-Dimensional Laboratory Equations be Cast Into Three-Dimensional Form?, *Proceedings, 29th U.S. Symposium on Rock Mechanics*, University of Minnesota, Minneapolis, MN.

7. **D. E. Munson, 1989.** Proposed New Structural Reference Stratigraphy, Law, and Properties, Sandia National Laboratories Internal Memorandum, Albuquerque, NM.

8. **R. D. Krieg, 1984.** Reference Stratigraphy and Rock Properties for the Waste Isolation Pilot Plant (WIPP) Project, SAND83-1908, January.

9. **D. E. Munson, A. F. Fossum, and P. E. Senseny, 1989.** Advances in Resolution of Discrepancies Between Predicted and Measured In Situ WIPP Room Closures, SAND88-2948, prepared by RE/SPEC Inc. for Sandia National Laboratories, Albuquerque, NM, March.

10. **H. S. Morgan, 1987.** Estimate of the Time Needed for TRU Storage Rooms to Close, Memorandum to D. E. Munson, Sandia National Laboratories, Division 6332, Albuquerque, NM, June.

11. **G. D. Callahan, 1990.** Presentation and Discussion of Properties Selected for Crushed Salt/Bentonite Backfill Material (Sandia Contract No. 40-2521), RE/SPEC Inc. External Memorandum to B. M. Butcher, Sandia National Laboratories, Division 6345, Albuquerque, NM, January.

PART IV

Scaling Approaches of Mechanics in Disordered Solids

SCALING THEORY OF ELASTICITY AND FRACTURE IN DISORDERED NETWORKS

P.M. Duxbury and S.G. Kim,
Dept. of Physics and Astronomy and
Center for Fundamental Materials Research,
Michigan State University.

Abstract

We discuss scaling theories for the elasticity and tensile fracture of random central force spring networks with bond dilution disorder. Effective medium theory works quite well for elasticity but needs very new ingredients to be even qualitatively correct for tensile fracture. A novel "extreme scaling theory" predicts a dilute limit singularity and a size effect in the tensile strength. These predictions are supported by numerical simulations.

We extend the above arguments to networks with distributions of bond disorder, and compare the central force network theories to models currently used in the study of rigid cellular materials.

I. INTRODUCTION

Effective medium theories and rigorous bounds, along with scaling theories and numerical simulations on lattice models provide a good understanding of the elastic properties [1-3] of random materials. In contrast, effective medium theories and bounds on fracture strength are less reliable [4], and it is only recently that numerical methods and scaling theories have been developed for random networks [5-13]. This is because theory for fracture in disordered materials must take into account extreme statistics and this requires new ideas in the formulation of analytic theories of tensile strength. In addition, numerical simulations must use an efficient yet sufficiently realistic method for the propagation of damage. Here we describe recent progress toward the resolution of these problems, and use as an illustrative example the central force spring model [14] with dilution disorder.

Cellular materials [15] are widely used in the packaging industry and have been recently suggested as frameworks for the production of "interpenetrating phase composites" [16], which are materials in which both matrix and reinforcing materials are connected in all three dimensions. A beautiful book by Gibson and Ashby [15] collects together a wide variety of information on cellular materials, and models the experimental data with unit cell models, that incorporate disorder through the average dimensions of the unit cell and cell walls or edges. We point out that effective medium theory provides a better method for including disorder into these models, and we use the central force spring network to illustrate this point. Unfortunately no effective medium theory exists for the beam model as yet so our discussion is restricted to rigid lattices.

II. RANDOMLY DILUTED NETWORKS

Consider a regular lattice containing N central force springs which connect the nearest neighbor sites of the lattice. Randomly remove volume fraction, f, of the springs. As f increases, the elastic moduli decrease, up to the rigidity percolation point [14], f_{cen} (~0.33 for the triangular lattice) at which the elastic moduli vanish. If the springs are allowed to break when strained beyond a threshold value, ε_c, the network may be used to study damage evolution. In this case we slowly increase the strain, allowing local spring rupture until the damage spreads across the network and catastrophic failure occurs. The total energy, E, of a spring configuration is given by,

$$E = \sum_{\langle ij\rangle} k_{ij}(\ell_{ij} - \ell_0)^2 \tag{1}$$

where k_{ij} is the spring constant between adjacent sites of the lattice, ℓ_{ij}, is the length of the spring between sites i and j, and ℓ_0 is the natural length of the springs (assumed constant). Disorder is introduced through k_{ij}, which in the case of dilution are either 0 or k.

Effective medium theory for elasticity

The simplest estimate of $p_{cen}=1-f_{cen}$ is found using a constraint counting argument [3,14]. When p is small the system consists of disconnected pieces and hence has many zero frequency modes, the number of such modes being approximated by the number of degrees of freedom (Nd) minus the number of constraints (zNp/2). Here, d is the spatial dimension and z is the nearest neighbor co-ordination number of the lattice. Thus the effective medium theory estimate of the fraction of zero frequency modes is,

$$x = (Nd - zNp/2)/Np = 1 - zp/2d \tag{2}$$

When $x \to 0$, the network is no longer "floppy", and this defines the rigidity threshold to be,

$$p_{cen} = 1 - f_{cen} \approx 2d/z \tag{3}$$

Effective medium theory has been used to find effective elastic properties on approach to f_{cen}. The spirit of the theory is to replace the random network with a regular lattice having an effective spring constant k_m on each bond. This assumes that the macroscopic symmetry is unaltered by disorder. Several different methods have been used to estimate k_m and they all lead to the result,

$$k_m/k = C^m{}_{ij}/C_{ij} = (p-p_{cen})/(1-p_{cen}) \tag{4}$$

where C_{ij} are the elastic constants of the network, and the subscript or superscript, m, denotes the effective medium value. Numerical simulations on triangular and fcc lattices show that equation (4) provides an excellent approximation for the elastic properties of random but rigid networks, for all but a very small region near the rigidity percolation point (see also Figure 3a for a comparison with the triangular lattice result).

The tensile fracture stress of spring networks has also been recently studied, and it is useful to first review the algorithm used in these calculations. It turns out that there are many possible algorithms, and the choice of algorithm strongly affects the final pattern of fractured springs. In contrast, the fracture strength appears to be robust, and this is due to the fact that these networks are brittle.

Numerical simulation of damage and fracture [11]

We apply an external strain to the networks and relax them using a conjugate gradient method, with a convergence criterion of 10^{-8} in the energy. When one of the springs is strained beyond a critical strain, $\varepsilon_c = (\ell_{ij}-\ell_0)/\ell_0 = 10^{-4}$, the spring breaks (we set its k_{ij} to zero). In the results to be described below, (following Beale and Srolovitz [11]) we take a diluted triangular spring network, and slowly increase the external tensile strain. At each external strain we relax the network to minimize (1) and check the strain in each bond against the spring rupture criterion. If at a given external strain, a spring breaks, we fix the external strain at that level, and continue relaxing the network and breaking the spring carrying the largest strain (provided it is greater than ε_c) until either catastrophic failure occurs, or until no more springs break (we call this the hottest bond algorithm [17]). We then slowly increase the external strain until it is just large enough to break another spring in the network and again iterate the hottest bond procedure described above. The stress strain relationship of a single spring, along with the stress strain behavior found from these simulations on a 50x50 triangular network are shown in Figure 1.

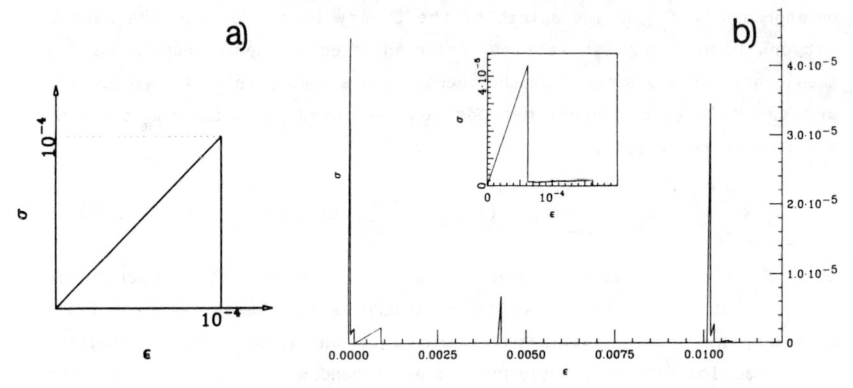

Figure 1 a) The stress strain behavior of a single spring b) The stress strain behavior of a L=50 triangular network with f=0.1. The inset shows the behavior at small external strains.

It is seen from Figure 1b that a catastrophic event usually occurs very soon after the first spring in the network breaks. We call the stress at which this event occurs $\sigma_1(f)$. A second important stress is the stress required to rupture the entire network. This is the maximum stress in the stress strain curve, and we label this $\sigma_b(f)$. In dilute networks (f small), $\sigma_1(f) \sim \sigma_b(f)$ while at larger dilutions this is not always the case. Here, as in previous work [11], we concentrate on $\sigma_1(f)$, as this models the brittle fracture of the networks. The large strain behavior involves a great deal of bond rotation, and is probably not relevant to either brittle or ductile fracture. To illustrate the model network response, we show in Figure 2 three snapshots of the microstructural evolution of damage and fracture in a 50x50 lattice. The initial catastrophic event (usually at $\sigma_1(f)$), leads to a path of broken bonds across the network, but not total fracture. Final fracture occurs at a much larger strain, though the final fracture path is often close to the path of the first catastrophic event, and is usually preceded by bond rotation along the fracture path.

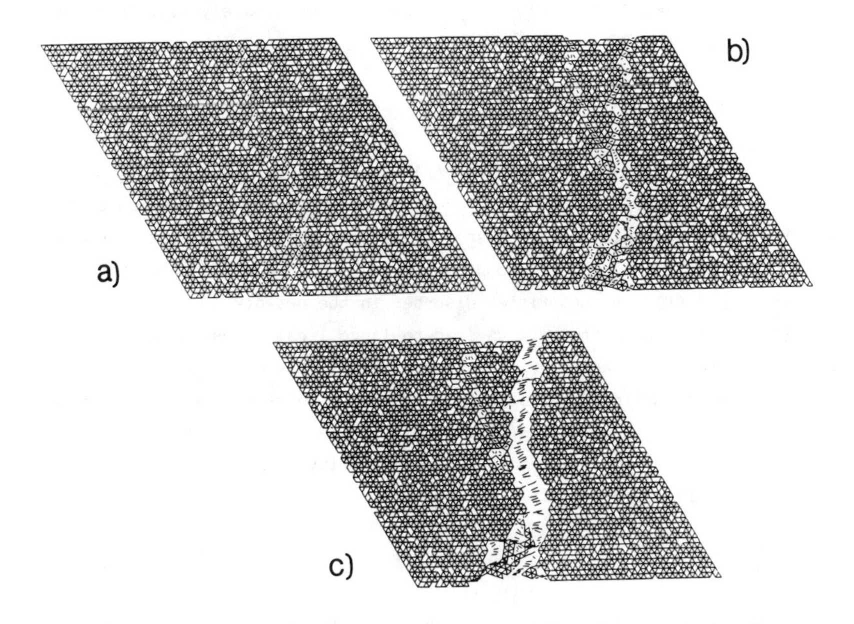

Figure 2 a) The network configuration after the first catastrophic event at $\sigma_1(f)$. b) An intermediate configuration just before the second major peak in Fig 1b. c) the final fracture configuration. (---) present bonds; (-) bonds that were initially present but which broke in the damage evolution process. The simulations were on 50x50 lattices with f (initially)=0.10.

There are exceptions to this rule however, and crack branching events do sometimes occur. As the initial f increases, the crack path becomes more tortuous, though the structure of the crack path is dependent on the model and damage evolution algorithm used. The fracture strength however is more robust and its scaling behavior appears to be fairly model independent [11]. This makes the use of simple models most valid for the calculation of strength and strength statistics, while the crack topologies found using these models are only valid for the specific loading condition and microstructure that applies to the model.

III. SCALING THEORY OF TENSILE FRACTURE STRESS OF DILUTE NETWORKS [8,11]

Since for small f, $\sigma_1(f) \sim \sigma_b(f)$ we can approximately calculate the fracture stress from the relationship,

$$\sigma_b(f)/\sigma_b(0) \sim \sigma_0/\sigma_{max}(f) \quad (5)$$

where σ_0 is the applied stress. $\sigma_{max}(f)$ <u>is the stress in the spring with the largest strain, prior to any bond fracture</u>. Equation (5) uses the (small strain) linearity of the system to extrapolate from a linear elasticity calculation of $\sigma_{max}(f)$ to deduce $\sigma_b(f)$. The problem now reduces to estimating $\sigma_{max}(f)$ for the initial disorder in the network.

It is well known that cracks lead to large local stress enhancements. In dilute networks, crack-like flaws are unlikely, but do occur with finite probability somewhere in the networks. It is these unlikely flaws that dominate the strength of the networks. The simplest crack-like flaw contains n adjacent removed bonds. The probability of occurrence of such a configuration is approximately

$$P(n) \sim L^d f^n \qquad \text{as } n,L\to\infty \text{ and } f\to 0 \quad (6)$$

The largest such configuration that occurs with finite probability is estimated from

$$L^d f^n \sim 1, \text{ which implies } \quad n_{max} \sim -d\ln L/\ln f \quad (7)$$

Such a crack-like defect configuration has a stress intensity at its tips that scales as (for $-\ln L/\ln f$ large);

$$\sigma_{max}(f) \sim \sigma_0 \, (1 + K_m(-d\ln L/\ln f)^{1/2}) \quad (8)$$

where K_m is an unknown constant that depends on the curvature of the crack tip and the size of a plastic zone (this is the continuum elasticity result for tensile loading of a through crack).

In two dimensions (and for cracks for fixed finite separation), it is known that two adjacent cracks enhance the stress between them by a factor approximately proportional to crack length. In this case the exponent in (8) is changed to 1 [11]. From these arguments and (5) we then find;

$$\sigma_b(f) \sim \sigma_b(0)/[1 + K_m(-d\ln L/\ln f)^{\beta}], \quad \text{for } -d\ln L/\ln f \to \infty,\ f\to 0 \quad (9)$$

where $1/2(d-1) \leqq \beta \leqq 1$. Equation (9) also applies to the tensile fracture stress of three dimensional systems, with only minor modifications of K_m and β. Equation (9) contains two interesting effects:

1. At fixed f, a logarithmic size effect in fracture stress occurs.
2. The fracture stress is rapidly decreasing for small f, and in fact is singular as $f \rightarrow 0$.

Equation (9) is only strictly valid for $-d\ln L/\ln f \rightarrow \infty$, which does not apply to the lattice sizes available for simulations. Despite this, the general trends outlined in 1. and 2. above are present in the numerical data presented in Figure 3. For comparison the Young's modulus which in this limit is linear in dilution and clearly non-singular.

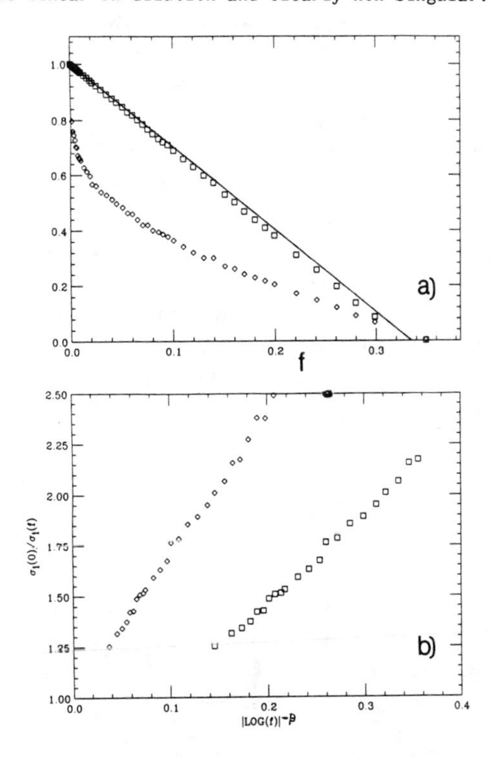

Figure 3 a) The Young's modulus, Y (□), and tensile fracture stress, $\sigma_1(f)$ (◇) of L=60 central force networks. Each point is an average over 40 configurations. The solid line is from equation (4) applied to the triangular lattice b) a test of the form (10) for $\sigma_1(f)$ for β=1.0 (◇) and β=1.7 (□).

The difference in dilute limit scaling behavior of the Young's modulus and the tensile fracture stress is clearly evident in Figure 3a. The qualitative similarity of $\sigma_1(f)$ in Fig. 3a with the theoretical prediction (9) suggests that the scaling behav. of small systems may be well represented by the form;

$$\sigma_b(f) \sim \sigma_b(0)/[1 + B\ |\ln f|^{-\beta}] \qquad (10)$$

where β and B are system size dependent parameters. A test of this form using the tensile fracture data of Fig. 3a is presented in Fig. 3b. It is seen from Fig. 3b that $\beta=1.7\pm0.1$ gives an excellent fit to the data in the singular region $f<0.10$. We thus suggest that (10) provides a useful representation of the strength of the dilute random networks for a wide range of system sizes and that β only lies in the range $1/2(d-1)<\beta<1$ in the infinite lattice limit. We expect that in the dilute limit, models with bond bending forces should show a behavior similar to the central force networks. This may be understood if we consider a lattice constructed of blobs of central force elements. Since the lattice is rigid on most length scales near the pure limit, most of these central force blobs are rigid. We can thus renormalise the problem on to an effective model with bond bending forces. In contrast, the behavior of the bond bending and central force problems are very different at higher f. The bond bending problems lose rigidity at f_c the connectivity percolation point, while the central force networks lose rigidity at f_{cen} ($f_{cen}<f_c$) [14].

In bond bending problems, $\sigma_b(f)$ exhibits the critical scaling;

$$\sigma_b(f\to f_c) \sim \sigma_b(0)\ (f_c-f)^x \qquad (11)$$

Several workers have placed bounds on x, though its precise value is as yet unknown. A nodes links and blobs picture [10] which states that the problem may be replaced by a network with effective lattice constant ξ (the percolation correlation length) implies the upper bound,

$$\sigma_b(f) < \sigma_b(0)\ (f_c-f)^{(d-1)\nu} \qquad (12)$$

If the effective lattice constant, ξ, acts as a rigid moment arm on a given bond, the fracture stress will be greatly reduced. This leads to the lower bound;

$$\sigma_b(f) > \sigma_b(0)\ (f_c - f)^{d\nu} \tag{13}$$

We thus deduce that x lies in the range $(d-1)\nu < x < d\nu$. We do not at present have any bounds on x for the central force problem, but it is evident from Fig. 3a, that x is nearly 1 for f quite close to f_{cen}. Some experiments have been attempted to find x. Using a 2mm thick aluminum sheet with holes drilled to form a triangular lattice, Sieradzki and Li [7] found $x=1.7\pm0.1$. This value lies within the bounds given above (with $\nu=4/3$ in 2-d). It must be remembered though that the experiment in not truely 2-d, the aluminum is ductile and the lattice size is small (20x20) so this agreement may be fortuitous. A second interesting feature of the Sieradzki and Li paper (see Fig. 2 of [7]) is that shows some experimental evidence of the qualitative difference between dilute limit scaling of the elastic modulus and that of the fracture stress, despite the small system size. A second experiment to find x was published by Benguigui et al [9] who randomly punched holes in a copper sheet. They used a square lattice and found $x\sim2.5\pm0.4$. This also lies within the bounds given above but is also subject to the caveats stated above.

IV THE STATISTICS OF TENSILE FRACTURE IN DILUTE NETWORKS

Sample to sample variability in fracture stress is usually much larger than that of elastic moduli. Often a statistical approach is necessary, especially in brittle materials, and the distribution most often used for the statistical analysis is the Weibull form [17],

$$C(\sigma) = 1 - \exp[-V(\sigma/\sigma_s)^m] \tag{14}$$

Here σ_s and m are fitting constants usually referred to as the scale parameter and Weibull modulus respectively. $C(\sigma)$ is the probability that on application of an external stress σ, the system will fracture. Clearly for finite σ_s and m, $C(\sigma)$ is zero for V infinite. A size effect is thus built into the Weibull statistic. Although tests are usually performed on at most 100 samples in fundamental research and 10 in industrial tests, reliabilities typically in the range 10^{-8} - 10^{-4} are required for applications. The form (14) is thus used to extrapolate from the test range to the applications range.

We have developed a statistic for dilute networks that is different than the Weibull form, and leads to design predictions which can differ by of order 30% at the 10^{-6} reliability level. The statistic is understood by noting that the statistics of $\sigma_b(f)$ are closely related to those of $\sigma_{max}(f)$ through equation (5). The probability that a network will fail when a stress σ is applied, is related to the probability that σ_{max} is greater that a value proportional to $1/\sigma$. We thus study the statistics of the springs carrying the largest stresses, and a numerical study of these springs in diluted central force lattice is presented in Fig. 4.

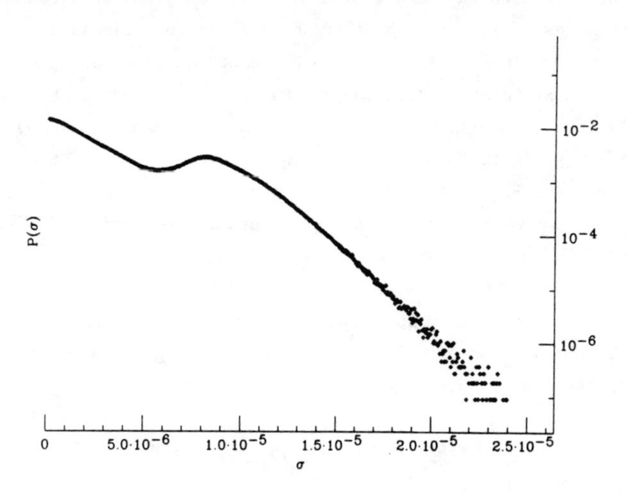

<u>Figure 4</u> A log-linear plot of the probability that a spring in the random spring network carries load σ. Found from simulations on 1000 configurations of 50x50 triangular spring networks with f=0.10.

It is evident from Fig. 4 that the tail of the distribution of bond stresses is exponential. Given this the cumulative distribution $C_N(\sigma_{max})$ [the probability that in a network with N springs the spring with the maximum stress carries stress less than σ_{max}] is given by (this is a standard result in the statistics of extremes - the Gumbel limiting form for maxima [20])

$$C_N(\sigma_{max}) = \exp(-cN\exp(-k\sigma_{max})) \qquad (15)$$

Using (5) and (15) we thus find the fracture statistic (with $N \sim V$)

$$C(\sigma) = 1 - \exp(-cV\exp(-k/\sigma^{\mu}) \qquad (16)$$

with $\mu = 1$. This form may also be derived using percolation cluster statistics in combination with the stress enhancements near cracks. This leads to the possibility $1<\mu<2$ [8]. Numerical simulations of the fracture statistics [11] indicate that the form (16) sometimes provides a better fit to the data than the Weibull form (14).

As a final comment on the statistics of spring networks, note that the reliability distribution (16) is derived from the statistics of large local stresses. In contrast, the usual derivation of the Weibull distribution or the usual Gumbel distribution of strengths are based upon weak link ideas [20]. In the conventional derivations, the stress distribution is implicitly assumed to be uniform, while the material has a distribution of weak links in its microstructure. Here we have taken the completely opposite view of fracture statistics, where the variability in fracture stress is produced by spatial variations in the local stress level. Fracture is initiated where the stress is largest, and so the statistics of fracture is determined by the statistics of the largest local stresses. It may be shown that the Weibull distribution applies whether the weak link or "hot spring" model is used. The Gumbel distribution applies to "weak link" situations while the modified Gumbel (16) applies to "hot spring" situations. In real materials it is probable that a combination of weak link and hot spring mechanisms apply, and it is apparent that the Weibull distribution is, to first approximation, robust to this mixing. In contrast, for the Gumbel type it is not obvious which of the Gumbel or the modified Gumbel is appropriate when both weak links and hot springs play a role in fracture.

V. SYSTEMS WITH A DISTRIBUTION OF BOND DISORDER

Consider now a network of central force springs in which the spring constants are drawn from a distribution p(k). In this case, the effective spring constant of the regular lattice is given by the solution to the equation [3];

$$\int p(k)(k-k_m)/[k_m/p_{cen} - k_m + k]\, dk = 0 \qquad (17)$$

The effective elastic constants, as in equation (3) have the same trends as k_m. Numerical simulations indicate that (17) is a good approximation for broad as well as narrow distributions of bond disorder [19].

In the case of fracture, we must choose the distribution of bond fracture thresholds as well as the distribution of bond spring constants. Most work has been carried out with the assumption that $k_{ij} = k$, a constant, while the bond fracture stress thresholds belong to a distribution $D(\sigma')$. In this case, a simple analysis of the fracture process is still possible

[21]. In the early stages of fracture, the stress is distributed almost uniformly throughout the sample, and failure occurs in the weakest bonds. These bond failures are almost uncorrelated, and correspond to an effective dilution process, where the fraction of removed bonds is given by,

$$f_{eff} = \int_0^{\sigma} D(\sigma')\, d\sigma' \qquad (18)$$

where σ is the applied stress. We develop a scaling theory for the fracture strength, σ_c, of these networks, by realizing that f_{eff} leads to the large local stress enhancements like those estimated in equation (8). When the largest of such stress enhancements is sufficient to break a typical bond in the network, i.e.

$$\sigma_{max} = \bar{\sigma} = \int_{-\infty}^{\infty} \sigma' D(\sigma')\, d\sigma' \qquad (19)$$

catastrophic network failure will occur. Equation (19) along with (18) and (8) leads to an implicit equation for the network tensile fracture strength. In the case of an constant distribution of bond fracture strengths between the two limits σ_1 and σ_2, the resulting equation is,

$$\sigma_1 + w/2 = \sigma_c[\ 1 + K_m(-d\ln L/\ln((\sigma_c - \sigma_1)/w))^{\alpha}] \qquad (20)$$

where $w = \sigma_2 - \sigma_1$. For small system sizes and strong disorder, the damage is distributed (labelled "ductile" by some workers). However, for large enough system size, there is a logarithmic size effect, and the fracture is more brittle [21]. At fixed sample size, and for increasing disorder, there is a transition from brittle to "ductile" behavior. This has been quantified by measuring the number of bonds that fail, $n_b = N_b/L^{d-1}$ before the whole network fractures. If cleavage fracture occurs, $n_b \sim 1$, while if "ductile" fracture occurs $n_b \sim L/(\ln L)^{\alpha}$. Simulations on random fuse networks are consistent with this theory [21]. However, it has also been shown that algebraic scaling forms provide a useful fit to numerical data on spring and fuse networks [12]. There is as yet no physical insight as to why the postulated algebraic forms should be preferred to the theory described above.

VI. COMPARISON WITH UNIT CELL MODELS FOR CELLULAR MATERIALS

In the previous sections, we have described theories for the elasticity and failure of random central force networks. In contrast, cellular materials are usually modelled using beams between the nodes of a regular lattice. Each beam has length, ℓ, and thickness, t, which are found from experiment. Disorder is incorporated through replacing t and ℓ by their by average values $\bar{t}$ and $\bar{\ell}$. Beams may support angular as well as central forces between lattice sites, so the theory developed in the previous sections must, in general, be extended before it is possible to apply them to cellular materials. Unfortunately, there has been little progress in developing effective medium theory for networks with angular forces, or to our knowledge, any attempt to develop effective medium theory for the beam model. However, there are some important problems where the response of the beam networks is dominated by local compressions or extensions, in which case the beams may be replaced by central force springs. This is true of the elastic properties of highly co-ordinated lattices such as the triangular or face-centered cubic lattices, and we also expect it to be true of the tensile fracture stress of these lattices. Consider then the triangular lattice beam model and define an effective spring constant between the sites of the lattice to be $k = E_s bt/\ell$ (here we have assumed a beam of cross-sectional area bt). With this identification, we compare the effective medium theory predictions with those of Gibson and Ashby [15] (called GA henceforth). For a triangular network of beams, from 4A.3 of GA we deduce that in the presence of disorder,

$$Y/Y_0 = (\bar{t}/t)\,(\bar{b}/b)\,(\ell/\bar{\ell}) \quad (21)$$

Where Y_0 is the Young's modulus in the absence of disorder. In contrast, the effective medium prediction is given by equation (17). To illustrate the difference between the predictions that these two theories make, consider the case of random dilution, in which volume fraction f=1-p of the beams is randomly removed from a regular triangular lattice. In this case $\bar{\ell}=\ell$, $\bar{b}=b$ and $\bar{t}=pt$, so that the GA theory reduces to,

$$Y/Y_0 = p \quad (22)$$

The effective medium theory for the same disorder is given by,

$$Y/Y_0 = (p-2d/z)/(1-2d/z) \quad (23)$$

From these equations, it is seen that the GA theory seriously <u>underestimates the effect of disorder</u> on the elastic properties of triangular lattices. In fact equations (22) and (23) are valid for <u>rigid lattices</u> in both two and three dimensions. Although the dilution case is one of the worst for the GA theory, the GA approach underestimates the effect of disorder on the elastic modulus for any reasonable distribution of bond disorder in central force lattices. Unfortunately, the extension of the above analysis to the more relevant case of honeycombs and other systems dominated by beam bending is, and at present, an outstanding problem.

For the case of tensile fracture strength, the analysis of Gibson and Ashby invokes an unidentified parameter c, which is the largest crack in the material. In the absence of a fabricated crack, the parameter c is related to the disorder, and in the case of dilution disorder, c is approximated by (applying GA variables to equation (7));

$$c \sim \ell(1-d\ln(L/\ell)/\ln(1-p)) \tag{24}$$

which when included, for example, in expressions (4.36) - (4.40) of Gibson and Ashby, predicts a size effect and "dilute limit singularity" in the tensile fracture strength of brittle cellular materials. Similarly, when a distribution of strut disorder is included in the analysis, the "ductile" to brittle transition characterized by the number of struts broken before fracture, is again predicted (see the analysis of Section V).

VI. CONCLUSIONS

A comparison of the elasticity and fracture of random networks illustrates the different scaling behavior of these properties in the presence of disorder. Equations (3) and (9) along with Figure 3a, show that the elastic constants are non-singular except near special points such as the rigidity or connectivity percolation points. In contrast, the brittle fracture stress exhibits the following features.

i) There is a dilute limit singularity in tensile strength (see Figure 3a), which is well represented by equation (10). A dilute limit singularity is present in some published experimental works [see e.g. Fig. 2 of ref 7], though the specific form (10) has not been tested experimentally. It should be emphasized that experiments on fracture strength trends must be performed on several samples, as sample to sample variations can mask the predicted trends in average strength. This is illustrated for example in the analogous problem of dielectric breakdown in metal loaded dielectrics where the trend analogous to equation (10) was observed, but about 10 samples at each value of f were required to reliably see the trend [22]. A logarithmic size effect also occurs in the dilute limit (see equation (9)).

ii) The statistics of fracture in the dilute limit obey a modified Gumbel form (16) rather than the usual Weibull distribution. Although many hundreds of samples are required to differentiate between Weibull and modified Gumbel statistics [8,11], extrapolations to reliabilities of order 10^{-6} leads to design predictions that can differ by of order 30%.

Effective medium theory applied to cellular materials shows that GA theory [15] sometimes seriously underestimates the reduction of elastic properties caused by disorder (see equations (22) and (23)). Strength scaling theory applied to the tensile fracture of brittle cellular materials shows that these materials should exhibit the features (i) and (ii) stated above.

Acknowledgements

Financial support by the DOE under grant number DE-FG02-90ER45418 and by the donors of The Petroleum Research Fund, administered by the ACS is gratefully acknowledged. Thanks to Karl Sieradzki and Lorna Gibson for the invitation to participate in this symposium. PMD also thanks Mike Thorpe for several useful discussions.

REFERENCES

1. Z. Hashin, "Analysis of Composite Materials - A Survey", J. Appl. Mech. **50**, 1983, pp. 481-505

2. S. Feng, M.F. Thorpe and E. Garboczi, "Effective-medium theory of percolation on central-force elastic networks", Phys. Rev.Rev **B31**, 1985, pp. 276-280.

3. S. Feng, B.I. Halperin and P.N. Sen, "Transport properties of continuum systems near the percolation threshold", Phys. Rev. **35**, 1987, pp.197-214.

4. D. Krajcinovic and M.A.G. Silva, "Statistical Aspects of the Continuous Damage Theory", Int. J. Solids Structures **18**, 1982, pp. 551-562

5. D.L. Turcotte, R.F. Smalley Jr. and Sara A. Solla, "Collapse of loaded fractal trees", Nature **313**, 1985, pp. 671-672; P. Ray and B.K. Chakrabarti, "A microscopic approach to the statistical fracture analysis of disordered brittle solids", Solid St. Comm. **53**, 1985, pp.477-479.

6. M. Sahimi and J.D. Goddard, "Elastic percolation models for cohesive failure in heterogeneous systems", Phys. Rev. **B33**, 1986, pp. 7848-7851.

7. K. Sieradzki and R. Li, "Fracture behavior of a solid with random porosity", Phys. Rev. Letts. **56**, 1986, pp. 2509-2511.

8. P.M. Duxbury, P.L. Leath and P.D. Beale, "Breakdown properties of quenched random systems-the random fuse network", Phys. Rev. **B36**, 1987, pp. 367-380.

9. L. Benguigui, P. Ron and D.J. Bergman, "Strain and Stress at the fracture of percolative media", J. de Phys. **48**, 1987, pp. 1547-1551.

10. E. Guyon, S. Roux and D.J. Bergman, "Critical Behavior of Elastic Failure Thresholds in Percolation", J. de Phys. **48**, 1987, pp. 903-904.

11. P.D. Beale and D.J. Srolovitz, "Elastic fracture in random materials", Phys. Rev. **B37**, 1988, pp. 5500-5507; G.N. Hassold and D.J. Srolovitz, "Brittle fracture in materials with random defects", Phys. Rev. B **39**, 1989, pp 9273-9281.

12. S. Roux, A. Hansen, H.J. Herrmann and E. Guyon, "Rupture of Heterogeneous Media in the limit of infinite disorder", J. Stat. Phys. **52**, 1988, pp. 237-244; H.J. Herrmann, A. Hansen and S. Roux, "Fracture of disordered elastic lattices in two dimensions", Phys. Rev. **39**, 1989, pp. 637-648; A. Hansen, S. Roux and H.J. Herrmann, "Rupture of central-force lattices", J. Phys. France **50**, 1989, pp. 733-744; W.A. Curtin and H. Scher, "Brittle fracture of disordered materials", J. Mater. Res. **5**, 1990, pp. 535-553. S. Arbabi and M. Sahimi, "Test of universality for three-dimensional models of mechanical breakdown of disorderd solids", Phys. Rev. B **41**, 1990, pp. 772-775.

13. H.J. Herrmann and S. Roux eds., "Statistical Models for the Fracture of Disordered Materials", Elsevier Science Publishers, NY, 1990.

14. M.F. Thorpe, "Rigidty Percolation in Glassy Structures", J. non-Cry. Sol. **76**, 1987, pp. 109-116.

15. L.J. Gibson and M.F. Ashby, "Cellular Solids", Pergamon Press, NY (1988).

16. David Clarke private communication.

17. see e.g. E. Castillo, "Extreme Value Theory in Engineering", Academic Press, NY, 1988.

18. L. de Arcangelis, S. Redner and H.J. Herrmann, "A random fuse model for breaking processes", J. de Physique Lett. **46**, 1985, L585-590.

19. E.J. Garboczi, "Effective froce constant for a central force random network", Phys. Rev. B **37**, 1988, pp. 318-320.

20. E.J. Gumbel, "Statistics of Extremes", Columbia Univ. Press, NY 1958.

21. B. Kahng, G.G. Bartrouni, S. Redner, L. de Arcangelis and H.J. Herrmann, "Electrical breakdown in a fuse network with continuously distributed breaking strengths", Phys. Rev. B **37**, 1988, pp. 7625-7637; K

Sieradski private commumication; Y.S. Li and P.M. Duxbury, "Crack arrest by residual bonding in spring and resistor networks", Phys. Rev. B **38**, 1988, pp. 9257-9260;

22. R.W. Coppard, L.A. Dissado, S.M. Rowland and R. Rakowski, "Dielectric breakdown of metal-loaded polyethylene", J. Phys. CM. **1**, 1989, pp. 3041-3045; P.M. Duxbury, P.D. Beale, H. Bak and P.A. Schroeder, "Capacitance and dielectric breakdown of metal loaded dielectrics", J. Phys. D, **23**, 1546 (1990).

A COMPARISON OF MECHANICAL PROPERTIES AND SCALING LAW RELATIONSHIPS FOR SILICA AEROGELS AND THEIR ORGANIC COUNTERPARTS

R.W. Pekala, L.W. Hrubesh, T.M. Tillotson, C.T. Alviso, J.F. Poco, and J.D. LeMay,
Chemistry & Materials Science Department, Lawrence Livermore National Laboratory, Livermore, CA 94550.

ABSTRACT

Aerogels are a unique type of ultrafine cell size, low density foam. Traditional aerogels are inorganic, but the synthesis of organic aerogels has also been reported. In all cases, solution chemistry can be used to tailor the structure and properties of the resultant aerogels. This study examines the microstructural dependence of the compressive mechanical properties of silica, resorcinol-formaldehyde, carbon, and melamine-formaldehyde aerogels.

INTRODUCTION

Aerogels are a special class of open-cell foams derived from the supercritical extraction of highly crosslinked, inorganic or organic gels. The resultant materials have ultrafine cell/pore sizes (< 100 nm), high surface areas (350-1000 m^2/g), and a microstructure composed of interconnected colloidal-like particles or polymeric chains with characteristic diameters of 10 nm. TEM and SAXS show that this microstructure is sensitive to variations in processing conditions that influence crosslinking chemistry and growth processes prior to gelation [1-4].

Traditional silica aerogels are prepared via the hydrolysis and condensation of tetramethoxy silane (TMOS) or tetraethoxy silane (TEOS). Factors such as pH and the $[H_2O]/[TMOS]$ ratio affect the microstructure of the dried aerogel. It is generally accepted that 'polymeric' silica aerogels result from acid catalysis while 'colloidal' silica aerogels result from base catlaysis [5]. Recently, Hrubesh and Tillotson developed a new 'condensed silica' procedure for obtaining silica aerogels with densities as low as 0.004 g/cc, i.e. only 3X the density of air [6-8]!

Organic aerogels are formed from the aqueous, polycondensation of (1) resorcinol/formaldehyde or (2) melamine/formaldehyde. The microstructure of the resorcinol-formaldehyde (RF) aerogels is dictated by the amount of base catalyst used in the sol-gel polymerization. In addition, these materials can be pyrolyzed in an inert atmosphere to form vitreous carbon aerogels [9-11]. Melamine-formaldehyde (MF) aerogels that are both colorless and transparent are only formed under acidic conditions (i.e. pH= 1-2) [12].

In this paper, the microstructural dependence and scaling law relationships for the compressive modulus of silica, carbon, RF, and MF aerogels will be discussed in detail.

EXPERIMENTAL

Synthesis

Aerogel preparation has been described elsewhere [10]. Briefly, all RF aerogels were synthesized from aqueous solutions containing various %solids at a constant [formaldehyde]/[resorcinol] ratio equals 2.0. Sodium carbonate was used as a base catalyst, and all formulations were referenced by their [Resorcinol]/[Catalyst] ratio. R/C ratios of 50-300 produced transparent gels. After curing the gels for up to 7 days at 95 °C, they were further crosslinked in a dilute acid solution, and then exchanged into an organic solvent (e.g. isopropanol). All RF gels were supercritically dried from carbon dioxide (T_c=31°C; P_c=7.6 MPa). The red-colored RF aerogels were pyrolyzed in an inert atmosphere at 1050 °C to produce vitreous carbon aerogels.

Melamine-formaldehyde aerogels were also synthesized from aqueous solutions containing various %solids with a constant [formaldehyde]/[melamine] ratio equals 3.7. Initially, the MF slurry was heated at 70 °C under basic conditions to give a clear polymer solution. The polymer solution was then cooled to 45 °C and acidified with HCl. This mixture was cast into glass vials and cured for up to 14 days at 95 °C. The resultant gels were neutralized and then exchanged into an organic solvent (e.g. acetone). All MF gels were supercritically dried from carbon dioxide to give aerogels that were both colorless and transparent.

Mat. Res. Soc. Symp. Proc. Vol. 207.

Silica aerogels were synthesized from tetramethoxy silane (TMOS). In the case of the acid (HBF_4) and base (NH_4OH) catalyzed gels, methanol was used as the diluent and a $[H_2O]/[Si]$ ratio of 4.0 was utilized. In the 'condensed silica' procedure, TMOS was refluxed under acidic (HCl) conditions with a substoichiometric amount of water. The resultant methanol was distilled off to give a viscous oil. This 'condensed silica' was dissolved in acetone, additional water was added to bring the overall $[H_2O]/[Si]$ ratio to 4.0, and the mixture was catalyzed under basic (NH_4OH) conditions to produce crosslinked gels [8]. All silica gels were dried in an autoclave above the critical point of their respective diluents (methanol T_c= 240°C P_c= 8.0 MPa; acetone T_c= 236°C P_c= 4.8 MPa).

Mechanical Testing

The modulus and strength of aerogels were measured in uniaxial compression with an Instron machine (model #1125). The tests were performed at an initial strain rate of 0.1%/sec. All measurements were made under ambient conditions at ~22 °C and 50-70% relative humidity. No precautions were taken to prevent moisture adsorption by the aerogels.

Test specimens were either machined into 1x1x1 cm^3 cubes or were 1-2 cm long cylinders cut from molded samples. Great care was exercised to ensure that specimens were machined with flat, smooth faces and plane-parallel opposing faces. The surfaces of as-produced aerogels were completely removed to eliminate possible contributions from surface skins. The density of each specimen was determined just prior to testing, and the compressive modulus was derived from the linear region of the stress-strain curve.

RESULTS AND DISCUSSION

As a result of their high porosity, aerogels exhibit elastic moduli many orders of magnitude smaller than their full density analogs. Figure 1 shows the compressive modulus of RF, carbon, MF, and silica aerogels as a function of density and catalyst conditions. As expected, the modulus increases as a function of bulk density. The linear log-log plot in each case demonstrates a power-law density dependence that has been observed in many other low density materials. This relationship is expressed as

$$E = c\rho^n \qquad (1)$$

where ρ is the bulk density, c is a prefactor (constant), and n is the scaling exponent. For most open-cell foams, the scaling exponent usually falls very close to 2.0 while ideal closed-cell foams give an exponent of 3.0. Foams with irregular, fractal type morphologies generally have values that exceed 3.0 [13-15].

In the case of RF aerogels, the above exponent equals 2.7±0.2 for all formulations [16,17]. Although the scaling exponents are the same, the prefactors scale inversely with the R/C ratio. Previous TEM studies have shown that the interconnected particle size of these aerogels depends upon the R/C ratio. At R/C=50, the particles have diameters of approximately 30Å and are joined together with large necks between particles. In fact, it is often difficult to visualize individual particles since these networks have a fibrous appearance. At R/C=300, the particles have diameters of 160-200 Å and are lightly fused together giving a "string of pearls" appearance. These RF aerogels have been described as being 'polymeric' and 'colloidal', respectively. Mechanical property data support these descriptions since the modulus increases due to an apparent improvement in particle interconnectivity as the R/C ratio and the particle size decrease.

The scaling exponents (2.7 ± 0.2) for the carbon aerogels are identical to their RF precursors. In addition, the same rank order is observed for the various formulations, i.e. carbon aerogels prepared at R/C=50 have higher moduli than those prepared at R/C=300. In general, carbon aerogels are ~10X stiffer than their RF analogs at equivalent densities and R/C ratios.

In the case of silica aerogels, the scaling exponents and prefactors both depend upon catalyst conditions [17]. For acid catalyzed and base catalyzed silica aerogels, the scaling exponents are 3.4±0.1 and 2.9±0.2, respectively. Over comparable density ranges, the 'polymeric' acid catalyzed silica aerogels are ~7X stiffer than their 'colloidal' base catalyzed counterparts. The aerogels produced from the 'condensed silica' procedure have a scaling exponent of 3.8±0.2. These materials have moduli that are between those of acid and base catalyzed silica aerogels. This result is not surprising since both acid and base catalysis are used in the 2-step, 'condensed silica'

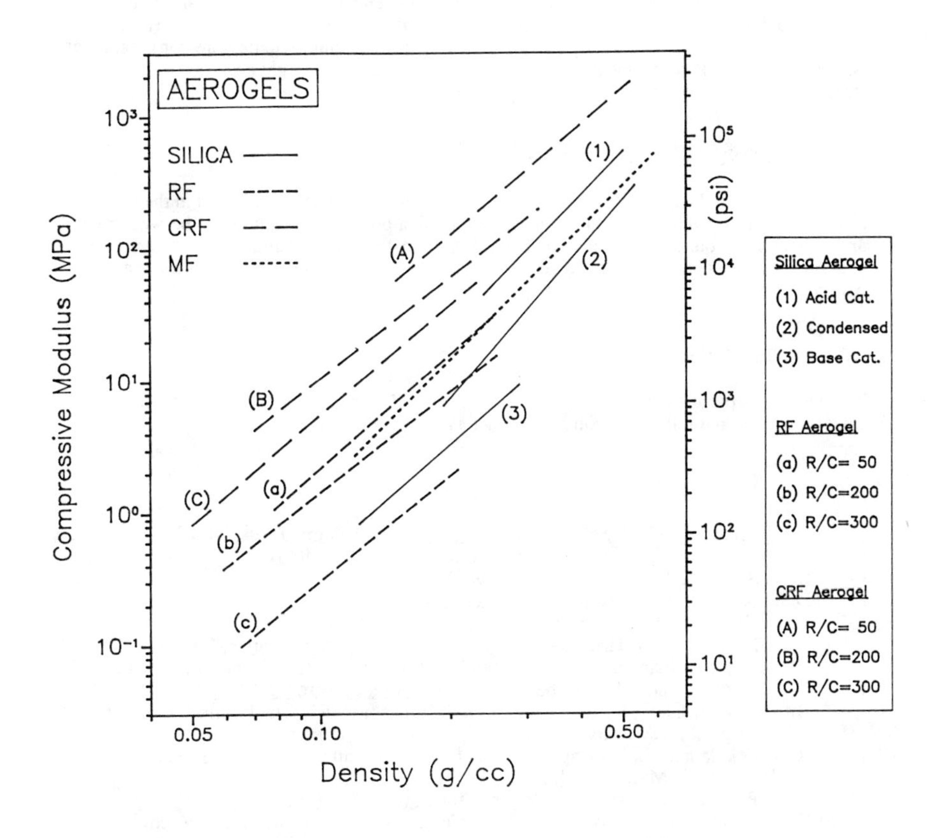

Figure 1. A log-log plot of compressive modulus vs. density for various inorganic and organic aerogels.

procedure.

Interestingly, MF aerogels have a scaling exponent of 3.3±0.1 and moduli that approximate those of acid catalyzed silica aerogels. In RF and carbon aerogels, a scaling exponent of 2.7±0.2 is observed that matches base catalyzed silica aerogels. Considering that in each comparable system, the crosslinking reactions and growth processes take place under similar conditions, similar microstructures might be expected. Mechanical property, TEM, and surface area data provide some evidence for this hypothesis. Thus, we are continuing to explore commonalities between inorganic and organic aerogels so that a universal model might be developed for the structure and properties of these unique materials.

CONCLUSIONS

Differences in the microstructure of organic and silica aerogels are reflected in their compressive properties. In all cases, aerogels exhibit a power-law relationship between modulus and bulk density, but this relationship depends upon matrix material and synthetic conditions. In general, 'polymeric' aerogels are stiffer than 'colloidal' aerogels, independent of the matrix material.

ACKNOWLEDGMENT

This work was performed under the auspices of the U.S. Department of Energy by Lawrence Livermore National Laboratory under contract #W-7405-ENG-48

REFERENCES

[1] C.J. Brinker and G.W. Scherer, Sol-Gel Science (Academic Press, New York, 1990).
[2] Aerogels, ed. by J. Fricke (Springer-Verlag, New York, 1986).
[3] D.W. Schaefer, Science 243, 1023 (1989).
[4] D.W. Schaefer, MRS Bulletin 13(2), 22 (1988).
[5] C.J. Brinker and G.W. Scherer, J. Non-Cryst. Solids 70, 301 (1985).
[6] T.M. Tillotson, L.W. Hrubesh, and I.M. Thomas, in Better Ceramics Through Chemistry III, ed. by C.J. Brinker, D.E. Clark, and D.R. Ulrich (MRS, Pittsburgh, 1988) p. 685.
[7] T.M. Tillotson and L.W. Hrubesh, in Better Ceramics Through Chemistry IV, ed. by C.J. Brinker, D.E. Clark, D.R. Ulrich, and B.J. Zelinski (MRS, Pittsburgh, 1990) in press.
[8] L.W. Hrubesh, T.M. Tillotson, and J.F. Poco, ibid., in press.
[9] R.W. Pekala and F.M. Kong, J. de Physique Coll. Suppl. 50(4), C4-33 (1989).
[10] R.W. Pekala, J. Mat. Sci. 24, 3221 (1989).
[11] R.W. Pekala and F.M. Kong, Polym. Prpts. 30(1), 221 (1989).
[12] R.W. Pekala and C.T. Alviso, in Better Ceramics Through Chemistry IV, ed. by C.J. Brinker, D.E. Clark, D.R. Ulrich, and B.J. Zelinski (MRS, Pittsburgh, 1990) in press.
[13] L.J. Gibson and M.R. Ashby, Proc. Royal Soc. Lond. 382(A), 43 (1982).
[14] T. Woignier, J. Phalippou, and R. Vacher, J. Mat. Res., 4(3), 688 (1989).
[15] T. Woignier, J. Phalippou, R. Sempere, and J. Pelous, J. Phys. Fr. 49, 289 (1988).
[16] R.W. Pekala, C.T. Alviso, and J.D. LeMay, J. Non-Cryst. Solids, accepted.
[17] J.D. LeMay, R.W. Pekala, and L.W. Hrubesh, Pacific Polym. Prpts. 1, 295 (1989).

This paper also appears in Mat. Res. Soc. E.A. 25

SCALING LAWS FOR TRANSPORT, MECHANICAL AND FRACTURE PROPERTIES OF DISORDERED MATERIALS

MUHAMMAD SAHIMI* AND SEPEHR ARBABI**
*Department of Chemical Engineering, University of Southern California, Los Angeles, CA 90089-1211
** Reservoir Engineering Research Institute, 845 Page Mill Road, Palo Alto, CA 94304.

ABSTRACT

We discuss scaling laws for scalar and vector transport properties of, and fracture processes in, disordered materials. Random resistor networks, and elastic and superelastic percolation networks are used to model the disordered material. While scalar transport properties of such systems (e.g. conductivity or diffusivity) obey universal scaling laws near the percolation threshold, vector transport properties (e.g. elastic moduli) may not follow such universal laws, and the critical exponents characterizing such scaling laws may depend on the microscopic force laws of the system. On the other hand, fracture processes in such systems appear to obey universal scaling laws. In particular, the external stress F for the fracture of the system scales with its linear size L as $F \sim L^{d-1}/(\ln L)^{\psi}$, where d is the dimensionality of the system and ψ is a small critical exponent ($\psi \simeq 0.1$). Moreover, as the macroscopic fracture point of the system is approached, the ratio of various elastic moduli of the system approaches a *universal* fixed point, *independent* of the microscopic details of the system. Finally, the distribution of fracture strength in a *randomly reinforced* system, or in a system near its percolation threshold with a broad distribution of elastic constants, is in the form of a Weibull distribution, rather than the recently-proposed Gumbel distribution.

1. INTRODUCTION

The relationships between the microstructure and effective properties of materials, e.g. their transport (electrical and thermal conductivities), mechanical (elastic moduli) and failure properties (the fracture strength distribution and toughness) have been studied for a long time. The effect of *disorder* on transport and mechanical properties has been studied and a better understanding has been developed over the last two decades by combining a variety of techniques and ideas, e.g. percolation and scaling concepts [1], effective medium approximation [2], renormalization group methods [3], as well as large scale numerical simulation [4-8] and well-controlled experiments on model systems [9]. Disordered morphology (i.e. geometry and topology) occurs in a wide variety of physical and technological systems, ranging from granular materials to composite solids such as ceramics, superconductors, glasses, polymers and porous catalysts that are essential for producing a wide variety of products in the petroleum, chemical and electronic industries. This paper attempts to summarize our current understanding of the linear and non-linear mechanical and molecular-transport properties of microscopically disordered materials. Generally speaking, one can distinguish three families of disordered materials. These are polymers, metals and rock-like materials. The last family includes materials such as concrete, glass, ceramics, granular superconductors, heavy clays and tectonic plates. Although each of these systems has been studied for quite some time, only recently have concerted attempts been made to bring the theoretical and computer simulation methods of modern statistical physics of disordered systems to bear on the prediction of mechanical fracture, electrical breakdown, and yield and flow of such disordered materials.

The strictly microscopic description of heterogeneous materials that we consider here has but limited applicability to the direct solution of boundary value problems in various scientific fields, since it is obviously impossible to describe, except in a statistical sense, the typical random microstructure. A more plausible approach is to replace the microscopically heterogeneous medium by an *effective continuum*, an idea which is standard in the statistical mechanics of molecular-kinetic theories and in the field of composite materials, where techniques for the requisite micro-structural averaging or *smoothing* (i.e. parametric *lumping*) are now well-established. This effective continuum provides an intermediate or *mesoscopic* level of description, interposed between the macroscopic boundary-value problem and the micromechanics, and a major theoretical problem is to predict properties of this effective continuum.

However, despite the many years of research on such systems, much remains to be understood. For example, in the area of particle dispersions, e.g. granular materials, most existing theories are limited to highly dilute systems, whereas most real granular materials and similar systems are characterized by high particle concentration and strong particle interaction. As with other branches of condensed-matter physics, one of the outstanding challenges is the developement of kinetic theories and constitutive models for such systems. One of the main goals of the present paper is to point out some of the outstanding issues regarding the transport properties, mechanics, rheology and fracture of disordered materials such as composite solids and granular systems.

In recent times, attractive analogies have become increasingly evident between phenomena such as thermodynamic phase transitions and critical points and mechanical processes, such as *bridging* or *arching* in granular media, or plastic yielding, strain localization and fracture [10-15] in more general materials. At the microscopic level, many of the latter involve statistical fluctuations and correlations which are not always well-described by the existing mean-field theories. While such approximations may be avoided by numerical particle-dynamic simulations, even these require a sound theoretical appreciation of statistical mechanics and critical phenomena.

Any fruitful attempt for studying of transport and mechanics of disordered media has to include, as a first step, a realistic model of the disordered system. Many models have been proposed in the past, most of which have been too simple to realistically represent *both* the geometry and the topology of disordered systems. These include the bundle of pores model for describing porous media, and the spatially-periodic model for describing disordered solids. None of these can represent the topology of the kind of disordered systems that we are interested in here, e.g. gel polymer networks, colloidal aggregates and various composite solids, which are made of networks of highly interconnected elements. Another approach has been to totally bypass this issue, and use average macroscopic continuum equations in which appear transport coefficients that have to be predetermined. This approach is phenomenological and provides no insight into how these coefficients depend on the morphology of the medium, unless careful experiments are carried out to determine this dependence.

In principle, any disordered system can be mapped onto an equivalent network of bonds and sites, in which bonds represent the elements of the system that transmit momentum, energy, stress, etc., and sites are (in most cases) massless points where bonds meet. That this can be done rigorously was shown by Mohanty [16]. The same type of networks can be obtained by discretization of the governing continuum equations of (scalar or vector) transport. Thus, network models represent a natural extension of continuum models. In the present paper, we use network models to represent a disordered material.

Suppose now that the bonds of the network have distributed properties, e.g. they represent channels of a porous medium, through which the flow of a fluid takes place, where effective sizes(radii and lengths)are distributed according to statistical distributions. Now, imagine that a *fraction* q of the bonds are *absent*, in the sense that they do not support the transport process, e.g. a fraction q of the channels of a porous medium have been plugged because of a chemical reaction that produces solid products which are deposited on the surface of the pores. Such a network is called a bond percolation network [1]. If q is large enough, no sample-spanning paths of *present* or intact bonds exists, and no macroscopic transport (of current, stress, etc.) takes place. On the other hand, if $q \approx 0$, the system is almost fully connected and macroscopic transport occurs. Thus, there is a critical value q_c of q, so that for $q > q_c$ no macroscopic transport takes place, whereas for $q \leq q_c$ the network can support macroscopic transport. In such a network, $p_c^B = 1 - q_c$ is called the bond percolation threshold of the system. Alternatively, one may imagine that if a fraction of the sites of the network are absent, then, all bonds that are connected to such sites are also absent. Then, one can define, in an analogous way, a site percolation threshold p_c^S such that for $p = 1 - q < p_c^S$ the network can not support macroscopic transport. It is clear that the passage from a regime of no macroscopic transport to one in which there is macroscopic transport is because of the structural or geometrical phase transition that takes place at the percolation threshold. Percolation networks have proven to be reasonable models for gel polymers [17], sintered powders [18], colloidal aggregates [19], porous media [20] and glasses [21].

So far, this description of percolation networks is abstract. Depending on a specific application, the bonds or sites of the network can take on specific physical meaning. For example, the bonds can represent linear or nonlinear resistors with distributed resistivities or conductances. Thus, if the network represents a disordered solid, one can establish a one-to-one correspondence between the flow of heat or electrical current in the solid and the current field in the network. We call this *scalar percolation*, since one is concerned with a scalar transport process in a percolation system. On the other hand, when bonds represent elastic elements (springs) that can be stretched and/or bent, we have an elastic percolation network (EPN) which, with various microscopic force laws, can take into account the vectorial nature of transport of stress in disordered solids. For the obvious reason, we call this *vector percolation.*

In this paper, we discuss the scaling behavior of the scalar and vector transport properties of percolation networks near the percolation threshold. We summarize our current understanding of such properties and, in particular, we discuss the relation between the scaling properties of scalar and vector percolation models. We also review and discuss scaling laws in fracture of EPNs. While percolation processes usually represent static and linear phenomena, the growth of cracks in spring networks is a dynamic and non-linear process. Despite this, we show that the two systems are related to one another, and propose that universal fixed points, similar to those in renormalization group theory of critical phenomena, can be used to classify all models of fracture of disordered solids. We provide a more detailed discussion of some of our results reported previously, and present a number of new results.

The plan of this paper is as follows. In the next section we describe in detail the EPNs that have been developed in the past few years, and discuss their percolation properties. Next, we discuss scaling laws for various properties of EPNs near the percolation threshold. In the last section, we describe the elastic percolation models of fracture and discuss their scaling properties.

2. TRANSPORT AND MECHANICAL PROPERTIES OF PERCOLATION NETWORKS

In this paper, the disordered system is represented by a two- or three-dimensional network in which each bond represents the elastic element (spring) with an elastic constant e, which can take on values selected from a probability density function $H(e)$. We consider here the simplest case in which

$$H(e) = p\delta(e - a) + (1 - p)\delta(e - b), \tag{1}$$

i.e. e takes the values a and b with the probabilities p and $1 - p$, respectively. In a large enough system, this is equivalent to having a network in which a randomly-selected fraction p of the springs have an elastic constant a, whereas the remaining springs have an elastic constant b. More generally, we can replace this distribution with, $H(e) = pf_1(e) + (1 - p)f_2(e)$, where f_1 and f_2 are two continuous (and normalized) probability density functions, but for the purpose of the present paper the above simple distribution suffices. If a is finite and $b = 0$, we obtain an EPN described above. In this case, as p decreases all elastic moduli of the network decrease and eventually vanish at $p = p_e^B$, which is the *elastic* bond percolation threshold of the network. The value of p_e^B is not necessarily the same as p_c^B, the bond percolation threshold of the network defined above. The reason for this is that one may form networks of such springs which, although they are macroscopically connected, do not support transport of stress, since the deformation of such networks might be done at no cost to the elastic energy of the system. Later in this section, we discuss condition(s) under which $p_e^B = p_c^B$. Another case of interest is when $a = \infty$ and b is finite, i.e. a fraction p of the springs are totally *rigid*, and the rest are *soft*. We call this a *superelastic percolation network* (SEPN) [22]. In this system, as p_e^B is approached from *below*, larger and larger islands of totally rigid springs are formed, as the result of which all elastic moduli of the network increase. At $p = p_e^B$, a sample-spanning cluster of totally rigid springs is formed and all elastic moduli of the network *diverge*. It has been argued that [23], this system my be relevant to modeling of rheological properties of a gelling solution below the gel point.

It remains to specify the microscopic force law that each spring of the system obeys. In principle, one can use any kind of microscopic law, but in the present paper we restrict our attention to a few microscopic models which we now describe. In the first model, the elastic energy of the network is given by

$$E = (\frac{1+\nu}{1-\nu})\sum_{ij}[(\mathbf{u}_i - \mathbf{u}_j).\mathbf{R}_{ij}]^2 e_{ij} + \frac{1-3\nu}{4(1-\nu)}\sum_{ij}(\mathbf{u}_i - \mathbf{u}_j)^2 e_{ij} \tag{2}$$

where $\mathbf{u}_i$ and $\mathbf{u}_j$ are the (infinitesimal) displacements of sites i and j, $\mathbf{R}_{ij}$ a unit vector from site i to site j, e_{ij} the elastic constant of the bond (spring) between sites i and j, and ν the Poisson's ratio. Theoretically, the Poisson's ratio is in the range $-1 < \nu < 1/3$, but for three-dimensional *isotropic* materials one usually has $0 < \nu < 1/3$. The first term in equation (2) represents the contribution of the stretching or central forces (CFs). Equation (2) describes the elastic energy of the so-called *Born model.* Jerauld [24] has shown that the above elastic energy can be obtained by discretizing the *Navier* equation. If a disordered system is described by the above elastic energy, then, it can be shown that near p_e^B the contribution of the second term of equation (2), which is essentially a *scalar* term (as opposed to the first term which is a vectorial contribution), dominates the elastic energy E. Therefore, as long as $\nu \neq 1/3$, the elastic percolation threshold of the system is the same as p_c^B, since the contribution of the scalar term, which is completely due to the *connectivity* of the system dominates E. The disadvantage of this model is that it *is not rotationally invariant* which is not realistic. This non-rotational invariance of the system is unexpected since the Navier equation, which is supposedly the continuum counterpart of the Born model, is of course rotationally invariant. Although the origin of this contradiction between the continuum model and its discrete counterpart is not completely clear yet, one may speculate that the discretization procedure can give rise to this unexpected behavior [22].

If $\nu = 1/3$, one obtains the CF model, which is essentially a network of simple springs that can be stretched. It is easy to see that the CF model *is* rotationally invariant, and since there is no scalar contribution to the elastic energy of the CF model, one has $p_e^B \neq p_c^B$. In certain EPNs with only CFs, the problem of determining the elastic moduli of the network is trivial. For example, it is easy to see that for the square and cubic networks with only CFs, $p_e^B = p_e^S = 1$. Thus, the CF model is restricted to certain networks whose connectivity and topological properties are complex enough that can support transmission of CFs. Computer simulations show that [25,26] for a triangular network (coordination number $Z = 6$), $p_e^B \simeq 0.642$, which should be compared with the exact value, $p_c^B = 2\sin(\pi/18) \simeq 0.347$, and [27] for a body-centered cubic (BCC) network, $p_e^B \simeq 0.737$, which should be compared with $p_c^B \simeq 0.1795$. Thus, an EPN with only CFs is not a very realistic model for many disordered materials, since for $p_c^B < p < p_e^B$, one has an elastic network which , although it is macroscopically connected (since $p > p_c^B$), it has no non-zero elastic modulus. Despite this shortcoming, the CF model is a useful system to study, since it is the simplest EPN and many of its properties are quantitatively similar to those of EPNs with more complex energy and microscopic force laws described below. Moreover, a system with only CFs is essentially identical with a granular packing of spherical particles in which there is no friction between the particles. We should point our that, for any given system determining p_e^B is much more difficult than p_c^B. Whereas numerical estimation of p_c^B only requires finding the smallest value of p for which a system is macroscopically connected (which is a purely topological problem), estimating p_e^B requires determining the smallest value of p for which the EPN is not only macroscopically connected, but it also has non-zero elastic moduli. Thus, in this latter case, one has to explicitly solve for the elastic moduli which requires large scale simulation (see below).

In deforming any EPN, one expects to not only stretch the bonds, but also change the angle between any pair of bonds that have one site in common. Thus, the elastic energy of the system should contain the contribution of both the CF and angle-changing or bond-bending (BB) forces. Hence, for such a system one has

$$E = \frac{\alpha}{2}\sum_{ij}[(\mathbf{u}_i - \mathbf{u}_j).\mathbf{R}_{ij}]^2 e_{ij} + \frac{\beta}{2}\sum_{jik}(\delta\theta_{jik})^2 e_{ij}e_{ik}, \tag{3}$$

where α and β are the stretching (CF) and BB force constants, respectively. The first term in equation (3) represents the usual contribution of the CFs, while the second term is the contribution of angle-changing or BB forces which, for two bonds ij and ik, is written in terms of the change in the angle $\delta\theta_{jik}$ between the two bonds. The two sums in equation (3) are, respectively, over all bonds and all pairs of bonds with a site i in common. The precise form of $\delta\theta_{jik}$ depends on how much microscopic details one would like to include in the model. If bending of bonds that make 180 degrees with one another is not allowed (i.e. colinear bonds do not change the angle between them), then,

$$\delta\theta_{jik} = (\mathbf{u}_i - \mathbf{u}_j).\mathbf{R}_{ik} + (\mathbf{u}_i - \mathbf{u}_k).\mathbf{R}_{ij}. \tag{4}$$

This model is essentially the same as that of Kirkwood [28] who studied the vibrational properties of rod-like molecules, and that of Keating [29] who investigated the elastic properties of covalent crystals. We refer to this as the Kirkwood-Keating (KK) model. If, however, the bending of colinear bonds is allowed, then,

$$\delta\theta_{jik} = \begin{cases} (\mathbf{u}_{ij} \times \mathbf{R}_{ij} - \mathbf{u}_{ik} \times \mathbf{R}_{ik}).(\mathbf{R}_{ij} \times \mathbf{R}_{ik})/|\mathbf{R}_{ij} \times \mathbf{R}_{ik}| & \mathbf{R}_{ij} \text{ not } \parallel \text{ to } \mathbf{R}_{ik}, \quad (5a) \\ |(\mathbf{u}_{ij} + \mathbf{u}_{ik}) \times \mathbf{R}_{ij}| & \mathbf{R}_{ij} \parallel \text{ to } \mathbf{R}_{ik}, \quad (5b) \end{cases}$$

where, $\mathbf{u}_{ij} = \mathbf{u}_i - \mathbf{u}_j$. In the special case of a cubic network in d dimensions, equations (5) are simplified to

$$\delta\theta_{jik} = (\mathbf{u}_i - \mathbf{u}_j) \times \mathbf{R}_{ij} - (\mathbf{u}_i - \mathbf{u}_k) \times \mathbf{R}_{ik}. \quad (6)$$

We refer to this system as the BB model.

Phillips and Thorpe [30] studied the percolation properties of the KK model. Using a mean-field argument, they proposed that for a d-dimensional KK model in a network of coordination number Z, one has

$$p_e^B \simeq \tfrac{1}{Z}[(d^2 + d)/(2d - 1)] \qquad \text{KK model} \quad (7)$$

On the other hand, for scalar percolation one has [31]

$$p_c^B \simeq \tfrac{1}{Z}[d/(d - 1)] \qquad \text{scalar model} \quad (8)$$

which means that for two-dimensional systems the elastic percolation threshold of the KK model and that of the scalar percolation threshold are identical ($p_c^B = p_e^B \simeq 2/Z$). For three-diemsional systems, equation (7) predicts that, $p_e^B \simeq 2.4/Z$, whereas for scalar percolation equation (8) yields, $p_c^B \simeq 1.5/Z$, so that the KK and scalar percolation systems have different percolation thresholds for $d = 3$. The BB model, on the other hand, has the same percolation threshold as the scalar percolation if each site of the system interacts with at least $d(d-1)/2$ of its nearest -neighbor nodes. For this reason, the BB model is the most general elastic percolation model since, by including enough microscopic details, one can force the elastic percolation threshold of the system to be identical with that of the scalar percolation. Jerauld [24] and Feng *et al.* [32] studied the CF model, and derived the following approximate formula

$$p_e^B \simeq 2d/Z \qquad \text{CF model} \quad (9)$$

which correctly predicts, $p_e^B = 1$ for a d-dimensional cubic network ($Z = 2d$). Equation (9) also predicts that $p_e^B = 2/3$ for the triangular network (Z=6), and $p_c^B = 3/4$ for the BCC network ($Z = 8$), which are in close agreement with the numerical estimates mentioned above.

We can provide a geometrical interpretation of equations (7)-(9). Since for any given p, the quantity pZ is the *average* coordination number $\bar{Z}$ of the network, equation (7) predicts that *at* the elastic percolation threshold of the KK model, one has

$$\bar{Z} \simeq (d^2 + d)/(2d - 1) \qquad \text{KK model} \quad (10)$$

whereas at the elastic threshold of the CF model one obtains

$$\bar{Z} \simeq 2d \qquad \text{CF model} \quad (11)$$

and at the percolation threshold of the scalar model, we have

$$\bar{Z} \simeq d/(d - 1) \qquad \text{scalar model} \quad (12)$$

Equation (11) explains clearly why the CF model is a trivial problem on certain networks since, according to this equation, the coordination number of a d-dimensional network must be *greater* than $2d$ in order for it to have non-zero elastic moduli. Therefore, honeycomb ($Z = 3$) and square ($Z = 4$) networks in two dimensions, and diamond ($Z = 4$) and cubic ($Z = 6$) networks in three dimensions do not have any non-zero shear modulus for any p, or Young's or bulk modulus for $p < 1$.

Let us now discuss a few properties of percolation networks. It is generally believed, based on computer simulations, an exactly-solved (the Bethe lattice) model, and renormalization group theory [33,34] that in an EPN above and near p_c, *all* elastic moduli G of the system obey a scaling law in the form

$$G \sim (p - p_c)^f, \tag{13}$$

where, depending on the model, p_c can be either the connectivity or elastic percolation threshold. The electrical or thermal conductivity σ of the same system (if the present or active bonds between sites are conducting elements) obeys

$$\sigma \sim (p - p_c)^t. \tag{14}$$

Similar to thermal systems near the critical temperature, one can also define a percolation correlation length ξ which obeys a universal scaling law near p_c

$$\xi \sim (p - p_c)^{-\nu_p} \tag{15}$$

and an order parameter $P(p)$ which obeys

$$P(p) \sim (p - p_c)^{\beta_p}. \tag{16}$$

The physical significance of ξ is that, only for length scales $L > \xi$ the system is macroscopically homogeneous. For $L < \xi_p$, the sample-spanning percolation cluster is a self similar fractal object, such that its total mass M (i.e. the total number of active bonds or sites in the cluster) scales with L as $M \sim L^{d_f}$, where

$$d_f = d - \beta_p/\nu_p, \tag{17}$$

for a d-dimensional system. The physical interpretation of P(p) is that it is the *fraction* of active bonds or sites in the sample-spanning cluster.

In a SEPN, all elastic moduli of the system diverge as p_c is approached from below according to the scaling law

$$G \sim (p_c - p)^{-\tau}. \tag{18}$$

In an analogous way, we may define a superconductive percolation network in which a fraction p of all bonds are superconductors (with zero resistance) and the rest are ordinary conductors. For such a network, one has

$$\sigma \sim (p_c - p)^{-s}, \tag{19}$$

as p approaches p_c from below. A duality argument [35] established that in two dimensions, $t = s$. A SEPN can be thought of as a model of *random reinforcement* of disordered materials in which rigid material is inserted in the matrix to increase its resistance to breakage. Various forms of such reinforcements are in fact used in practice in order to toughen the material.

In the past few years, the exponents f, t, τ and s have been determined very accurately. The main tool of determining such critical exponents has been the finite-size scaling analysis (FSSA). If, for example, we combine equations (13) and (15), we can write

$$G \sim \xi^{-\delta} \tag{20}$$

where $\delta = f/\nu_p$. In principle, scaling laws (13)-(19) are valid for infinitely large systems, whereas in practice one deals with a finite system. In a finite system, the correlation length ξ cannot exceed L, the linear size of the system, so that if, for any value of p, ξ (of the infinitely large system) is larger than L, then, L is the dominant length scale of the system, and equation (20) must be replaced with

$$G \sim L^{-\delta}. \tag{21}$$

However, equation (21) is also supposed to be valid for large values of L (but such that $L < \xi$). In practice, one cannot use very large networks, so that correction-to-scaling (CTS) terms become important. In such a case equation (21) is rewritten as

$$G \sim L^{-\delta}[1 + a_1 g_1(L) + a_2 g_2(L)]. \tag{22}$$

Here $g_1(L)$ and $g_2(L)$ represent, respectively, the leading non-analytical and analytical CTS, and a_1 and a_2 are constant. In a similar way, for a SEPN we can write

$$G \sim L^{x}[1 + b_1 h_1(L) + b_2 h_2(L)] \tag{23}$$

where $x = \tau/\nu_p$. Similar equations can be used for σ. We recently showed that [36], $h_1(L) = g_1(L)$ and $h_2(L) = g_2(L)$. Moreover, we showed that for *all* transport properties of percolation networks (elasticity, superelasticity, conductivity, diffusivity, etc.) one has

$$g_1(L) = (\ln L)^{-1}, \tag{24}$$
$$g_2(L) = L^{-1}. \tag{25}$$

Since equations (22) and (23) are supposed to be valid for a system in which $L < \xi$, calculations are done at $p = p_c$, where ξ is divergent and thus exceeds any L that can be used in the simulations.

In Table 1, we summarize the most accurate estimates of t and s for $d = 2$ and 3 currently available, along with their mean-field values which are supposed to be valid for $d \geq 6$. The estimate of $t(d = 2)$ was obtained by Normand *et al.* [4], while Gingold and Lobb [37] obtained $t(d = 3)$. Normand and Herrmann [38] estimated $s(d = 3)$ using FSSA and the transfer matrix method. These exponents have been found to be largely universal, independent of the microscopic details of the system.

Table I
Currently-accepted values of the critical exponents of scalar percolation in d-dimensions.

d	ν_p	β_p	t	s
2	4/3	5/36	1.2993 ± 0.0015	1.2993 ± 0.0015
3	0.88	0.41	2.003 ± 0.047	0.735 ± 0.005
≥ 6	1/2	1	3	0

Unlike t and s, the values of f and τ have been controversial. Originally, we argued that [39] the CF model may not have a unique universality class, i.e. the values of f may be different for site, bond and correlated percolation in the CF model. It also seemed that the value of f for two-dimensional bond percolation in the CF model may not be the same as that of the BB model. More recently, very accurate simulations by us [25,40] and others [41] seem to provide the following picture. The value of f for two-dimensional BB model is universal and is the same as that of the CF model in the *bond percolation* model. However, the value of f for two-dimensional *site percolation* in the CF is still very different from that of the BB and the CF models in bond percolation. For three-dimensional systems, we have estimated f for both the BB and the CF models. As in the case of two-dimensional systems, the BB model seems to have a unique universality class, in which bond percolation [42], and site and correlated percolation [43] have the same value of f. However, for three-dimensional CF model, f seems to have a value distinctly different [44] from that of the BB model. We should also mention that it appears that the estimation of critical exponents for the CF model by FSSA is extremely sensitive to the value of elastic percolation thresholds p_{ce}. For example, for bond percolation on the triangular lattice in the CF model, one obtains [39], $f/\nu_p \simeq 1.45$ if $p_e^B \simeq 0.65$, whereas $f/\nu_p \simeq 3$ if $p_e^B \simeq 0.642$ [25,40,41]. Therefore, it appears that the CF model is not characterized by a unique universality class. We have also estimated [23] the value of τ for the BB model in both two and three dimensions, using the FSSA. Table 2 summarizes the estimates of f and τ for various elastic and superelastic percolation networks, along with their mean-field values if available. The two-dimensional value of f for the BB model was obtained by Zabolitzky *et al.* [6]. For the Born model one always finds that $f = t$ and $\tau = s$. As typical examples, figure 1 shows the variation of the Young's modulus K with L for three-dimensional BB model at $p = p_c$, while figure 2 presents the variations of the Young's modulus for the CF model in the BCC network at $p_e^B \simeq 0.737$, from which f/ν_p is estimated. The value of f for three-dimensional BB model is in excellent agreement with the critical exponent of the elastic moduli of gel polymers and sintered materials [42], while the estimated τ for three-dimensional BB model is consistent with the critical exponent of the viscosity of a gelling solution near the gel point [23]. Thus, the models can be used to model mechanical and rheological properties of disordered materials.

Table II

Currently-accepted values of the critical exponents of vector percolation in d-dimensions. Values of f and τ for the CF model refer to bond percolation.

d	f/ν_p	τ/ν_p	Model
	2.95 ± 0.25	1.05 ± 0.05	CF
2	2.97 ± 0.03	0.92 ± 0.03	BB
	1.2993 ± 0.0015	1.2993 ± 0.0015	Born
	2.02 ± 0.10	?	CF
3	4.3 ± 0.1	0.74 ± 0.04	BB
	2.31 ± 0.07	0.835 ± 0.005	Born
	?	0	CF
≥ 6	8	0	BB
	6	0	Born

Are there any relations between the transport exponents f, t, τ and s and the static exponents ν_p, β_p and so on? Many relations have been proposed relating t to ν_p and β_p, a summary of which is given by Sahimi [45], none of which is exact. Some of them do not agree with the ϵ-expansion [46] of t ($\epsilon = 6 - d$), while others show deviations from numerical estimates of t. However, for two-dimensional percolation, we proposed that [36]

$$s = t = \nu_p - \beta_p/4 = 187/144 \simeq 1.2986... \quad (26)$$

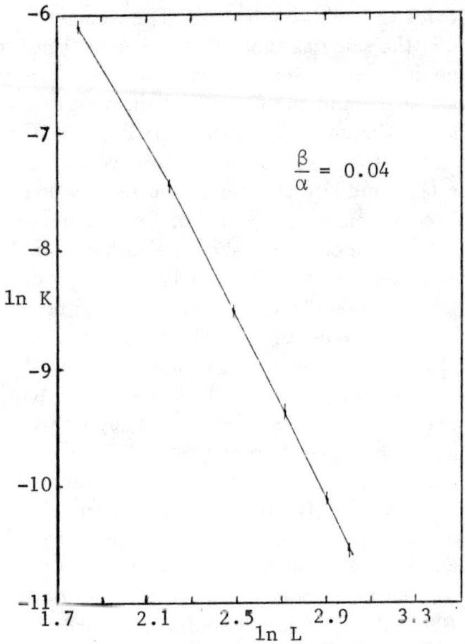

Figure 1: Variations of the Young's modulus with the linear size L at the bond percolation threshold of a simple-cubic network with the BB model.

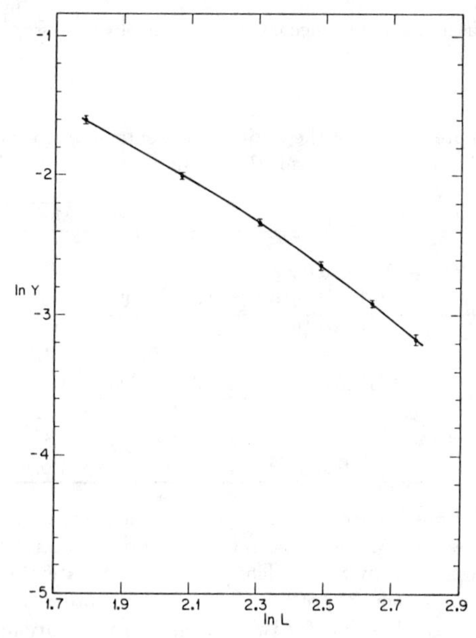

Figure 2: Variations of the bulk modulus with the linear size L at the bond percolation threshold of a BCC network with the CF model.

which agrees completely with the numerical estimate given in Table 1. For the BB model, we also proposed [47] a relation between f, t and ν_p

$$f = t + 2\nu_p \tag{27}$$

which agrees completely with the numerical estimates (see Tables 1 and 2), and is likely to be exact. For the exponent τ in the BB model, we proposed that [23]

$$\tau = \nu_p - \beta_p/2 \tag{28}$$

which also agrees with the numerical estimates given in Tables 1 and 2.

3. SCALING AND UNIVERSALITY IN FRACTURE OF PERCOLATION NETWORKS

Mechanical failure in disordered materials is of immense technological and economic importance. An in-depth understanding of the macroscopic and microscopic details of the mechanical failure of real materials is critical to the engineering of structural components and the design of advanced structural materials. Microscopic failure plays a fundamental role in many systems of industrial importance ranging from aircraft structures and pressurized nuclear reactors to the propagation of cracks in underground oil reservoirs and in ceramics, fibrous composites, glasses and high temperature superconducting materials.

There exists an extensive literature on the general problem of mechanical failure of disordered solids [48]. The traditional fracture mechanics approach to failure has certainly provided the framework for analyzing wide variety of problems without considering the details of microscopic disorder. The basis of the analysis of brittle failure is a very important criterion which was proposed by Griffith [49] in 1921. The key idea is to formulate an energy balance. Griffith proposed that a *single* crack becomes unstable to extension when the elastic energy released in crack extension by a small length $\triangle l$ becomes equal to the surface energy required to create a length $\triangle l$ of crack surface. In other words, the equilibrium extension of a *single* flaw is governed by a balance between the mechanical energy released and the surface energy gained as the flaw propagates. However, such an approach cannot take into account the effect of disorder.

In real engineering materials, presence of large number of flaws, with various sizes, shapes and orientations makes the situation far more complex. Disorder comes into play in many ways during a fracture process. Fracture generally enhances an initially present disorder, through the nucleation of new cracks, or simply due to the heterogeneity of the stress field that results from the complex geometrical arrangements of the existing cracks. Even small, initially present disorder can be enormously amplified during fracture. Therefore , fracture is a collective phenomenon in which disorder plays a fundamental role. In fact, due to disorder and existence of many random cracks, brittle materials generally exhibit wide statistical fluctuations in fracture strengths, when nominally identical samples are tested under nominally identical loading conditions and, therefore, one should generally talk about the *average* strength of such materials. Besides the statistical fluctuations in the fracture strengths, the mean fracture strength depends on the sample size and in general it decreases with an increase in the sample size. Similarly, if the sample size remains constant but the characteristic size of the microstructure changes, the mean fracture strength will be affected. This has been experimentally observed in aluminum and ceramics. These indicate the importance of sample size in fracture processes. Finally, consider a typical stress-strain curve during fracture. It is usually characterized by an initial increase in the stress as the result of an increase in the strain, reaching a maximum and, then, decreasing again with increasing the strain. It has been shown by Bažant [50] that, in the decreasing part of the curve, called *strain softening*, the material develops *instabilities* and damage will concentrate in local zones. If we use a finite element or finite difference technique to numerically model this behavior, then, we will see that the localization of damage is very dependent on the mesh size used, that is, the finer the mesh, the more localized the damage. In addition, the mechanical response depends very much on the mesh size. A homogeneous solid having this kind of behavior at a local scale will not have the same behavior at the macroscopic scale. Its behavior is *not* scale invariant. Since this kind of stress-strain relation is observed at a macroscopic scale for real materials, a stabilizing factor must exist. Disorder indeed provides such a stabilization. Thus, it is not sufficient, and indeed inappropriate, to represent the behavior of a material by its average (as done in the mean-field approaches) for fracture processes. Fluctuations are fundamental, not only because of the necessary statistical analysis of failure in many practical instances, but because these fluctuations at a local scale provide a necessary mechanism to prevent the localization of damage.

Recently, several network models have been introduced for both electrical [51-53] and mechanical [54] failure of disordered solids. In these models each bond of the network is supposed to describe the disordered system on a microscopic level, with elasticity and failure characteristics described by a few control parameters. For example, the bonds represent the microscopic elements of a disordered material that follow the laws of linear elasticity (e.g. Hooke's law) up to a critical threshold (e.g., in their length), beyond which they can break irreversibly and create a microscopic crack in the system. If a large number of bonds break, a macroscopic crack is formed and the material suffers mechanical breakdown. Alternatively, the bonds can represent the microscopic elements of a macroscopically insulating material, having otherwise a few isolated conducting regions, such that on an application of an external potential, some of the bonds can suffer a large potential drop, break down and become conducting. If a large enough number of bonds break down, the system suffers dielectric breakdown and becomes conducting. In any event, once such a model is generated on a computer, an external potential, strain or stress is applied to the network and gradually increased, as a result of which the individual bonds will break in a certain manner until the system fails macroscopically. The sequence of breaking bonds and the spatial patterns they form are supposed to represent a real breaking process. Various properties of such failure processes have recently been investigated [55-62] and several important features of their behavior have been discussed (see below). It is straightforward to incorporate almost any kind of disorder in such a system and study its effect on the fracture properties of the system. In particular, percolation effects, which represent the presence of cracks of various shapes and sizes, can easily be included in such a model.

Following Sahimi and Goddard [54], who first introduced percolation models of mechanical breakdown in disordered solids, four general classes of disorder can be introduced:

(i) deletion or suppression of a fraction of bonds at random or in a prescribed manner. This represents a percolation effect. Alternatively, one can make, at random or in a prescribed fashion, a fraction of bonds to be totally rigid. This would represent another form of percolation effect which corresponds to the SEPNs and, as discussed above, can be thought of as a *reinforcement* of the material to make it resist mechanical or electrical breakdown.

(ii) Random (or prescribed) distribution of the elastic constants e or conductances g of the bonds. The idea is that in real materials the shapes and sizes of the microscopic elements that allow transport of stress or current are different, resulting in different e or

g for each bond.

(iii) Random (or prescribed) distribution of the critical threshold. For example, each elastic element (spring) can be characterized by a critical displacement or length ℓ_c, such that if it is stretched beyond ℓ_c, it breaks irreversibly. Alternatively, each bond can be characterized by a critical force or stress f_c, such that if it suffers a stress $f > f_c$, it breaks irreversibly. In the context of dielectric problem, each bond can be characterized by a critical potential v_c, such that if a bond suffers a voltage drop $v > v_c$, it breaks down and becomes conductive. The idea is that a composite solid made up intrinsically of the same material (same e or g everywhere), may contain regions having different resistances to breakage under an imposed external stress or potential, as refelected in different critical thresholds ℓ_c or v_c, because of defects in the manufacturing process.

(iv) One important source of disorder in stressed materials is the so-called residual stress variations. Residual stresss arises in a variety of disordered materials such as polycrystalline $A\ell_2O_3$ and ZrO_3-toughened $A\ell_2O_3$. Among other factors, residual stress may be caused by thermal expansion mismatch. That is, when there is a mismatch between the coefficients of thermal expansion of the matrix material and second-phase particles. This gives rise to residual stresses upon cooling the material by an amount ΔT from high processing temperatures to room or use temperature. The simplest way of introducing this effect is to allow for variations in the equilibrium element or spring length ℓ_0 with no variations in the elasticity or failure strain. Isolated from the network, each spring or element has no stress on it, it is only upon imposing the constraint of the surrounding network elements that a spring has a residual stress. This is the fourth class of disorder that one can introduce into the model. In the present paper, we mainly use types (i) and (ii) of disorder.

Of course we may combine any combination of the above kinds of disorder, depending on the intented application. One may also start with a *fractal* network, instead of a fully-connected, macroscopically homogeneous system, and study its fracture properties. Such systems usually contain large porosities and are relevant to a wide variety of materials. For example, they are relevant to the transgranular stress corrosion cracking of ductile metal alloys such as stainless steel and brass [63]. In this phenomenon a stressed A_xB_{1-x} alloy immersed in an aggregation aqueous solution undergoing a process in which the less noble element, A say, is selectively dissolved out. The resulting porous structure, which has a fractal structure, is extremely brittle, and under stress starts to nucleate cleavage cracks. By appropriate choice of the alloy-environment system and variation of the alloy composition, it is possible to develop a range of random porous structures with fractal characters [63]. A study of fracture in percolating systems near or at the percolation threshold is also relevant to the propagation of fractures in weakly connected granular systems which provide a reasonable description of sedimentary rocks.

In any case, once the particular form of disorder is introduced into the model, one can study the breakage of the bonds and the formation of macroscopic cracks. However, the sequence of breaking bonds depends on the type of disorder and its statistical distribution. It may also depend on the microscopic laws that govern the electrical or elastic behavior of the individual bonds and their interaction with one another. We already saw in section 2 that even for linear mechanical properties different microscopic force laws can give rise to different macroscopic behavior. Moreover, these models represent non-equilibrium and highly non-linear systems, and as such are very different from their linear and static counterparts which are usually represented by random resistor networks or EPNs in which a fraction of the bonds of the network has been cut at *random*. The macroscopic transport and geometrical properties of percolation networks obey well-defined scaling laws, whereas until recently, such universality concepts had not been tested for fracture properties.

In this section, we study three issues that are fundamental to the fracture of disordered solids. We first study the distribution of fracture strength (DFS) in EPNs and SEPNs, especially near the percolation threshold. Next, we test the idea of universality in the models of fracture. In particular, we obtain the stress-strain curves during fracture of a variety of two- and three- dimensional elastic networks to see whether they follow universal scaling laws. Finally, we propose a new method of classifying various models of fracture according to universal fixed points, much like those found in renormalization group theory of critical phenomena.

3.1 Distribution of Fracture Strengths

The failure stress of a solid sample is usually determined in a tensile test. There are several valid definitions of the failure stress, depending on the nature of the tensile test. In a *stress-controlled* test the stress is incremented, and the strain is the dependent variable. In this case the sample fails at the highest value of stress in the stress-strain curve. In the context of a network model, this usually occurs at the point where the first bond breaks. In a *strain-controlled* test, on the other hand, the strain is incremented and the stress is the dependent variable. As the stress is finite for all strains, the failure stress in this case corresponds to the point where the stress first drops to zero. Thus, the stress in a strain-controlled failure is always less than or equal to that in a stress-controlled failure. Similarly, the strain at which failure occurs in a stress-controlled test is always less than or equal to that for a strain-controlled test.

We define fracture strength σ_f of a system as the lowest externally applied stress at which the system breaks down. We use the hypothesis that the eventual failure of the system is governed by the most critical flaw in the system. That is, the weakest part of the system fails first. Hence, calculation of the full distribution function of fracture strenghts σ_f reduces to the calculation of the distribution functions of the most critical flaws in the system. It is shown [59] that this is an excellent approximation for the failure stress of the system in a *stress-controlled* tensile test.

The Weibull distribution has been traditionally used in fitting the fracture strength data. Recently, however, Duxbury and Leath [57] formulated a new distribution which is usually referred to as the Gumbel distribution, and argued that such a distribution fits the fracture strength data much better than the Weibull distribution. They demonstrated the superiority of the Gumbel distribution over the Weibull distribution in a random resistor network model of electrical breakdown. However, their distribution is only intended for percolation networks far above p_c.

If we define the distribution function $F_L(\sigma_f)$ as the probability that a sample of size L will experience elastic failure if an external stress σ_f is applied to the sample, then, the classical Weibull distribution is given by

$$F_L(\sigma_f) = 1 - exp(-cL^d\sigma_f^m), \tag{29}$$

whereas the Duxbury-Leath [57] or the Gumbel distribution is given by

$$F_L(\sigma_f) = 1 - exp[-cL^d exp(\frac{-k}{\sigma_f^\eta})], \tag{30}$$

where c, k, m and η are constant. One can use equations (29) and (30) directly to see which distribution fits the fracture strength data better. However, a better and more sensitive test of the validity of these two distributions can be carried out if we rewrite equations (29) and (30) in alternative forms. Equation (29) can be rewritten in the form

$$A_W = -\ln\left\{\frac{-\ln[1-F(\sigma_f)]}{L^d}\right\} = m\eta\ln(1/\sigma_f) - \ln c, \qquad \text{Weibull} \qquad (31)$$

while equation (30) can be rewritten in the form

$$A_G = -\ln\left\{\frac{-\ln[1-F(\sigma_f)]}{L^d}\right\} = k\frac{1}{\sigma_f^\eta} - \ln c. \qquad \text{Gumbel.} \qquad (32)$$

A glance at these two equations shows that the left hand sides of the two equations are the same, $A_w = A_G = A$. The Weibull distribution is of the form

$$A = b_1\ln(\frac{1}{\sigma_f}) + c_1, \qquad \text{Weibull} \qquad (33)$$

while the Gumbel distribution is of the form

$$A = b_2\frac{1}{\sigma_f^\eta} + c_2. \qquad \text{Gumbel} \qquad (34)$$

Both of these equations predict linear variations of A with $\ln(\frac{1}{\sigma_f})$ or with $\sigma_f^{-\eta}$. The precise value of η has not been determined yet, but lower and upper bounds for it have been proposed.

We used a triangular EPN of size $L = 40$ in all of our simulations of the DFS. Both the CF and BB models were used. We generated an EPN for a given p, and applied a very small macroscopic strain to the network in the transverse direction as in a shear modulus test. All bonds of the network have the same threshold value l_c for their lengths. We, then, solved for the equilibrium state of the network at the given p and found the bond whose length l is the maximum and is larger than l_c. If such a bond could not be found, then, we increased the macroscopic strain by a small amount and repeated the process. For each simulation realization we only need to find the first (the most stretched) critical bond. Then, we can easily find the external stress σ_f necessary to break this most critical bond. One repeats the process many times from which the cumulative distribution function $F_L(\sigma_f)$ can be determined. Once F_L is known, we can test the applicability of the Weibull or Gumbel distributions for fitting the simulation data.

The case of the CF model has already been studied by Beale and Srolovitz [58] for $p \gg p_c$. They found that the Gumbel distribution fits the DFS data much better than the Weibull distribution at $p = 0.9$, far away from $p_{ce} \simeq 0.642$ for a triangular EPN. We studied the DFS both far from and near p_c and much more extensively. Here, we present some of our results for p closer to p_{ce}. The complete details can be found elsewhere [64]. Fracture in a CF network is not only of theoretical interest, it is also of direct relevance to the mechanics of granular materials, if there is no friction between the particles in the granular packing.

We took $p = 0.8$ and simulated 1000 realizations. Figure 3 shows our results for $F(\sigma_f)$ where it is clear that it does not resemble (the expected) S shape, but has a long tail. We then tested the applicability of the Gumbel and Weibull distributions. Figure 4 presents the fit with the Gumbel distribution using $\eta = 1$. We clearly see that the data do not lie on a straight line. The solid line in the figure is just a guide to the eye which only connects the data points to each other. Using $\eta = 2$ did not improve the quality of the fit. The fit with the Weibull distribution is shown in the Fig. 5 which is again not a straight line. Therefore, neither Gumbel nor Weibull distributions provide good fits to

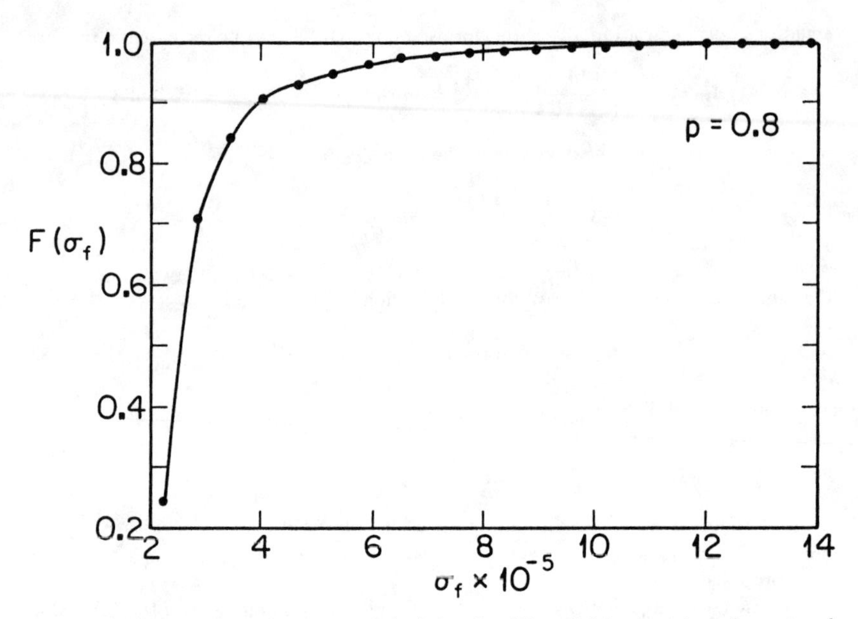

Figure 3: The distribution function $F_L(\sigma_f)$ for the CF model in the triangular network at $p = 0.8$.

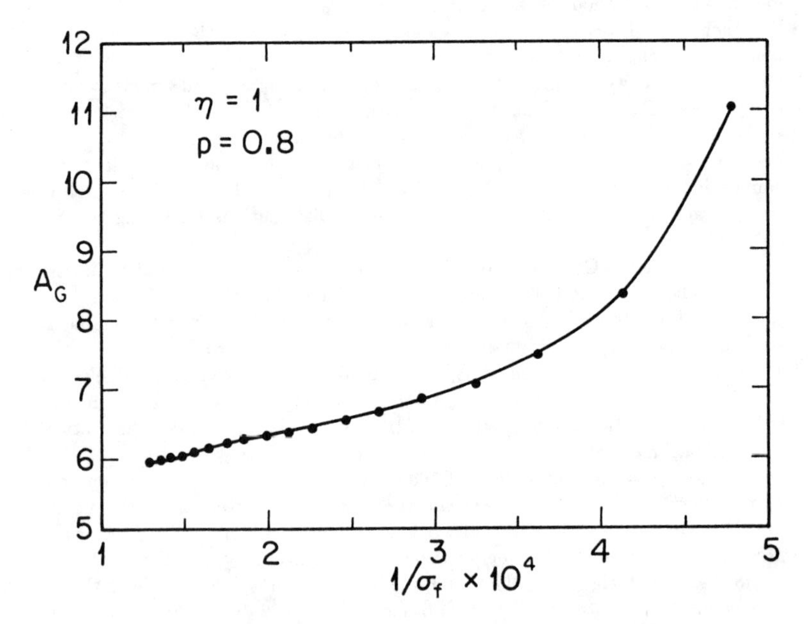

Figure 4: The Gumbel distribution fit of the data of figure 3 with $\eta = 1$.

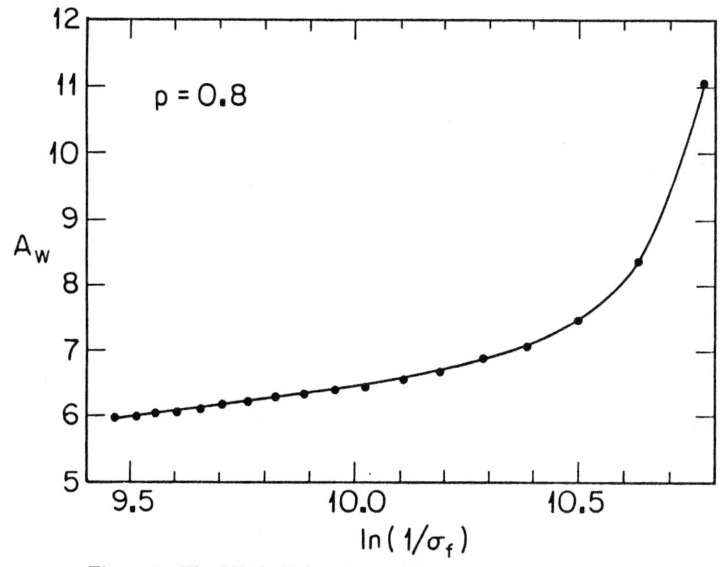

Figure 5: The Weibull distribution fit of the data of figure 3.

the DFS, as the data have distinct curvature and do not lie on straight lines, whereas for $p \geq 0.9$ the Gumbel distribution with $\eta = 2$ provides a very good fit to the data [58,64].

Before we present the results with the BB model, we should try to explain the above result. Consider a percolation cluster far from p_c. One can find many connected paths which support transport of stress, in the form of many macro-links of various sizes $\mathcal{L}$. In such a system, the DFS may appear as a result of two major factors: (1) fluctuations of the macro-link sizes $\mathcal{L}$ around the percolation correlation length ξ, and (2) fluctuations of the individual strengths of the bonds in the network. Since we use discrete percolation networks here, all the individual bonds of the network have been assigned the same strength and elastic constant, and, hence the second factor does not come into play. (We have shown elsewhere [65] that if we assign a different elastic constant e or conductance g to each bond, then, depending on the distribution of e or g, the DFS can be neither the Weibull nor the Gumbel distribution, even at $p = 1$.) As we move toward p_e^B, two changes take place: First, one has fewer macro-links and, secondly the contributions of the shorter macro-links to the transport process become negligible compared to those of the fewer longer macro-links. This has the direct effect of reducing the fluctuations and making the DFS narrower. At p_e^B, there is only one huge macro-link which supports the transport of stress. Therefore, there are no macro-link to macro-link fluctuations. In other words, in the critical region of discrete percolation the notion of a largest defective cluster has no meaning. This means that in a discrete lattice at p_e^B in which all bonds have the same e, failure occurs with no fluctuations and the distribution of fracture strength is a Dirac function. Some of this argument was given by Sornette [66] who also argued that the DFS approaches a Weibull-like distribution in the critical region, if one assigns a different e to each bond. We also studied the DFS with the BB model, the results of which are given elsewhere [64]. We found that in most cases the Gumbel distribution provides a reasonable fit to the data. We should mention here the recent experimental work of van den Born *et al.* [67]. These authors measured the mechanical strength of highly porous ceramics (with porosities ranging from 0.71 to 0.79), and found

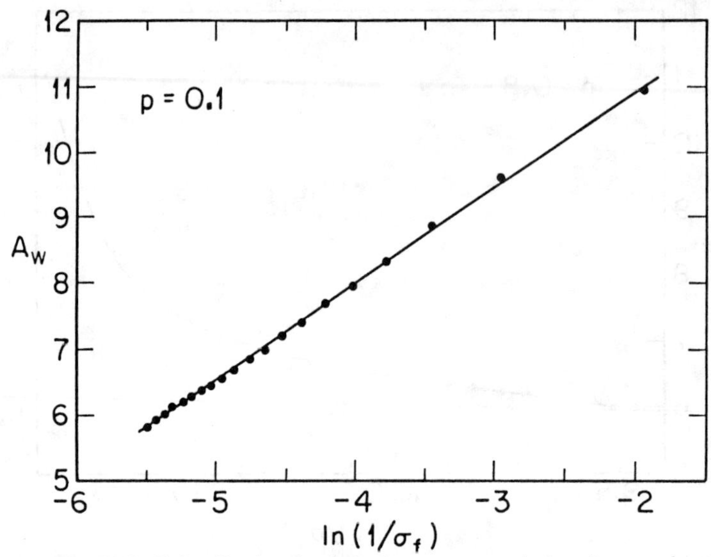

Figure 6: The Weibull distribution fit of the fracture strength data in a randomly reinforced triangular network with the BB model. The fraction of rigid bonds is $p = 0.1$.

the Gumbel distribution with $\eta = 1$ to be more accurate than the Weibull distribution in representing the data for the failure pressure distribution. Evidently, these systems were not near their percolation thresholds, even at such high porosities.

We next considered the DFS in SEPNs, or the randomly reinforced model, with the BB model and determined this distribution for several values of p, where p now refers to the fraction of rigid bonds in the network. We again used $L = 40$ with $\beta/\alpha = 0.1$ and made 1000 realizations for each case. We found that the Gumbel distribution cannot provide a good fit to the data for any $1 \leq \eta \leq 2$, but as Figure 6 indicates, the Weibull distribution provides an excellent fit to the data for $p = 0.1$. Therefore, the fracture process in a randomly reinforced system is very different from that in EPNs. We found that the Gumbel distribution can provide a satisfactory fit to the data if we take $\eta \simeq 0.1$. However, in the Gumbel distribution, σ_f^η with $\eta \simeq 0.1$ is essentially equivalent to $\ln \sigma_f$, i.e. the Weibull distribution.

Unlike the case of fracture in EPNs, one can define and calculate the DFS at $p_c \simeq 0.347$ in the randomly reinforced systems, since there are large variations in the elastic constants of the bonds. We found that [64] the most accurate fit of the data with the Gumbel distribution is obtained for $\eta \simeq 0.5$, whereas the Weibull distribution cannot provide an accurate fit to the data, which is again the opposite of fracture in EPNs discussed below.

3.2 Test of Universality of the Models of Mechanical Breakdown

In this section we investigate the universality of the models of mechanical breakdown for disordered solids represented by two- and three-dimensional EPNs. If such a universality exists, we may then use these models to study fracture of disordered solids without worrying much about which particular model to use or how the microscopic details of the system could affect the predictions of a particular model.

We used a BCC network to study failure phenomena with the CFs only, whereas a simple-cubic network was used to study the BB model. This allows us to test the universality of the scaling laws with respect to both the types of the network and the microscopic force laws that govern the behavior of the bonds in the network. In two dimensions we used a triangular network.

As discussed above, we introduce a threshold value ℓ_c for the length of a bond which is selected according to the probability density function

$$P(\ell_c) = (1-\gamma)\ell_c^{-\gamma}, \tag{35}$$

where we use two values of γ, $\gamma = 0.80$ and 0 [a uniform distribution in (0,1)]. These two values of γ allow us to investigate the effect of the statistical distribution of ℓ_c on the universal properties of the failure phenomena. We use this power-law distribution, because, unlike a uniform distribution (the limit $\gamma = 0$), such distributions can give rise to unusual properties for percolation networks and affect their universal properties and, therefore, we would like to see to what extent such extreme distributions can affect failure phenomena studied here. We then initiate the failure process by applying a fixed external strain on a fully-connected network in a given direction, and determine the nodal displacements $\mathbf{u}_i$ by minimizing the elastic energy of the system with respect to $\mathbf{u}_i$ for all nodes i of the network. Two different methods were used to initiate the failure process. In the first method, we selected that spring for which the ratio $\rho = \ell_m\ell/\ell_c$ is maximum (that is, ℓ is maximum), where ℓ is the current length of the spring in the strained network and ℓ_m is the maximum microscopic length of a bond in the network, and removed the spring from the system (broke it). In the second method, we selected that bond for which the ratio $\lambda = f_m\ell_c/f$ is minimum (that is, f is maximum), where f is the total microscopic force that the spring suffers, and f_m is the maximum microscopic force on a bond of the network, and removed the spring (in the case of the BB model, both f_m and f include the angle-changing forces). This second method of breaking a bond is somewhat similar to Tresca's or von Mises's classical yielding criterion for an elastic beam or spring. Breaking one bond at a time is equivalent to the assumption that the rate at which the elastic forces relax through the network is much faster than the breaking of a spring. These two methods allow us to investigate the effect of yielding criterion on the universal properties of the failure process. One can also remove all the bonds whose lengths have exceeded their thresholds (see below).

After a spring was broken, we recalculated the nodal displacements $\mathbf{u}_i$ for the new configuration of the network, selected the next spring that was to be broken, and so on. This process continued until the network finally became macroscopically disconnected. For three-dimensional systems, we used system sizes ranging from L=4 to 12 and averaged our results over many independent realizations of the network. Up to 600 independent realizations were used. Use of larger networks is currently not possible, because it would require an enormous amount of computer time.

We now address the question of universality in these models. Here, we are interested in the scaling behavior of the external stress or force and its variations with the size of the system, since this can be easily measured. To study this, we calculate the external force F that must be applied to break a bond. This force is proportional to ρY and λY in the first and second method of bond breaking, respectively, where Y is the Young's modulus. Thus, a plot of F versus ρ or λ would be similar to the traditional stress-strain curves that have been measured experimentally for many composite systems. Instead of showing the results for each model and network size L separately, we collapse the data for all values of L. Figures 7 and 8 represent the results for the microcracked CF and BB models with $\gamma = 0.80$, respectively. In these figures the bonds are broken

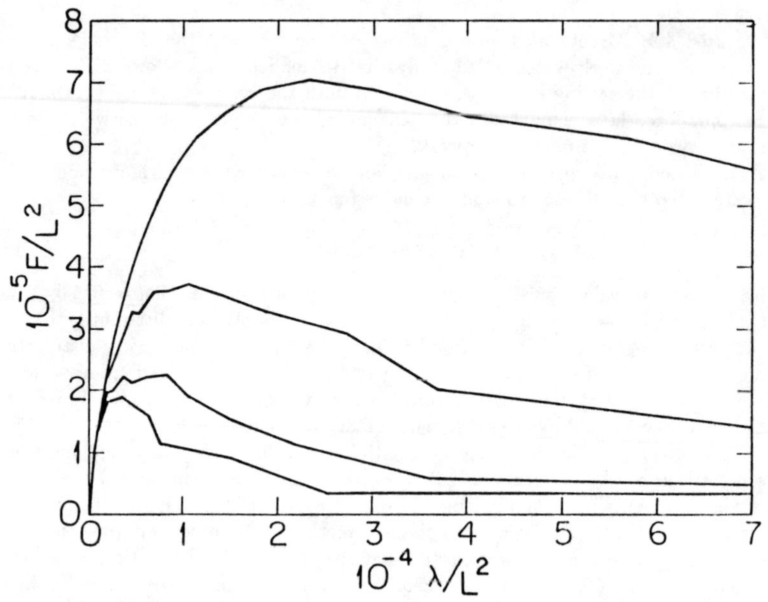

Figure 7: The collapse of the data (stress F versus strain) for the microcracked BB model in a simple-cubic network with $\gamma = 0.8$.

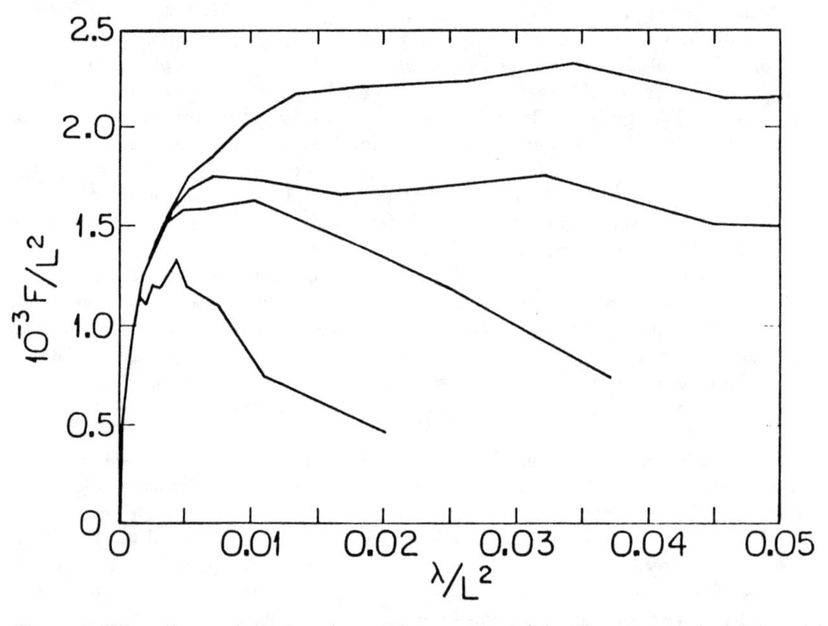

Figure 8: The collapse of the data (stress F versus strain) for the microcracked CF model in a BCC network with $\gamma = 0.8$.

according to the second method of bond breaking. However, the data collapsing is not complete and, as can be seen, there are three distinct regimes. In the first regime, which is before the maximum has been reached, and is far from it, microcracking propagates at a relatively slow rate (this is the regime of linear elasticity). As microcracking proceeds, one arrives in the second regime, which is in the vicinity of the maximum, in which intense microcraking takes place and the system is close to macroscopic failure. Beyond the maximum, the system is in the so-called post-failure regime, and is highly sensitive to small variations in λ (or ρ). These qualitative features are in agreement with direct experimental measurements and observations [68]. The shape of the curves in figures 7 and 8 are also in excellent agreement with the stress-strain curves measured by van Mier [69] for various kinds of concrete, which again indicates the usefulness of these models for investigating real systems. Moreover, as figures 7 and 8 indicate, the regimes before the maximum can be described well by the scaling law

$$F \sim L^{\Omega} h(\lambda / L^{\Omega}), \tag{36}$$

where $h(x)$ is a scaling function and $\Omega \simeq 2 \pm 0.1$. The estimated errors are only statistical, and systematic errors due to the finite size of the lattices may be significantly higher. We also considered a more complete version of the above scaling equation in order to estimate Ω more accurately. We considered

$$F \sim L^{\Omega} h(x) / (\ln L)^{\psi}, \tag{37}$$

which is suggested by theoretical arguments [56,60] and used the data to estimate Ω and ψ. We varied both Ω and ψ and found that the best data collapse was again obtained for $\Omega \simeq 2$ and $\psi \simeq 0.1$. We find the values of Ω and ψ to be insensitive to γ, network type (BCC or simple cubic), the microscopic force law (with or without BB forces) or the bond breaking method. More details can be found elsewhere [70].

To further test these results, we also calculated the stress-strain diagram with the BB model in the triangular network. In figure 9 we show the data collapse with $\Omega = 1$ for the two sizes $L = 50$ and 70 of a triangular network with $\beta/\alpha = 0.01$. It is clear that the data collapsing is very good up to the maximum in the curves. We also looked at the variation of F with N_c, the number of broken bonds during fracture, and tried to collapse the data for F and N_c for the two sizes $L = 50$ and 70. We used a similar scaling relation as in equation (36) [or (37) with a small ψ],

$$F \sim L^{\Omega_1} h(N_c / L^{\Omega_2}). \tag{38}$$

We found that the best data collapse is provided by $\Omega_1 \simeq 1$ and $\Omega_2 \simeq 1.7$, consistent with our results discussed above. Thus, we propose that $\Omega = d - 1$ for a d-dimensional system, with the possibility of a small value for ψ, as suggested by the theory [56,60]. The value $\Omega_2 \simeq 1.7$ represents the fractal dimension of the macroscopic crack. For the three-dimensional system, we find [70], $\Omega_2 \simeq 2.7$, i.e., $\Omega_2(d) = \Omega_2(d-1) + 1$, which is intuitively expected.

We should mention here the results of de Arcangelis *et al.* [71] who also studied the fracture of two-dimensional elastic media. These authors suggested that $\Omega \simeq \Omega_1 \simeq 0.75$ and $\Omega_2 \simeq 1.7$. Their value of Ω_2 is in complete agreement with ours. Although these authors claim that they find the same value of Ω and Ω_1 for the scalar and vector (elastic) models and quote their previous results [61], de Arcangelis and Herrmann [61] give $\Omega \simeq \Omega_1 \simeq 0.9$ which, within the estimated error, *is* consistent with our results, but *not* with $\Omega \simeq \Omega_1 \simeq 0.75$. Moreover, their results can be fitted to equation (37) with $\Omega \simeq 1$ with essentially the same accuracy, or better. Thus, simulations with much larger systems are necessary to obtain a more accurate estimate of Ω.

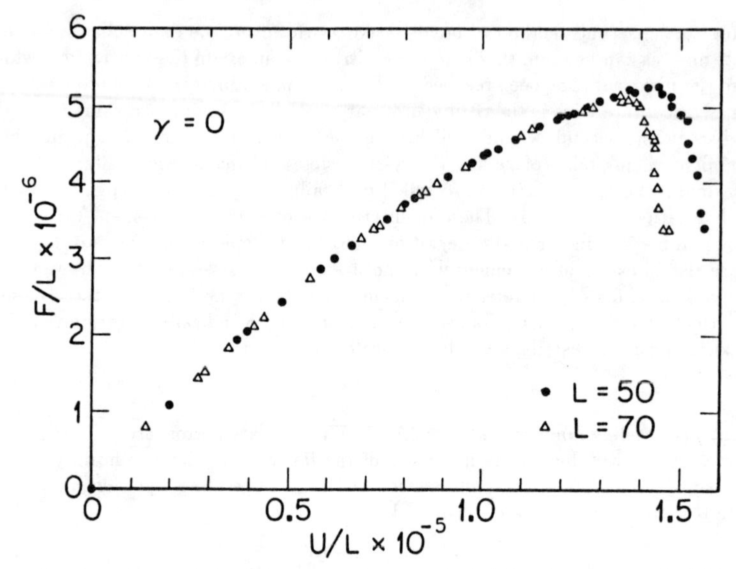

Figure 9: The collapse of the data (stress F versus strain) for the microcracked BB model in a simple-cubic network with $\gamma = 0$.

3.3 Universal Fixed Points in Fracture Processes

In this section we address four questions which we believe are of fundamental importance to the failure of disordered systems. (i) What is the *signature* (or a distinct property) of a failing system close to its macroscopic failure point? If such a signature does exist, it may help one to detect and prevent the catastrophic failure of a system. On the other hand, in some systems we would like to make sure that macroscopic cracks *are* developed. For example, in order to increase oil production artificial fractures are created in oil reservoirs to increase the permeability of the system, and it is highly desirable to know whether such macroscopic fractures have been created. (ii) If there does exist a signature of a failing system, how *universal* is it? Does it depend on the microscopic properties of the system or the dynamics of failure and is material dependent? (iii) In general, the growth of cracks in a disordered system is a non-linear phenomenon. On the other hand, linear properties of disordered systems are usually modeled by percolation networks of resistors or elastic bonds, in which the bonds are cut *at random*. Percolation phenomena represent second-order phase transitions (i.e. all effective properties vanish or diverge continuously at p_c), whereas at least some of the fracture phenomena studied by our models resemble first-order (discontinuous) phase transitions. However, under certain experimental conditions the accumulation of damage and the growth of cracks can be essentially at random as in, e.g. a system which is under rapid thermal cycling. In such a situation, a percolation model may be appropriate for describing the damage process. But, in the models of mechanical breakdown that we have studied so far, the failing system is under an external uniform *load* (stress or strain), and the breakdown of the bonds is not random, but depends on the stress or strain field in the network. Therefore, it is important to know the extent of similarities between the properties of such networks and those of a percolation process. (iv) How can we classify the universality classes of fracture phenomena, and how many universality classes are there?

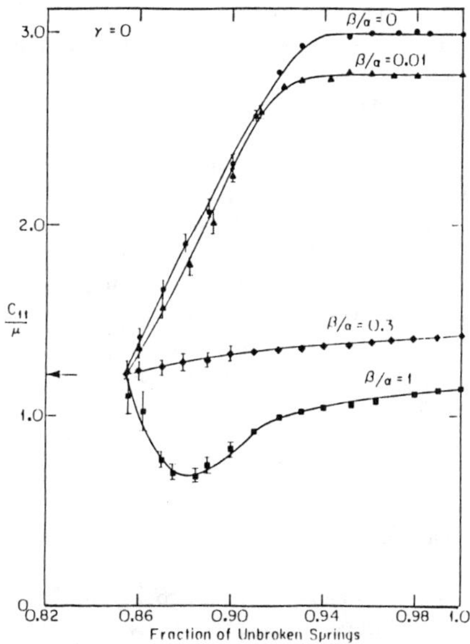

Figure 10: The ratio C_{11}/μ versus the fraction of unbroken springs with model 1 of fracture.

In order to provide at least partial answers to such questions, we used a triangular network with the BB model. We then introduced a threshold value ℓ_c for the length of a bond, which was selected according to the probability density function given by the equation (35), where we again used $\gamma = 0$ and 0.80, to investigate the effect of the statistical distribution of ℓ_c on the possible universality of the properties that we study. Two different methods were used to model the failure process. In the first method, we selected that spring for which the difference $\rho = \ell - \ell_c$ is maximum, where ℓ is the current length of the spring in the strained network, and removed the spring from the system (broke it). We call this model one (M1). In the second method, we removed *all* the bonds whose lengths have exceeded their thresholds, and refer to it as model two (M2). During the fracture process, we calculated three properties of the network. We first distributed the threshold values ℓ_c and measured the elastic modulus C_{11} of the network during the fracture process. We then used the *same* fully connected network (with the *same* values of ℓ_c), and calculated the shear modulus μ of the network during the fracture process. This is equivalent to using two *identical* samples for measuring C_{11} and μ. The results presented below are for $L = 70$, for which we used a large number of different realizations (to reduce the fluctuations) and averaged the results. We also used network sizes $L = 40$, 50 and 90, in order to assess the effect of sample size on the results.

In figure 10 we present the ratio $r = C_{11}/\mu$ as a function of the fraction of unbroken springs, for various values of β/α. These results were obtained by using M1 of fracture. The last points on these curves represent C_{11}/μ right before the system fails macroscopically. We refer to this as the incipient fracture point (IFP). As can be seen, even though the initial states of the systems are different, they all approach the same value as the IFP is approached. Note that initially r remains essentially constant, i.e. it is *not* sensitive to a few cracks or a collection of localized cracks. However, as damage accumulates and the cracks grow, r changes drastically. Because $\beta/\alpha = 0$ corresponds to a system in which only CFs are present, figure 10 indicates that this behavior is independent of the microscopic force laws of the system. The behavior of the system for $\beta/\alpha = 1$ is particularly interesting. Initially, r remains essentially constant. However, as damage accumulates a turning point (TP) appears beyond which r decreases and reaches a minimum. But near the IFP, r rises again and approaches the value at the IFP which appears to be the same for all values of β/α.

To check whether this behavior depends on the dynamics of failure in systems that are under an external load, or the distribution of the threshold values, similar simulations with M2 were carried out for $\gamma = 0$ and 0.8. In all cases, we found r to approach the same value as that shown in figure 10 (the details are given elsewhere [72]). We should point out that because in M2 one breaks several bonds at each step of the simulation, the fluctuations among the various realizations are also larger than those in M1, especially when $\gamma > 0$. Moreover, the geometry of the macroscopic fracture is very different in M1 and M2, particularly for $\gamma = 0$ and 0.8. Thus, we concluded that for *all* values of γ and β/α, and regardless of the dynamics of fracture, one has

$$\frac{C_{11}}{\mu} \simeq 1.25 \tag{39}$$

That is, as the IFP is approached, one has a *universal* fixed point.

The appearance of a universal fixed point means that for a broad class of materials (including some ceramics, concretes and sedimentry rocks) in which fracture instability occurs from a *non-local* regions of the materials, the approach of r to the fixed point value can be interpreted as the *signature* of a failing system. Although figure 11 indicates that for certain values of β/α one may have a non-monotonic variation of r with the accumulated damage, for most real systems one has [73] $\beta/\alpha \leq 0.3$, and for such values of β/α the approach of r to the fixed point is *always* monotonic. For the triangular lattice used here it can be shown that for the uniform network with no bonds broken, $r = (3 + 12\beta/\alpha)/(1 + 12\beta/\alpha)$ so that for $\beta/\alpha > 1/12 \simeq 0.083$ the Poisson's ratio ν_p will be negative, and since for *isotropic* systems $\nu_p > 0$ (although in theory one can have $\nu_p < 0$), only systems with $0 \leq \beta/\alpha \leq 1/12$ are of interest and for this range of β/α the approach of r to the fixed point is always monotonic.

What is the theoretical explanation for this apparent universality of r? Similar to the elastic moduli of EPNs, one expects C_{11} and μ to follow the same type of law as the IFP is approached (e.g. they both obey scaling laws with the *same* critical exponents in M2). As such r represents an *amplitude ratio.* It is known [74] from statistical mechanics and percolation that the amplitude ratios are *universal.* The fact that we obtain a fixed value of r for the fracture problem indicates that, not only the amplitude ratios in the fracture problem are also universal, they are also indicative of the possibility that one may be able to map this problem onto an equivalent statistical mechanical system.

Let us now discuss the possible similarities between the models simulated here and percolation networks. To begin with, we recall that for EPNs near p_c, C_{11}/μ approaches a fixed point, *independent of the microscopic details of the system* [42,75]. For two-dimensional isotropic systems near p_c one has [75] $C_{11}/\mu \simeq 3$. Although for fractured networks considered here, the value of C_{11}/μ at the IFP is different from that of percolation networks at p_c, the fact that in both systems C_{11}/μ approaches a universal value indicates that the two systems share some common features. This qualitative similarity also supports the idea that, under certain conditions, EPNs may describe some fracture processes.

The existence of a fixed point in a fractured system can be directly tested by experimental measurements. For example, for granular porous media, Schwartz, Johnson and Feng [76] proposed a model that can predict many experimental features of such systems. The percolation properties of this model were shown [75] to be very similar to those of EPNs with the BB model and, in particular, as p_c of the granular medium is neared, C_{11}/μ approaches a fixed point. Therefore, we expect a fracturing granular porous medium to show fixed point behavior similar to what we find here. Experimentally, this can be tested, since C_{11}/μ is directly related to $(V_c/V_s)^2$, where V_c and V_s are the velocities of the compressional and shear waves in the medium, respectively, which can be measured by established experimental techniques.

Based on our study, we propose that the value of r at the fixed point can be used to classify various universality classes of fractured systems. Specifically, we propose that there are *two* distinct universality classes. One is for systems that are under a uniform external load (stress or strain) in which the growth of a crack at a point depends on the stress field in the system and, therefore, the damage accumulation is *not* random. Such systems are described by the fixed point found here. The second one is for systems in which damage accumulates essentially at random. Such systems are described by the fixed point of EPNs at p_c.

4. SUMMARY

While scalar transport properties of disordered materials, modelled by percolation networks, mostly follow universal scaling laws near the percolation threshold, vector transport properties may not possess such universal behavior and may be dependent on the microscopic force laws of the system. On the other hand, fracture properties of materials may obey universal scaling laws, and show fixed point behavior, much like those in the renormalization group theory of critical phenomena.

ACKNOWLEDGMENTS

The authors are grateful to the National Science Foundation and the Air Force Office of Scientific Research for partial support of the research reported here. We are also grateful to the San Diego Supercomputer Center for providing much needed computer time on Crays X-MP and Y-MP.

REFERENCES

[1] D. Stauffer, *Introduction to Percolation Theory*, Taylor and Francis, London, (1985).

[2] S. Kirkpatrick, *Rev. Mod. Phys.*, **45**, 574 (1973).

[3] G. Ord, B. Payandeh and M. Robert, *Phys. Rev.*, **B 37**, 467 (1988).

[4] J. -M. Normand, H. J. Herrmann and M. Hajjar, *J. Stat. Phys.*, **52**, 441 (1988).

[5] J. G. Zabolitzky, *Phys. Rev.*, **B 30**, 4077 (1984).

[6] J. G. Zabolitzky, D. J. Bergmann and D. Stauffer, *J. Stat. Phys.*, **44**, 211 (1986).

[7] R. B. Pandey, D. Stauffer, A. Margolina and J. G. Zabolitzky, *J. Stat. Phys.*, **34**, 877 (1984).

[8] M. Sahimi, B. D. Hughes, L. E. Scriven and H. T. Davis, *J. Phys. C*, **16**, L521 (1983).

[9] C. J. Lobb and M. Forrester, *Phys. Rev.*, **B 35**, 1899 (1987).

[10] F. Drove, in *Mechanics of Engineering Materials* (C. S. Desai and Gallagher, eds.), John Wiley and Sons (1984), p. 179.

[11] J. W. Rudnicki and J. R. Rice, *J. Mech. Phys. Solids*, **23**, 371 (1975).

[12] J. K. Knowles and E. Steinberg, *J. Elasticity*, **8**, 329 (1978).

[13] P. V. Lade, *IUTAM Conference on Deformation and Failure of Granular Materials*, Delft (1982), p. 641.

[14] R. D. Mindlin, *Proc. 2nd U.S. National Congr. Appl. Mech.*, Ann Arbor, Michigan (1954), p. 8.

[15] J. D. Goddard, *Proc. R. Soc. London A*, **430**, 105 (1990).

[16] K. K. Mohanty, Ph.D. Thesis, University of Minnesota, Minneapolis (1981).

[17] D. Stauffer, *J. Chem. Soc. Faraday Trans. II*, **72**, 1354 (1976).

[18] D. Deptuck, J. P. Harrison and P. Zawadzki, *Phys. Rev. Lett.*, **54**, 913 (1985).

[19] S. Mall and W. B. Russel, *J. Rheol.*, **31**, 651 (1987).

[20] M. Sahimi, G. R. Gavalas and T. T. Tsotsis, *Chem. Eng. Sci.*, **45**, 1443 (1990).

[21] H. He and M. F. Thorpe, *Phys. Rev. Lett.*, **54**, 2107 (1985).

[22] M. Sahimi and J. D. Goddard, *Phys. Rev.*, **B 32**, 1869 (1985).

[23] S. Arbabi and M. Sahimi, *Phys. Rev. Lett.*, **65**, 725 (1990).

[24] G. R. Jerauld, Ph.D. Thesis, University of Minnesota, Minneapolis (1985).

[25] M. Sahimi and S. Arbabi, *Phys. Rev.*, **B 40**, 4975 (1989).

[26] S. Roux and A. Hansen, *Europhys. Lett.*, **6**, 301 (1988).

[27] S. Arbabi and M. Sahimi, *Phys. Rev. B*, to be published (1991).

[28] J. G. Kirkwood, *J. Chem. Phys.*, **7**, 506 (1939).

[29] P. N. Keating, *Phys. Rev.*, **152**, 774 (1966).

[30] J. C. Phillips and M. F. Thorpe, *Solid State Commun.*, **53**, 699 (1985).

[31] V.K.S. Shante and S. Kirkpatrick, *Adv. Phys.*, **20**, 325 (1971).

[32] S. Feng, M. F. Thorpe and E. J. Garboczi, *Phys. Rev.*, **B 31**, 276 (1985).

[33] S. Feng and M. Sahimi, *Phys. Rev.*, **B 31**, 1671 (1985).

[34] M. Knackstedt, M. Robert and B. Payendeh, *preprint* (1990).

[35] J. P. Straley, *Phys. Rev.*, **B 15**, 5733 (1977).

[36] M. Sahimi and S. Arbabi, *J. Stat. Phys.* , **62**, 453 (1991).

[37] D. B. Gingold and C. J. Lobb, *Phys. Rev.*, **B 42**, 8220 (1990).

[**38**] J. -M. Normand and H. J. Herrmann, *Inter. J. Mod. Phys. C*, to be published (1991).

[**39**] S. Arbabi and M. Sahimi, *J. Phys. A*, **21**, L863 (1988).

[**40**] S. Arbabi and M. Sahimi, *Phys. Rev. B*, to be published (1991).

[**41**] A. Hansen and S. Roux, *Phys. Rev.*, **B 40**, 749 (1989).

[**42**] S. Arbabi and M. Sahimi, *Phys. Rev.*, **B 38**, 7173 (1988).

[**43**] S. Arbabi and M. Sahimi, *Macromolecules*, to be published (1991).

[**44**] S. Arbabi and M. Sahimi, *Phys. Rev. B*, to be published (1991).

[**45**] M. Sahimi, *J. Phys. A*, **17**, L601 (1984).

[**46**] A. B. Harris, S. Kim and T. C. Lubensky, *Phys. Rev. Lett.*, **53**, 743 (1984).

[**47**] M. Sahimi, *J. Phys. C*, **19**, L79 (1986).

[**48**] H. J. Herrmann and S. Roux (eds.), *Statistical Models for the Fracture of Disordered Media*, North-Holland, Amsterdam (1990).

[**49**] A. A. Griffith, *Phil. Trans. R. Soc. Lond.*, **221**, 163 (1921).

[**50**] Z. P. Bažant, *ASME Appl. Mech. Rev.*, **39**, 675 (1986).

[**51**] L. de Arcangelis, S. Redner and H. J. Herrmann, *J. Physique*, **46**, L585 (1985).

[**52**] L. Niemeyer, L. Pietronero and H. J. Weismann, *Phys. Rev. Lett.*, **52**, 1023 (1984).

[**53**] H. Takayasu, *Phys. Rev. Lett.*, **54**, 1099 (1985).

[**54**] M. Sahimi and J. D. Goddard, *Phys. Rev.* , **B 33**, 7848 (1986).

[**55**] P. M. Duxbury, P. L. Leath and P. D. Beale, *Phys. Rev. Lett.*, **57**, 1052 (1986).

[**56**] P. M. Duxbury, P. D. Beale and P. L. Leath, *Phys. Rev.*, **B36**, 367 (1987).

[**57**] P. M. Duxbury and P. L. Leath, *J. Phys. A*, **20**, L411 (1987).

[**58**] P. D. Beale and D. J. Srolovitz, *Phys. Rev.*, **B 37**, 5500 (1988).

[**59**] G. N. Hassold and D. J. Srolovitz, *Phys. Rev.*, **B 39**, 9273 (1989).

[**60**] B. Kahng, G. G. Batrouni, S. Redner, L. de Arcangelis and H. J. Herrmann, *Phys. Rev.*, **B 37**, 7625 (1988).

[**61**] L. de Arcangelis and H. J. Herrmann, *Phys. Rev.*, **B 39**, 2678 (1989).

[**62**] H. J. Herrmann, A. Hansen and S. Roux, *Phys. Rev.*, **B 39**, 637 (1989).

[**63**] K. Sieradzki and R. C. Newman, *Phil. Mag. A*, **51**, 95 (1985).

[**64**] M. Sahimi and S. Arbabi, *Phys. Rev. B*, to be published (1991).

[**65**] M. D. Stephens and M. Sahimi, *Phys. Rev.*, **B 36**, 8656 (1987).

[**66**] D. Sornette, *J. Physique*, **48**, 1843 (1987); **49**, 889 (1988).

[67] I. C. van den Born, A. Santen, H. D. Hoekstra and J. Th. M. De Hosson, *Phys. Rev.*, **B43**, 3794 (1991).

[68] W. F. Brace and A. S. Orange, *J. Geophys. Res.*, **73**, 1433 (1968).

[69] J. G. M. van Mier, *Matériaux et Constructious*, **19**, 179 (1986).

[70] S. Arbabi and M. Sahimi, *Phys. Rev.*, **B 41**, 772 (1990).

[71] L. de Arcangelis, A. Hansen, H. J. Herrmann and S. Roux, *Phys. Rev.*, **B 40**, 877 (1989).

[72] M. Sahimi and S. Arbabi, *Phys. Rev. Lett.*, to be published (1991).

[73] J. L. Martin and A. Zunger, *Phys. Rev.*, **B 30**, 6217 (1984).

[74] A. Aharony, *Phys. Rev.*, **B 22**, 400 (1980).

[75] L. M. Schwartz, S. Feng, M. F. Thorpe and P. N. Sen, *Phys. Rev.*, **B 32**, 4607 (1985).

[76] L. M. Schwartz, D. L. Johnson and S. Feng, *Phys. Rev. Lett.*, **52**, 831 (1984).

THE POROSITY DEPENDENCE OF MECHANICAL AND OTHER PROPERTIES OF MATERIALS

Roy W. Rice, W. R. Grace and Co.-Conn., 7379 Route 32, Columbia, MD 21044

ABSTRACT

Though the volume fraction (or %) porosity is commonly used as the exclusive porosity variable to describe the porosity dependence of physical properties, it is not sufficient. The average minimum solid area, i.e., of the bond or sintered area between sintering particles, or the minimum web or strut cross-sections is the most appropriate second porosity parameter. Pore shape-stress concentration effects have, at best, limited direct effect on mechanical properties, but generally correlate with minimum solid area effects. Methods of combining effects of different types of porosity within the same body are important and result in quite reasonable descriptions of mechanical properties across a broad range of porosities.

INTRODUCTION

Despite extensive study, understanding of the porosity dependence of material properties is incomplete. Key issues and neglects have been 1) Identifying parameters necessary to adequately characterize porosity-property effects, 2) Combining effects of different porosities within a given body, and 3) Adequate determination and use of porosity characterization. This paper summarizes recent efforts to address these issues.

POROSITY PARAMETERS

The volume fraction (or %) porosity (P) is appropriately, universally used in evaluating porosity-property relations, often as the only porosity parameter. A second porosity parameter; e.g., pore shape, whether the porosity is open or closed, or bond area between sintering particles, has been used, but whether one is needed, and if so what is most appropriate has not been systematically evaluated. P by itself is not a complete characterization of the porosity for property dependance. A second parameter, most appropriately the average minimum solid area, is needed.[1]

There are two basic methods of making a porous body; i.e. by: 1) partially compacting and sintering or bonding particles, or 2) introducing isolated or interconnected pores, e.g., using gas bubbles (as for a foam) or fugitive particles (as for refractory bricks) to hold particles of the body in place till bonding occurs. The first method gives pores similar to or smaller than the size of the particles, while the second one has no specific relationship between the microstructure of the webs defining the individual pores and the pores. These two methods can be combined, e.g., packing and bonding of small balloons or porous (or foam) particles.

The above two basic methods can be idealized as stacking uniformly shaped and sized cells (containing a central solid particle or pore) in various arrays with varying degrees of intersection of the particles (i.e. sintering or bonding) or of the bubbles (i.e. various degrees of interconnection of the porosity). Both of these methods of constructing idealized porous bodies follow the same S-shaped curve of volume fraction (or %) porosity versus the normalized pore to surrounding solid size ratios.[1] Since there is a unique one-to-one relationship between P and this ratio, the latter is not an independent variable. On the other hand, pore shape and the degree of interconnectiveness are a function not only of P, but also of the initial stacking of the particles or pores. Since they do not uniquely correlate with P, they thus represent some independent aspects of the porosity. However, the average minimum solid area (i.e., bond area between sintering particles or minimum web or strut area) combines these remaining variables. Further, this area is generally the most determinable parameter by physical and mathematical means, and should most broadly correlate with various physical properties, and is thus the most appropriate second parameter.[1] Further, most, if not all, minimum solid area models can be closely approximated over most of their range of applicability by an exponential function; i.e., e^{-bP} where b is the parameter relating to the minimum solid area.[2,3] Thus, this offers a convenient and effective mechanism of handling and comparing data.

PORE SHAPE-STRESS CONCENTRATION VERSUS MINIMAL SOLID AREA

While various mechanical (as well as other properties, e.g., electrical or thermal connectivities) obviously correlate with the minimal solid area, many have used pore shape-stress concentration effects, usually maximum ones, for mechanical property-porosity models. However, recent evaluations seriously question whether such effects fundamentally determine mechanical properties. Consider first elastic properties for which stress concentration models have been frequently derived and used. Introducing porosity into an otherwise homogeneous, isotropic body results in two changes in strain under stress relative to the dense body of the same material; namely: 1) a net increase in the average, i.e., macro, strain of the body, and 2) a range of micro strains depending upon the character of porosity in the body instead of a uniform micro strain equal to the macro strain. Maximum pore stress concentrations clearly determine the micro strain range within the body, but have, at best, an indirect effect on average body response, e.g., macro strain, due to: 1) varying spacial orientations and distributions of these concentrations and 2) the maximum being only a small part of the continuum of stress concentrations about each pore. Further, minimum solid area models generally predict similar, but more accurate, description of elastic behavior than stress concentration effects.[2] For example, one of the most widely used stress concentration models predicts elastic properties going to 0 as pores become cracks, while both a variety of recent theoretical developments[4] and experimental determinations show this not to be the case, but instead to lead to values generally consistent with those predicted by common minimal solid area models.[5]

There are two other key cases of important stress concentration effects in solid bodies showing that they generally play no role in determining elastic properties. While almost all crystalline materials have substantial elastic anisotropy, resulting in significant stress concentrations (and, hence, variable strains) at or near grain boundaries in a stressed body, such effects play no significant role in determining the elastic properties of such polycrystalline solids. These are, instead, simply given by a spatial averaging of the single crystal elastic properties. Similarly, built-in stresses due to thermal expansion anisotropy, or phase transformation, in solids result in built-in boundary strains, but have no effect on elastic behavior until such stresses result in cracks.

Fracture energy is closely related to Young's modulus, so they have very similar P dependance, as does fracture toughness. Thus, tensile strength also typically has very similar porosity dependance to Young's modulus, since the only other major parameter determining it besides fracture toughness is flaw size (C),[6] and C generally doesn't depend on P, e.g., at high P pores are typically much finer than C (and stress concentrations are again not significant). When large pores act as failure sources, they usually do so in conjunction with surrounding flaws, thus significantly moderating any stress concentration effects from the pore itself.[7] The only mechanical properties that are probably significantly dependent on pore stress concentrations are with significant compressive loading, e.g., compressive strength, hardness, and wear since failure in these is often a cumulative damage process, or may involve slip, enhanced by stress concentrations.[2,3]

COMBINATIONS OF POROSITY

A serious limitation of past property-porosity models is their directly or indirectly assuming significant constraints on the amount or character of porosity, or both, generally assuming one particular type of pore.[2] However, most bodies are made up of two or more different types of porosity, e.g., most bodies made by partially sintering compacted particles commonly have two or more pore structures from either (or both) variable particle sizes and stacking. Further, the highest porosity achievable by either random or uniform stacking of particles is ~ 50%,[1] so any higher porosities achieved by loosely bonding particles together must involve pores between, and smaller than, the particles, as well as pores comparable to, or larger than, the size of the particles being sustained in the forming process, e.g., by fugitive particles or gas pressure.

Another important reason for combining effects of different pore structures is the inherent porosity changes with changes in processing to change porosity levels. In generating porosity by changing the degree of bonding between packed particles, there is almost always an inherent increase in the packing density of the particles, e.g., from particle coordination numbers about 6-8 at high P, and ~ 12-14 at P~0. Past minimum solid area (as well as other) models were often based on a given stacking, i.e., coordination number, so changing combinations of these as densification occurs must be accounted for to accurately describe the porosity dependance of properties as sintering occurs. Similarly, for higher P in foams, the structure becomes progressively more open and pores typically progress from an essentially ideal spherical shape to a prismatic shape, which is commonly approximated by a cylinder (or even more accurately by a cylinder with hemispherical or polyhedral end caps).

Effects of different porosities can be combined by the rule of mixtures (i.e., a Voight approach for elastic elements in parallel) or a Reuss approach, e.g., for elastic objects in series. Both give similar results if not very dissimilar effects are being combined, in which case the rule of mixtures is the easier method. However, when there are large disparities in properties, especially where one type of porosity gives a zero property value, then the rule of mixtures generally gives erroneous results. A third possible combination method is to treat effects of the two porosities as the product of the two since the porosity dependance is commonly given as the property of the matrix material times a function of porosity, hence, this is treating one porosity as part of the matrix.

Recent evaluation shows substantial support for such pore structure changes and modeling of these via combinations of different porosity models.[3] Thus, Atkin's Young's modulus data for gel-derived silica bodies[8] cuts across individual models for stacking particles with decreasing coordination numbers, subsequently approaches, and crosses over a model for stacked spherical pores, i.e., approaching a cylindrical pore model in the fashion expected as porosity increases.[3] Similarly, data of Walsh, et al,[9] for pores formed by sintering glass particles agrees over much of its range with minimal solid area models for spherical pores,[3] then passes beyond this and approaches minimum solid area models based on cylindrical pores,[3] as expected.

CHARACTERIZATION AND MEASUREMENT NEEDS

Most porosity-property studies are limited in their utility by the absence of important information, much of which is not necessarily difficult to obtain. A simple and fundamental need for porosities derived by partially bonding of packed powders is the green (or alternatively bisque fired) densities. These define, at least approximately, the initial particle coordination number and, hence, what minimum solid area the system starts from. Particle size distributions and compaction pressures can also be of value. Similarly, for foam-type materials the degree of interconnectiveness and sizes and shapes of the pores, can be very valuable. In either type of system, but probably most important for porosity in partially bonded powder compacts, is homogeneity. While typically difficult to directly characterize, indications of inhomogeneity, and its approximate degree, can often be readily obtained by measuring physical properties in different areas of a given sample, or by measurements on progressively smaller samples cut from larger samples. Measuring more than one related physical property, e.g., Young's and shear modulus, as well as different physical properties is also important.

Unfortunately, such information has often not been used in evaluating data, e.g., in Dean and Lopez[10] fit of various elastic property-porosity models to ceramic data. Their model evaluation is seriously questioned since: 1) one study providing some of the most significant discrimination between different models had shown significant effects of porosity inhomogeneity at higher porosities and 2) one of the next most significantly discriminating studies showed Poison's ratio increasing with increasing porosity, an effect not generally seen in ceramic bodies made by powder compaction, and not predicted by models for such bodies.

SUMMARY AND CONCLUSIONS

Besides volume fraction (or %) of P, a second parameter is needed to describe porosity-property effects. Average minimum solid area is the most appropriate one. Stress concentration models are generally not appropriate for even most mechanical properties. Minimum solid area models for different porosity combinations show good correlation with available studies. Better characterization and comparison of different measurements are important.

REFERENCES

1. R. W. Rice, "Evaluating Porosity Parameters for Property-Porosity Relations", Am. Ceram. Soc. Annual Meeting (1990).

2. R. W. Rice, "Comparison of Stress Concentration Versus Minimum Solid Area Based Mechanical Property-Porosity Relations", submitted for publication to J. Am. Ceram. Soc.

3. R. W. Rice, "Microstructure Dependance of Mechanical Behavior of Ceramics" in Treatise on Materials Science and Technology, Vol. II, Academic Press (1977).

4. N. Laws and J. R. Brockenbrough, Int. J. Solids Structures, 23, 9, 1247-1268 (1987).

5. R. W. Rice, "Evaluation and Extension of Physical Property-Porosity Models Based on Minimum Solid Areas", Am. Ceram. Soc. Annual Meeting (1990).

6. R. W. Rice, Mat. Sci. and Eng., A112, 215-224 (1989).

7. R. W. Rice, J. Mat. Sci., 19, 895-914 (1984).

8. D. Ashkin, "Properties of Bulk Microporous Ceramics", Ph.D. thesis, Rutgers University (1990).

9. J. B. Walsh, W. F. Brace, and A. W. England, J. Am. Ceram. Soc., 48, 12, 605-608 (1965).

10. E. A. Dean and J. A. Lopez, J. Am. Ceram. Soc., 66, 5, 366-370.

OPTIMAL SELECTION OF FOAMS AND HONEYCOMBS IN PACKAGING DESIGN

J. Zhang[1] and M. F. Ashby[2]
1. Thayer School of Engineering, Dartmouth College, Hanover, NH 03755.
2. Cambridge University Engineering Department, Trumpington St., Cambridge CB2 1PZ, U.K.

ABSTRACT

The volume of foams used in packaging is enormous. Proper design requires identifying the right material and selecting the right density for each particular application. A new approach to package design is presented in the form of "the Packaging Selection Diagram", from which the optimal density of a cellular material can be obtained once the maximum permitted stress of the packaging is known. This approach offers greater generality and simplicity than the existing methods such as the Janssen factor or the Energy Absorption Diagram.

1. Introduction

The most common use of foams is in packaging. Packaging surrounds most commodities ranging from missiles to food. The essence of packaging is the ability to absorb energy while keeping the peak force on the packaged object below the limit which will cause damage or injury. Cellular materials (such as foams) are especially good at this. Their energy absorption capacity is compared to that of the solid in Fig. 1a [after Gibson and Ashby, 1988]: for a given energy absorption, the cellular material often generates a lower peak force than the solid. The energy absorbed by the foam per unit volume to strain ε is simply the area under the stress-strain curve up to ε (shown as shaded area in Fig. 1a). As the figure shows, it is the long plateau of the stress-strain curve, arising from cell collapse by elastic buckling, plastic yielding or brittle crushing, which allows large energy absorption at near-constant load. To absorb energy at a near-constant load, one needs to choose the right cell-wall material and the right relative density for the foam. Selecting the cell wall material is relatively simple. One needs to consider whether the packaging material carries static or repeated loading or is subjected to severe environmental conditions such as high temperature. An elastomeric cell wall material is needed for packaging which will be subjected to repeated loading. If the protection is needed only once, a plastic or brittle material is better because a cellular material made out of either one is more efficient (details given later). Choosing the right density for a given package is a more difficult area where this paper tries to add some understanding. If the density is too low, the cells are crushed before enough energy has been absorbed. If the density is too high, the stress exceeds the critical value before enough energy has been absorbed. This is best explained in Fig. 1b (taken from Gibson and Ashby, 1988), which compares the performance of foams with three different densities.

Four main ways of characterizing the energy absorption of cellular materials have been proposed. These are the Janssen Factor, J; the Cushion Factor, C; the Rusch Curve; and the Energy Absorption Diagram. In this paper, we first review all existing methods. Then we look at the physics behind the normalization of the mechanical properties. In section 4, "Packaging Selection Diagrams" for elastic foams, plastic foams, elastic honeycombs and plastic honeycombs are presented and compared with experimental data. A comparison is made between the energy-absorbing capacities of foams and honeycombs by plotting their packaging-selection diagrams together.

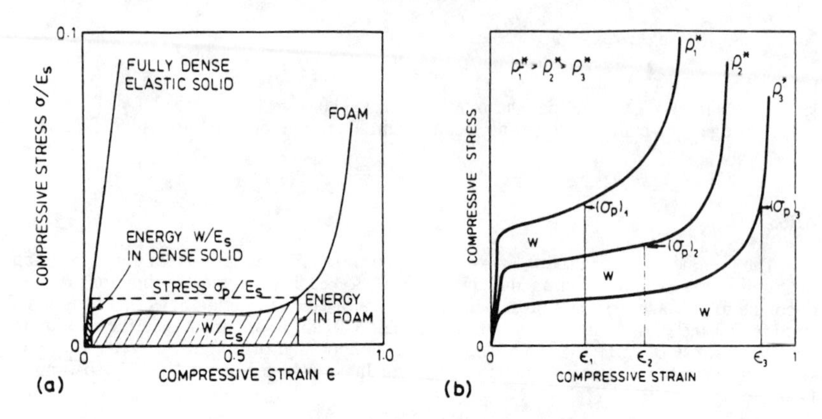

Fig. 1 (a) Stress-strain curves for an elastic solid and a foam made from the same solid, showing the energy per unit volume absorbed at a peak stress σ_p; (b) The peak stresses generated in foams of three densities in absorbing the same energy W are given by $(\sigma_p)_1$, $(\sigma_p)_2$ and $(\sigma_p)_3$. The lowest-density foam "bottoms out" before absorbing the energy W, generating a high peak stress. The highest-density foam also generates a peak stress before absorbing the energy W. Between these two extremes, there exists an optimal density, which absorb the energy W, at the lowest peak stress. (after Gibson and Ashby, 1988)

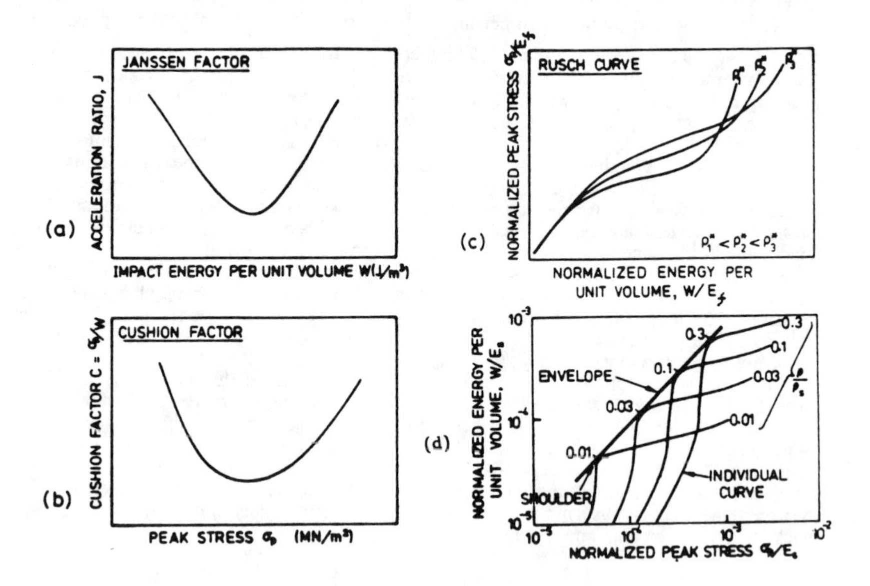

Fig. 2 Four Diagrams which are used to characterize energy absorption
(a) the Janssen Factor, J; (b) the Cushion Factor, C;
(c) the Rusch Curve (d) the Energy Absorption Diagram

2. Literature Review

There is considerable literature on the use of foams for cushioning and packaging. The interested reader might wish to consult the books by Mustin (1968), by Hilyard (1982) and by Gibson and Ashby (1988), the series of papers by Lockett, Cousins and Dawson (Cousins, 1976a, b; and Lockett et al, 1981), the papers by Green et al (1969), by Rusch (1970, 1971), by Lee and Williams (1971), by Melvin and Roberts (1971), by Schwaber and his co-workers (Meinecke and Schwaber, 1970; Meinecke et al, 1971; Schwaber and Meinecke, 1971; Schwaber, 1973) and by Maiti et al, (1984). We summarize their approaches in the following paragraphs.

(a) The Janssen Factor, J

According to Woolam (1968), the efficiency of a cushioning material at absorbing energy, J, can be defined as the ratio of the maximum acceleration experienced by the material, a_m, to the acceleration which would be experienced by an "ideal" absorber, a_i:

$$J = a_m/a_i \tag{1}$$

This "ideal" absorber can absorb energy at constant force, F, and can deform completely, so its acceleration is

$$a_i = F/m = mv^2/2hm = v^2/2h \tag{2}$$

where v and m are the initial speed and the mass of the cushioned object respectively, and h is the height of the cushioning material. The Janssen factor, J, is often plotted against the impact energy, W, per unit volume of the cushioning material, as shown in Fig. 2a.

(b) The Cushion Factor, C

The cushion factor, C, is the ratio of peak stress developed in the cushion to the energy stored per unit volume of the cushion. Gordon (1974) plots this factor against the peak stress by using data on the uniaxial stress-strain behavior of foams (Fig. 2b). One point is worth noting: this factor is equivalent to the Janssen factor, J, in dynamic testing such as drop weight testing

$$C = \sigma_p/W = (ma_m/A)/(mv^2/2Ah) = J \tag{3}$$

where A is the area of the cushion.

(c) The Rusch Curve

Rusch (1970, 1971) notes that the shape of the stress-strain curve can be defined by an empirical shape factor, $\psi(\varepsilon)$, in the form

$$\sigma = E_f \varepsilon \psi(\varepsilon) \tag{4}$$

where σ is the compressive stress; ε is the strain; and E_f is the Young's modulus of the foam. Rusch further defines K, the energy-absorbing efficiency, as the maximum deceleration of a material packaged by an ideal material to that packaged by the the foam under investigation, d_m:

$$K = v^2/2hd_m = 1/J \tag{5}$$

Another dimensionless quantity I is defined as the impact energy per unit volume of foam divided by E_f. The optimum foam for energy absorption for a given peak stress can be found by plotting I/K against I (Fig. 2c).

(d) Energy Absorption Diagrams

Maiti et al (1984) have further improved the Rusch method. The normalization of the peak stress and energy per unit volume (Fig. 2d) is done against the Young's modulus of the solid, which makes it more general than the Rusch method. However, the interpretation of the results from all the methods mentioned above is not really straightforward for design engineers. In this paper, we introduce a different approach — the Packaging Selection Diagram, which allows one to determine the optimal density and the energy absorbed once the maximum permitted stress is known. The diagrams can also incorporate dynamic results.

3. Post-Collapse Behavior in Foams

At first sight, it is hard to understand why the post-collapse behavior looks so much alike in some foams. What role do the cell dimensions, such as the cell edge length and the cell edge thickness, play in the overall stress-strain response? Can a scaling law be applied to cellular materials? This section addresses these issues.

First, we introduce some terminology and assumptions. When the "shape" of a cell is mentioned, we mean the scale-independent geometrical properties of the cell, so the thicknesses and lengths of the edges are irrelevant here. Cells with geometric similarity are thought to be of the same shape no matter how large the cells or how thick the cell edges. Although each cell within a foam sample can be geometrically different, foams with varying densities and cell sizes are visualized as related to each other by a simple geometric scaling. A perfect open cell is also assumed. An open-cell foam can be modelled as an array of polyhedral cells. A representative unit cell is assumed to be made of beams with equal length l and cross section t^2. The foam is considered as an aggregate of these representative units. The force P acting on each of these beams (Gibson and Ashby, 1988) can be related to the global stress on the foam:

$$P \propto \sigma l^2 \qquad (6)$$

First consider the post-collapse behavior of flexible foams under uniaxial compression. If we look at a typical beam AB (shown in Fig. 3), its curvature, $d\theta/ds$, is given by

$$E_s I d\theta/ds = -M \propto Pl \propto \sigma l^3 \qquad (7)$$

where $E_s I$ represents the flexural rigidity of the beam in the plane of bending; θ is the angle between the tangent to the cell edge and the x axis; and s is measured along the edge.

If the stress is normalized by the initial collapse stress of the foam ($\sigma^* \propto E_s I/l^4$, see Gibson and Ashby, 1988) and the dimensionless length $\bar{s}$ is used, that is, $\sigma/\sigma^* = \bar{\sigma}$ and $s/l = \bar{s}$, the above equation becomes:

$$d\theta/d\bar{s} \propto \sigma/\sigma^* = \bar{\sigma} \qquad (8)$$

Under the same normalized post-collapse stress $\bar{\sigma}$, all the corresponding positions in the frameworks of the foams of varying densities and materials, which have the same $\bar{s}$, will have the same angle or the same deformed geometrical shape. As a result, the same degree of post-collapse $\bar{\sigma}$ is related to the same strain ε. A single stress-strain curve would give a good description for all flexible foams.

Similarly, the post-collapse curves of flexible honeycombs under uniaxial compression can be expressed by a single function if the same normalization procedure is followed. In the case of plastic foams or plastic honeycombs, the upper bounds and lower bounds of the post-collapse curves can be normalized in a similar way to give a single curve for each of them. Therefore, the true post-collapse behavior of plastic foams or honeycombs can be approximated by a single curve in dimensionless stress-strain space.

From a load-bearing point of view, the cell corners in the foam structures contribute little to the deformation. A refinement of the scaling law (Zhang, 1989 for details) for the stress-strain behavior can be obtained by incorporating these corners or the "dead volume" into consideration.

4. Optimal Selection of Foams and Honeycombs as Energy Absorbers

In packaging, the aim is to absorb as much as possible the energy of the packaged object while at the same time keeping the force on the object below the limit which will cause damage. In mathematical terms, we are seeking a maximum in absorbed energy, subject to a given constraint on the stress:

$$\begin{cases} W = \int \sigma \, d\varepsilon = \int_0^{\varepsilon} f(\rho^*) k(\varepsilon) \, d\varepsilon = f(\rho^*) k_1(\varepsilon) \\ \sigma = f(\rho^*) k(\varepsilon) = \sigma_p \end{cases} \tag{9}$$

where ρ^* is the relative density, the ratio of the density of the foam, ρ, to that of the base solid, ρ_s; $k(\varepsilon)$ is the shape function of the stress-strain curve; $k_1(\varepsilon)$ is the integral of $k(\varepsilon)$ over strain; $f(\rho^*)$ is the initial collapse stress, σ^*; and σ_p is the maximum permitted stress set by a given application.

A Lagrange function is constructed as follows:

$$L(\rho^*, \varepsilon, \lambda) = W + \lambda\, (\sigma - \sigma_p) = f(\rho^*) k_1(\varepsilon) + \lambda [f(\rho^*)\, k(\varepsilon) - \sigma_p] \tag{10}$$

The partial derivatives of the constructed function must all be zero when the conditional maximum is reached, leading to:

$$k_1(\varepsilon)\, k'(\varepsilon) = k^2(\varepsilon) \tag{11}$$

From the above equation the strain ε_m, at which the maximum W is reached, can be obtained once the shape function $k(\varepsilon)$ is known. Then the optimal density and the energy absorbed can be obtained by substituting ε_m into Eq. (9). The shape functions for various types of isotropic foams and honeycombs of regular hexagonal cells, and solutions of the optimal density and the energy absorbed for these materials can be found in Table 1.

Table 1 The Optimal Selection of EF (elastic foams), PF (plastic foams), EH (elastic honeycombs) and PH (plastic honeycombs)as Energy Absorbers

Type	Shape Function $k(\varepsilon)$	Initial Collapse Stress $f(\rho^*)$	Optimal Density ρ^*	Energy W
EF	$0.95/(1-\varepsilon)$ (Ref. 1)	$0.05E_s(\rho^*)^2$ (Ref. 2)	$2.77(\sigma_p/E_s)^{1/2}$	$0.38\ \sigma_p$
PF	1 $\varepsilon \leq 0.5$; $5-[16-100(\varepsilon-0.5)^2]^{0.5}$ $\varepsilon > 0.5$ (Ref. 1)	$0.32\sigma_{ys}(\rho^*)^{3/2}$ (Ref. 1)	$2.03(\sigma_p/\sigma_{ys})^{2/3}$	$0.50\ \sigma_p$
Out-of-plane: PH	1 $\varepsilon \leq 0.75$; ∞ $\varepsilon > 0.75$ (Ref. 1)	$3.2\sigma_{ys}(\rho^*)^{5/3}$ (Ref. 1)	$0.50(\sigma_p/\sigma_{ys})^{3/5}$	$0.75\ \sigma_p$
In-plane: PH	1 $\varepsilon \leq 0.5$; ∞ $\varepsilon > 0.5$ (Ref. 1)	$0.28\sigma_{ys}(\rho^*)^2$ (Ref. 1)	$1.89(\sigma_p/\sigma_{ys})^{1/2}$	$0.50\ \sigma_p$
Out-of-plane: EH	1 $\varepsilon \leq 0.5$; ∞ $\varepsilon > 0.5$ (Ref. 1)	$7.2E_s(\rho^*)^3$ (Ref. 1)	$0.52(\sigma_p/E_s)^{1/3}$	$0.50\ \sigma_p$
In-plane: EH	1 $\varepsilon \leq 0.5$; ∞ $\varepsilon > 0.5$ (Ref. 1)	$0.14E_s(\rho^*)^3$ (Ref. 2)	$1.93(\sigma_p/E_s)^{1/3}$	$0.50\ \sigma_p$

References

1. Zhang, J. (1989) Ph.D. Thesis, Cambridge University Engineering Department, Cambridge, U.K.
2. Gibson, L. J. and Ashby, M. F. (1988) "Cellular Solids: Structure and properties", Pergamon Press.

A large amount of data from both static and dynamic tests are available for optimal design of packaging. They are presented in U shape curves in the case of the Janssen factor or the cushion factor. These plots contribute to the understanding of maximizing energy absorption under a given critical stress. However, to design engineers, the optimal points are of most interest. In the case of U shape curves, this is the bottom point, while in the case of energy absorption diagrams and Rusch's curves it is the point which touches the envelope. We replot these optimal points in our package selection diagrams and compare them with our theoretical modelling. One thing is worth noting: the normalization of each stress or energy in the following plots is done against the solid properties ρ_s, E_s and σ_{ys} measured at the strain-rate and the temperature at which the test was performed. The normalizing properties ρ_s, E_s and σ_{ys} for low strain-rate at room temperature are given in Table 2. A typical dynamic test at the strain rate around 50/s will bring up the solid Young's modulus E_s and yield strength σ_{ys} by a factor of around 2.5. As a result of our normalization, the data of the same foam tested at different strain-rates or drop heights lie on top of each other.

Table 2 Cell Wall Properties

Materials	ρ_s (Mg/m^3)	E_s (GN/m^2)	σ_{ys} (MN/m^2)
ABS (1)	1.07	2.6	70
Phenolic (1)	1.28	0.038	100
Polyethylene (2)	0.92	0.2	—
Polymethacrylimid (3)	1.2	3.6	360
Polyurethane, rigid (2)	1.2	1.6	127
Polyurethane, flexible (2)	1.2	0.045	—
Polystyrene, flexible (1)	1.05	3.0	80
Styrene-acrylonitrile (1)	1.05	3.0	80

Source of Data:
1. Handbook of Industrial Materials, First Edition, Trade and Technical Press LTD, Surrey, U.K.
2. Gibson and Ashby, 1988.
3. Maiti et al, 1984.

The data and the prediction for elastic foams are plotted in Fig. 4. The agreement for low relative density range is good. There is more discrepancy when the density becomes larger. The data and the prediction for plastic foams are plotted in Fig. 5. The agreement is good for both low and high relative density ranges.

There is considerable discrepancy in the measurements of solid properties σ_{ys} and E_s. Also, the initial collapse stress of a foam could vary considerably from the theoretical prediction. A more realistic design methodology is to identify a few possible cell wall materials and their corresponding optimal densities from the packaging selection diagrams and then to conduct simple compression tests to measure their initial collapse stresses. As shown in our earlier analysis, the best plastic foams will be the ones that have initial collapse stresses around 93% of the critical stress σ_p, whereas the best elastic foams will be the ones with initial stresses around 38% of σ_p.

The advantage of the plastic foams or honeycombs is that, compared with the elastic ones, they provide a better energy absorption capacity under the same critical stress σ_p. However, the plastic ones absorb much less energy after the first impact. The elastic ones can absorb energy repeatedly and their energy absorption capacity is not reduced appreciably after the first loading. In comparing honeycombs with foams (shown in Figs. 6 and 7), we find that honeycombs when loaded in the out-of-plane direction are better than their counterparts in energy absorption. In addition, the optimal density of honeycombs for a given critical stress is generally smaller than that of foams, which makes them very attractive when weight savings are crucial. On the other hand, honeycombs loaded in the in-plane directions are inferior to foams as far as weight savings are concerned.

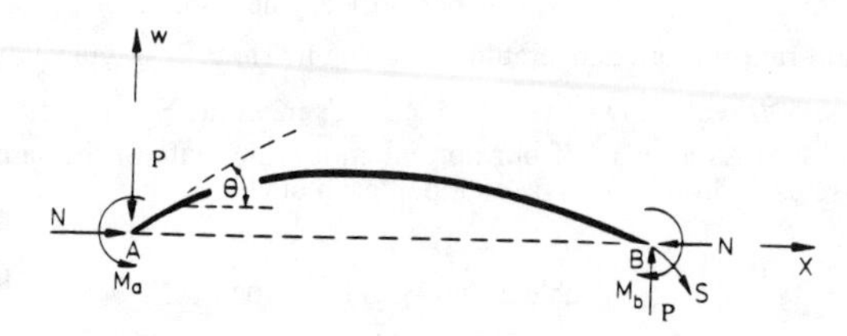

Fig. 3 A diagram for the analysis of bending deformation in a beam

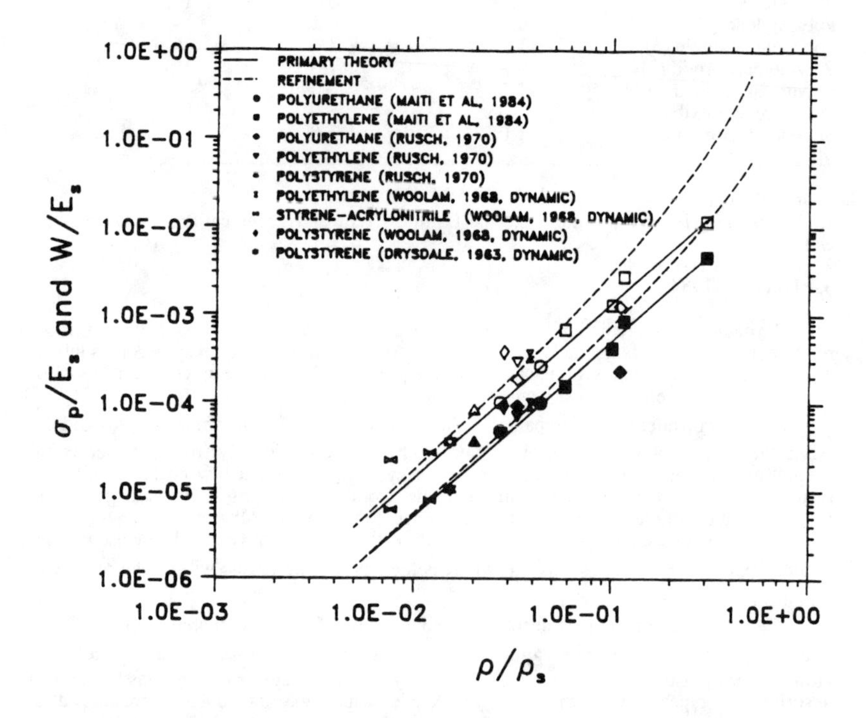

Fig. 4 The packaging selection diagram for elastic foams. Both the theoretical lines and the experimental data appear in pairs with the stress lying above the energy absorbed. A peak stress is plotted as an open symbol while the corresponding energy absorbed is shown as the same symbol but shaded.

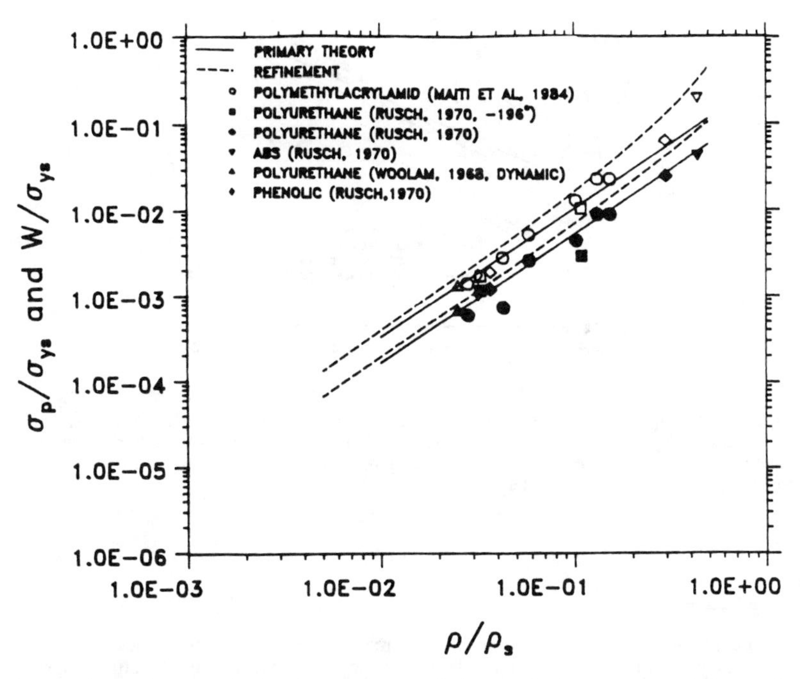

Fig. 5 The packaging selection diagram for plastic foams. Both the theoretical lines and the experimental data appear in pairs with the stress lying above the energy absorbed. A peak stress is plotted as an open symbol while the corresponding energy absorbed is shown as the same symbol but shaded.

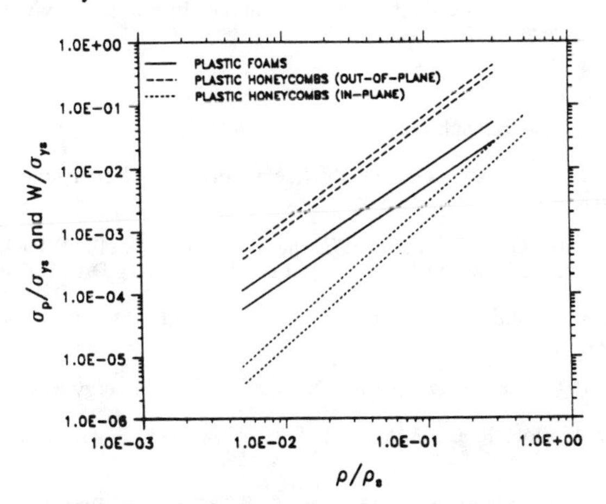

Fig. 6 The figure shows the packaging selection diagrams for plastic foams and honeycombs together. Plastic honeycombs loaded out-of-plane demonstrate better energy-absorption capacity than plastic foams. For an given application, the optimal honeycomb is much lighter than its foam counterpart.

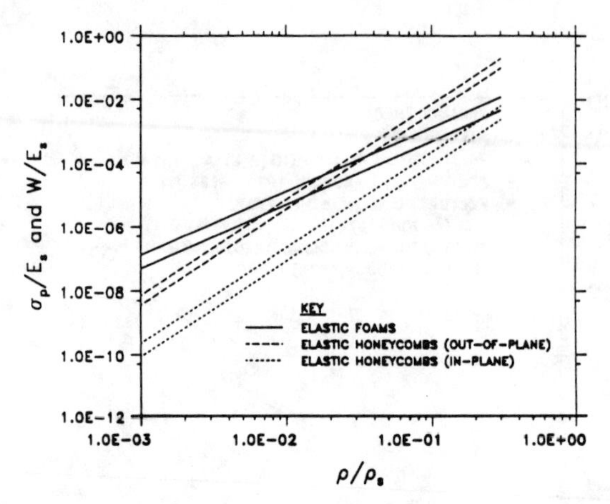

Fig. 7 The figure shows the packaging selection diagrams for elastic foams and honeycombs together. Elastic honeycombs loaded out-of-plane demonstrate better energy-absorption capacity than elastic foams. For an given application, the optimal honeycomb is much lighter than its foam counterpart.

5. Conclusions

Packaging systems employing cellular materials are traditionally designed with an experimental database, requiring a large number of impacts tests (Mustin, 1968). The Packaging Selection Diagrams in this paper allow empiricism to be combined with physical modeling. If properly used, the amount of experiment needed in the design process can be significantly reduced. The diagrams are adequate for the broad comparisons required in conceptual design, and, in most cases, for the rough calculations of embodiment design. They may not be appropriate for detailed design calculations. For those, it may be necessary to conduct a few selected experiments.

References

Cousins, R. R. (1976a) J. Appl. Polymer Sci., 20, 2893.

Cousins, R. R. (1976b) "Design Guide to the Use of Foams for Crash Padding", NPL Report DMA 237, London.

Drysdale, J., Gordon, G. A., Wheeler, E. E. and Marsden, P. D. (1963) Package, 34, (396, 399, and 400), (March, June, and July) (Memoir No. 6, Packaging Div., PATRA.)

Gibson, L. J. and Ashby, M. F. (1988) "Cellular Solids: Structure and Properties", Pergamon Press, Oxford.

Gordon, G. A. (1974) "Testing & Approval, Impact Strength & Energy Absorption", PIRA.

Green, S. J., Schierloh, F. L., Perkins, R. D. and Babcock, S. G. (1969) Exp. Mech., March, 103.

Handbook of Industrial Materials, First Edition, Trade & Technical Press LTD, Surrey, U.K.

Hilyard, N. C. (ed.) (1982) "Mechanics of Cellular Plastics", Applied Science Publishers, London.

Lee, W. M. and Williams, B. M. (1971) J. Cell. Plast., 7, 72.

Lockett, F. J., Cousins, R. R. and Dawson, D. (1981) Plast. Rubber Proc. Appl., 1, 25.

Maiti, S. K., Gibson, L. J. and Ashby, M. F. (1984) Acta Metal., 32, 1963.

Meinecke, E. A. and Schwaber, D. M. (1970) J. Appl. Polymer Sci., 14, 2239.

Meinecke, E. A., Schwaber, D. M. and Chiang, R. R. (1971) J. Elastoplast., 3, 19.

Melvin, J. W. and Roberts, V. L. (1971) J. Cell. Plast., 7, 97.

Mustin, G. S. (1968) "Theory and Practice of Cushion Design", US Government Printing Office, Washington, DC.

Rusch, K. C. (1970) J. Appl. Polymer Sci., 14, 1263 and 1433.

Rusch, K. C. (1971) J. Cell. Plast., 7, 78.

Schwaber, D. M. (1973) Polymer-Plast. Technol. Eng., 2, 231.

Schwaber, D. M. and Meinecke, E. A. (1971) J. Appl. Polymer Sci., 157, 2381.

Woolam, W. E. (1968) J. Cell. Plast., 4, 79.

Zhang, J. (1989) Ph.D. Thesis, Cambridge University Engineering Dept., Cambridge, U.K.

Author Index

Aksay, I.A., 151
Alviso, C.T., 197
Arbabi, Sepehr, 201
Ashby, M.F., 235
Aubert, J.H., 15, 117

Brezny, Rasto, 3
Butcher, B.M., 169

Callahan, G.D., 169
Cone, K., 163
Covino, Josephine, 129
Cowin, Stephen C., 83

Day, A.R., 95
Delarosa, Mark J., 141
Donald, S., 163
Donovan, J.A., 157
Duva, J.M., 109
Duxbury, P.M., 179

Elzey, D.M., 109

Fortes, M.A., 41

Garboczi, E.J., 95
Gehris, Jr., Allen P., 129
Gibson, L.J., 9
Green, David J., 3, 27, 35

Haggerty, J.S., 71
Hrubesh, L.W., 197

Ishizaki, Kozo, 135

Jin, Wei, 103

Khan, Ali A., 65
Kim, S.G., 179
Kunze, J.M., 109

Laferla, Raffaele, 141
Lannutti, J.J., 151
LeMay, James D., 21, 197
Lightfoot, A., 71

Mahanti, S.D., 103
Maji, A.K., 163
Mark, J.E., 15

Nair, S.V., 71

Okada, Shojiro, 135
Orenstein, Robert M., 27
Ozkul, M.H., 15

Pekala, R.W., 197
Pinto, J., 77
Poco, J.F., 197

Ritter, J.E., 71
Rice, Roy W., 229

Sahimi, Muhammad, 201
Schilling, C.H., 151
Segall, Albert E., 27
Sherman, Andrew J., 141
Shih, W.-H., 151
Snyder, K.A., 95
Stupak, P.R., 157

Takata, Atsushi, 135
Thorpe, M.F., 95, 103
Tillotson, T.M., 197
Triantafillou, T.C., 9

Van Voorhees, Eric J., 35
Vaz, M. Fátima, 41
Vincent, Julian F.V., 61, 65

Wadley, H.N.G., 109
Wagner, Ralph A., 169
Wiegand, Donald A., 77
Williams, Brian E., 141

Zhang, J., 235

Subject Index

Al_2O_3, 35
acoustic emission, 6
aerogels, 21, 197
alumina, 6
  foams, 27
aluminum honeycombs, 163
anisotropic, 83
anisotropy, 61, 65
aspect ratio, 65

backfill consolidation model, 169
bead fusion, 157
bending strength, 135
biological cellular materials, 61
blunt indenters, 41
bond dilution disorder, 179
bridge enhancement, 135

carbon foams, 5, 141
cask design, 163
cellular
  ceramics, 3
  foams, 9
  materials, 41, 61
  SiO_2 materials, 129
central force network theory, 179
ceramic, 141
  sandwich, 35
chemical vapor deposition, 141
collapse stress, 17
composites, 77
compressive
  fracture strength, 82
  modulus, 22
  strength, 6, 22, 28, 77
computer simulation, 95
consolidation, 151
constitutive models, 109
continuum
  composite, 95
  mechanical models, 83
cortex, 65
crack extension, 29
cracks, 77
crushed salt, 169
  -bentonite, 169
cyclotrimethylene trinitramine, 77

damage, 27
  accumulation, 7
defects, 18
demixed, 117
densification models, 109
density measurements, 117
deviatoric creep, 169
dielectric constant, 117
digital image, 95

effective medium theory, 95, 103, 179
elastic
  collapse, 9
  constants, 83
  modulii, 95, 201
  modulus, 27
  strain energy, 76
energy absorption, 157
expanded polystyrene, 157

failure, 35
flesh of apples, 61
flow zones, 67
foaming sol-gel process, 117
foams, 3
fracture 9, 157
  properties, 65
  toughness, 3, 65

gallery expansion, 103
  height, 103
gamma-ray densitometry, 151

hot isostatic pressing (HIP), 109, 135

image analysis, 95
indentation, 41
intercellular spaces, 65
internal friction, 28
isostatic polystyrene, 16
isotropic, 65

layered random alloys, 103
lightweight ceramics, 117
low density
  foams, 21, 197
  microcellular
    foams, 117
    materials, 21
Lycra, 16

mechanical
  behavior, 3
  properties, 21
melamine formaldehyde, 197
microcellular foams, 15, 21

micromechanical models, 83
microporous systems, 103
microstructure, 117
morphological characterization, 117
multiaxial loads, 9

nailing, 41
non-linear elastic creep, 169

open cell ceramic foam, 27

parenchyma, 65
particle rearrangement, 151
percolation, 95
  threshold, 201
phase separation, 23
plasma sprayed metal matrix composite, 109
plastic
  deformation, 67
  flow, 77
plasticity, 9
Poisson ratio, 95
poly(4-methyl-1)pentene, 16
polyacrylonitrile, 16
polymer foaming concept, 117
polysiloxanes, 151
polystyrene foams, 16
polyurethane, 16
  foams, 163
porosity, 77
processing conditions, 81

radial spaces, 65
random resistor networks, 201
refractory foams, 141
resorcinol formaldehyde, 197
rhenium, 141
rigid cellular materials, 179

scaling
  exponent, 22
  laws, 201
  theory of elasticity, 179
sediments, 151
sharp indenters, 41
silica, 197
  reinforcement, 17
soft impact limiters, 163
sol-gel routes, 23
stereological measures, 83
strain, 77
strength, 3
stress, 35
  concentrators, 67
structural
  efficiency, 24
  -property relationships, 22
super-critical extraction, 117
superelastic percolation, 201

tensile
  fracture, 179
  strength, 6
texture, 65
thermal
  diffusivity, 117
  gradients, 27
  shock, 27
    severity, 28
thermally induced phase separation, 15, 117
Ti-14Al-21Nb intermetallic matrix composite, 109
toughness, 3
triaxial confined compression, 77
trinitrotoluene, 77

uniaxial compression, 77

vascular tissue, 65
vector transport properties, 201
viscosity measurements, 117
vitreous carbon foams, 6
void, 77
volumetric creep, 169

wedge penetration test, 66
Weibull
  distribution, 201
  strength distribution, 29

x-ray crystallography, 117

yield, 77
  strength, 77
Young's modulus, 15, 27, 77

ISSN 0272 - 9172

Volume 1—Laser and Electron-Beam Solid Interactions and Materials Processing, J. F. Gibbons, L. D. Hess, T. W. Sigmon, 1981, ISBN 0-444-00595-1

Volume 2—Defects in Semiconductors, J. Narayan, T. Y. Tan, 1981, ISBN 0-444-00596-X

Volume 3—Nuclear and Electron Resonance Spectroscopies Applied to Materials Science, E. N. Kaufmann, G. K. Shenoy, 1981, ISBN 0-444-00597-8

Volume 4—Laser and Electron-Beam Interactions with Solids, B. R. Appleton, G. K. Celler, 1982, ISBN 0-444-00693-1

Volume 5—Grain Boundaries in Semiconductors, H. J. Leamy, G. E. Pike, C. H. Seager, 1982, ISBN 0-444-00697-4

Volume 6—Scientific Basis for Nuclear Waste Management IV, S. V. Topp, 1982, ISBN 0-444-00699-0

Volume 7—Metastable Materials Formation by Ion Implantation, S. T. Picraux, W. J. Choyke, 1982, ISBN 0-444-00692-3

Volume 8—Rapidly Solidified Amorphous and Crystalline Alloys, B. H. Kear, B. C. Giessen, M. Cohen, 1982, ISBN 0-444-00698-2

Volume 9—Materials Processing in the Reduced Gravity Environment of Space, G. E. Rindone, 1982, ISBN 0-444-00691-5

Volume 10—Thin Films and Interfaces, P. S. Ho, K.-N. Tu, 1982, ISBN 0-444-00774-1

Volume 11—Scientific Basis for Nuclear Waste Management V, W. Lutze, 1982, ISBN 0-444-00725-3

Volume 12—In Situ Composites IV, F. D. Lemkey, H. E. Cline, M. McLean, 1982, ISBN 0-444-00726-1

Volume 13—Laser-Solid Interactions and Transient Thermal Processing of Materials, J. Narayan, W. L. Brown, R. A. Lemons, 1983, ISBN 0-444-00788-1

Volume 14—Defects in Semiconductors II, S. Mahajan, J. W. Corbett, 1983, ISBN 0-444-00812-8

Volume 15—Scientific Basis for Nuclear Waste Management VI, D. G. Brookins, 1983, ISBN 0-444-00780-6

Volume 16—Nuclear Radiation Detector Materials, E. E. Haller, H. W. Kraner, W. A. Higinbotham, 1983, ISBN 0-444-00787-3

Volume 17—Laser Diagnostics and Photochemical Processing for Semiconductor Devices, R. M. Osgood, S. R. J. Brueck, H. R. Schlossberg, 1983, ISBN 0-444-00782-2

Volume 18—Interfaces and Contacts, R. Ludeke, K. Rose, 1983, ISBN 0-444-00820-9

Volume 19—Alloy Phase Diagrams, L. H. Bennett, T. B. Massalski, B. C. Giessen, 1983, ISBN 0-444-00809-8

Volume 20—Intercalated Graphite, M. S. Dresselhaus, G. Dresselhaus, J. E. Fischer, M. J. Moran, 1983, ISBN 0-444-00781-4

Volume 21—Phase Transformations in Solids, T. Tsakalakos, 1984, ISBN 0-444-00901-9

Volume 22—High Pressure in Science and Technology, C. Homan, R. K. MacCrone, E. Whalley, 1984, ISBN 0-444-00932-9 (3 part set)

Volume 23—Energy Beam-Solid Interactions and Transient Thermal Processing, J. C. C. Fan, N. M. Johnson, 1984, ISBN 0-444-00903-5

Volume 24—Defect Properties and Processing of High-Technology Nonmetallic Materials, J. H. Crawford, Jr., Y. Chen, W. A. Sibley, 1984, ISBN 0-444-00904-3

Volume 25—Thin Films and Interfaces II, J. E. E. Baglin, D. R. Campbell, W. K. Chu, 1984, ISBN 0-444-00905-1

Volume 26—Scientific Basis for Nuclear Waste Management VII, G. L. McVay, 1984, ISBN 0-444-00906-X

Volume 27—Ion Implantation and Ion Beam Processing of Materials, G. K. Hubler, O. W. Holland, C. R. Clayton, C. W. White, 1984, ISBN 0-444-00869-1

Volume 28—Rapidly Solidified Metastable Materials, B. H. Kear, B. C. Giessen, 1984, ISBN 0-444-00935-3

Volume 29—Laser-Controlled Chemical Processing of Surfaces, A. W. Johnson, D. J. Ehrlich, H. R. Schlossberg, 1984, ISBN 0-444-00894-2

Volume 30—Plasma Processing and Synthesis of Materials, J. Szekely, D. Apelian, 1984, ISBN 0-444-00895-0

Volume 31—Electron Microscopy of Materials, W. Krakow, D. A. Smith, L. W. Hobbs, 1984, ISBN 0-444-00898-7

Volume 32—Better Ceramics Through Chemistry, C. J. Brinker, D. E. Clark, D. R. Ulrich, 1984, ISBN 0-444-00898-5

Volume 33—Comparison of Thin Film Transistor and SOI Technologies, H. W. Lam, M. J. Thompson, 1984, ISBN 0-444-00899-3

Volume 34—Physical Metallurgy of Cast Iron, H. Fredriksson, M. Hillerts, 1985, ISBN 0-444-00938-8

Volume 35—Energy Beam-Solid Interactions and Transient Thermal Processing/1984, D. K. Biegelsen, G. A. Rozgonyi, C. V. Shank, 1985, ISBN 0-931837-00-6

Volume 36—Impurity Diffusion and Gettering in Silicon, R. B. Fair, C. W. Pearce, J. Washburn, 1985, ISBN 0-931837-01-4

Volume 37—Layered Structures, Epitaxy, and Interfaces, J. M. Gibson, L. R. Dawson, 1985, ISBN 0-931837-02-2

Volume 38—Plasma Synthesis and Etching of Electronic Materials, R. P. H. Chang, B. Abeles, 1985, ISBN 0-931837-03-0

Volume 39—High-Temperature Ordered Intermetallic Alloys, C. C. Koch, C. T. Liu, N. S. Stoloff, 1985, ISBN 0-931837-04-9

Volume 40—Electronic Packaging Materials Science, E. A. Giess, K.-N. Tu, D. R. Uhlmann, 1985, ISBN 0-931837-05-7

Volume 41—Advanced Photon and Particle Techniques for the Characterization of Defects in Solids, J. B. Roberto, R. W. Carpenter, M. C. Wittels, 1985, ISBN 0-931837-06-5

Volume 42—Very High Strength Cement-Based Materials, J. F. Young, 1985, ISBN 0-931837-07-3

Volume 43—Fly Ash and Coal Conversion By-Products: Characterization, Utilization, and Disposal I, G. J. McCarthy, R. J. Lauf, 1985, ISBN 0-931837-08-1

Volume 44—Scientific Basis for Nuclear Waste Management VIII, C. M. Jantzen, J. A. Stone, R. C. Ewing, 1985, ISBN 0-931837-09-X

Volume 45—Ion Beam Processes in Advanced Electronic Materials and Device Technology, B. R. Appleton, F. H. Eisen, T. W. Sigmon, 1985, ISBN 0-931837-10-3

Volume 46—Microscopic Identification of Electronic Defects in Semiconductors, N. M. Johnson, S. G. Bishop, G. D. Watkins, 1985, ISBN 0-931837-11-1

Volume 47—Thin Films: The Relationship of Structure to Properties, C. R. Aita, K. S. SreeHarsha, 1985, ISBN 0-931837-12-X

Volume 48—Applied Materials Characterization, W. Katz, P. Williams, 1985, ISBN 0-931837-13-8

Volume 49—Materials Issues in Applications of Amorphous Silicon Technology, D. Adler, A. Madan, M. J. Thompson, 1985, ISBN 0-931837-14-6

Volume 50—Scientific Basis for Nuclear Waste Management IX, L. O. Werme, 1986, ISBN 0-931837-15-4

Volume 51—Beam-Solid Interactions and Phase Transformations, H. Kurz, G. L. Olson, J. M. Poate, 1986, ISBN 0-931837-16-2

Volume 52—Rapid Thermal Processing, T. O. Sedgwick, T. E. Seidel, B.-Y. Tsaur, 1986, ISBN 0-931837-17-0

Volume 53—Semiconductor-on-Insulator and Thin Film Transistor Technology, A. Chiang. M. W. Geis, L. Pfeiffer, 1986, ISBN 0-931837-18-9

Volume 54—Thin Films—Interfaces and Phenomena, R. J. Nemanich, P. S. Ho, S. S. Lau, 1986, ISBN 0-931837-19-7

Volume 55—Biomedical Materials, J. M. Williams, M. F. Nichols, W. Zingg, 1986, ISBN 0-931837-20-0

Volume 56—Layered Structures and Epitaxy, J. M. Gibson, G. C. Osbourn, R. M. Tromp, 1986, ISBN 0-931837-21-9

Volume 57—Phase Transitions in Condensed Systems—Experiments and Theory, G. S. Cargill III, F. Spaepen, K.-N. Tu, 1987, ISBN 0-931837-22-7

Volume 58—Rapidly Solidified Alloys and Their Mechanical and Magnetic Properties, B. C. Giessen, D. E. Polk, A. I. Taub, 1986, ISBN 0-931837-23-5

Volume 59—Oxygen, Carbon, Hydrogen, and Nitrogen in Crystalline Silicon, J. C. Mikkelsen, Jr., S. J. Pearton, J. W. Corbett, S. J. Pennycook, 1986, ISBN 0-931837-24-3

Volume 60—Defect Properties and Processing of High-Technology Nonmetallic Materials, Y. Chen, W. D. Kingery, R. J. Stokes, 1986, ISBN 0-931837-25-1

Volume 61—Defects in Glasses, F. L. Galeener, D. L. Griscom, M. J. Weber, 1986, ISBN 0-931837-26-X

Volume 62—Materials Problem Solving with the Transmission Electron Microscope, L. W. Hobbs, K. H. Westmacott, D. B. Williams, 1986, ISBN 0-931837-27-8

Volume 63—Computer-Based Microscopic Description of the Structure and Properties of Materials, J. Broughton, W. Krakow, S. T. Pantelides, 1986, ISBN 0-931837-28-6

Volume 64—Cement-Based Composites: Strain Rate Effects on Fracture, S. Mindess, S. P. Shah, 1986, ISBN 0-931837-29-4

Volume 65—Fly Ash and Coal Conversion By-Products: Characterization, Utilization and Disposal II, G. J. McCarthy, F. P. Glasser, D. M. Roy, 1986, ISBN 0-931837-30-8

Volume 66—Frontiers in Materials Education, L. W. Hobbs, G. L. Liedl, 1986, ISBN 0-931837-31-6

Volume 67—Heteroepitaxy on Silicon, J. C. C. Fan, J. M. Poate, 1986, ISBN 0-931837-33-2

Volume 68—Plasma Processing, J. W. Coburn, R. A. Gottscho, D. W. Hess, 1986, ISBN 0-931837-34-0

Volume 69—Materials Characterization, N. W. Cheung, M.-A. Nicolet, 1986, ISBN 0-931837-35-9

Volume 70—Materials Issues in Amorphous-Semiconductor Technology, D. Adler, Y. Hamakawa, A. Madan, 1986, ISBN 0-931837-36-7

Volume 71—Materials Issues in Silicon Integrated Circuit Processing, M. Wittmer, J. Stimmell, M. Strathman, 1986, ISBN 0-931837-37-5

Volume 72—Electronic Packaging Materials Science II, K. A. Jackson, R. C. Pohanka, D. R. Uhlmann, D. R. Ulrich, 1986, ISBN 0-931837-38-3

Volume 73—Better Ceramics Through Chemistry II, C. J. Brinker, D. E. Clark, D. R. Ulrich, 1986, ISBN 0-931837-39-1

Volume 74—Beam-Solid Interactions and Transient Processes, M. O. Thompson, S. T. Picraux, J. S. Williams, 1987, ISBN 0-931837-40-5

Volume 75—Photon, Beam and Plasma Stimulated Chemical Processes at Surfaces, V. M. Donnelly, I. P. Herman, M. Hirose, 1987, ISBN 0-931837-41-3

Volume 76—Science and Technology of Microfabrication, R. E. Howard, E. L. Hu, S. Namba, S. Pang, 1987, ISBN 0-931837-42-1

Volume 77—Interfaces, Superlattices, and Thin Films, J. D. Dow, I. K. Schuller, 1987, ISBN 0-931837-56-1

Volume 78—Advances in Structural Ceramics, P. F. Becher, M. V. Swain, S. Sōmiya, 1987, ISBN 0-931837-43-X

Volume 79—Scattering, Deformation and Fracture in Polymers, G. D. Wignall, B. Crist, T. P. Russell, E. L. Thomas, 1987, ISBN 0-931837-44-8

Volume 80—Science and Technology of Rapidly Quenched Alloys, M. Tenhover, W. L. Johnson, L. E. Tanner, 1987, ISBN 0-931837-45-6

Volume 81—High Temperature Ordered Intermetallic Alloys, II, N. S. Stoloff, C. C. Koch, C. T. Liu, O. Izumi, 1987, ISBN 0-931837-46-4

Volume 82—Characterization of Defects in Materials, R. W. Siegel, J. R. Weertman, R. Sinclair, 1987, ISBN 0-931837-47-2

Volume 83—Physical and Chemical Properties of Thin Metal Overlayers and Alloy Surfaces, D. M. Zehner, D. W. Goodman, 1987, ISBN 0-931837-48-0

Volume 84—Scientific Basis for Nuclear Waste Management X, J. K. Bates, W. B. Seefeldt, 1987, ISBN 0-931837-49-9

Volume 85—Microstructural Development During the Hydration of Cement, L. Struble, P. Brown, 1987, ISBN 0-931837-50-2

Volume 86—Fly Ash and Coal Conversion By-Products Characterization, Utilization and Disposal III, G. J. McCarthy, F. P. Glasser, D. M. Roy, S. Diamond, 1987, ISBN 0-931837-51-0

Volume 87—Materials Processing in the Reduced Gravity Environment of Space, R. H. Doremus, P. C. Nordine, 1987, ISBN 0-931837-52-9

Volume 88—Optical Fiber Materials and Properties, S. R. Nagel, J. W. Fleming, G. Sigel, D. A. Thompson, 1987, ISBN 0-931837-53-7

Volume 89—Diluted Magnetic (Semimagnetic) Semiconductors, R. L. Aggarwal, J. K. Furdyna, S. von Molnar, 1987, ISBN 0-931837-54-5

Volume 90—Materials for Infrared Detectors and Sources, R. F. C. Farrow, J. F. Schetzina, J. T. Cheung, 1987, ISBN 0-931837-55-3

Volume 91—Heteroepitaxy on Silicon II, J. C. C. Fan, J. M. Phillips, B.-Y. Tsaur, 1987, ISBN 0-931837-58-8

Volume 92—Rapid Thermal Processing of Electronic Materials, S. R. Wilson, R. A. Powell, D. E. Davies, 1987, ISBN 0-931837-59-6

Volume 93—Materials Modification and Growth Using Ion Beams, U. Gibson, A. E. White, P. P. Pronko, 1987, ISBN 0-931837-60-X

Volume 94—Initial Stages of Epitaxial Growth, R. Hull, J. M. Gibson, David A. Smith, 1987, ISBN 0-931837-61-8

Volume 95—Amorphous Silicon Semiconductors—Pure and Hydrogenated, A. Madan, M. Thompson, D. Adler, Y. Hamakawa, 1987, ISBN 0-931837-62-6

Volume 96—Permanent Magnet Materials, S. G. Sankar, J. F. Herbst, N. C. Koon, 1987, ISBN 0-931837-63-4

Volume 97—Novel Refractory Semiconductors, D. Emin, T. Aselage, C. Wood, 1987, ISBN 0-931837-64-2

Volume 98—Plasma Processing and Synthesis of Materials, D. Apelian, J. Szekely, 1987, ISBN 0-931837-65-0

Volume 99—High-Temperature Superconductors, M. B. Brodsky, R. C. Dynes, K. Kitazawa, H. L. Tuller, 1988, ISBN 0-931837-67-7

Volume 100—Fundamentals of Beam-Solid Interactions and Transient Thermal Processing, M. J. Aziz, L. E. Rehn, B. Stritzker, 1988, ISBN 0-931837-68-5

Volume 101—Laser and Particle-Beam Chemical Processing for Microelectronics, D.J. Ehrlich, G.S. Higashi, M.M. Oprysko, 1988, ISBN 0-931837-69-3

Volume 102—Epitaxy of Semiconductor Layered Structures, R. T. Tung, L. R. Dawson, R. L. Gunshor, 1988, ISBN 0-931837-70-7

Volume 103—Multilayers: Synthesis, Properties, and Nonelectronic Applications, T. W. Barbee Jr., F. Spaepen, L. Greer, 1988, ISBN 0-931837-71-5

Volume 104—Defects in Electronic Materials, M. Stavola, S. J. Pearton, G. Davies, 1988, ISBN 0-931837-72-3

Volume 105—SiO_2 and Its Interfaces, G. Lucovsky, S. T. Pantelides, 1988, ISBN 0-931837-73-1

Volume 106—Polysilicon Films and Interfaces, C.Y. Wong, C.V. Thompson, K-N. Tu, 1988, ISBN 0-931837-74-X

Volume 107—Silicon-on-Insulator and Buried Metals in Semiconductors, J. C. Sturm, C. K. Chen, L. Pfeiffer, P. L. F. Hemment, 1988, ISBN 0-931837-75-8

Volume 108—Electronic Packaging Materials Science II, R. C. Sundahl, R. Jaccodine, K. A. Jackson, 1988, ISBN 0-931837-76-6

Volume 109—Nonlinear Optical Properties of Polymers, A. J. Heeger, J. Orenstein, D. R. Ulrich, 1988, ISBN 0-931837-77-4

Volume 110—Biomedical Materials and Devices, J. S. Hanker, B. L. Giammara, 1988, ISBN 0-931837-78-2

Volume 111—Microstructure and Properties of Catalysts, M. M. J. Treacy, J. M. Thomas, J. M. White, 1988, ISBN 0-931837-79-0

Volume 112—Scientific Basis for Nuclear Waste Management XI, M. J. Apted, R. E. Westerman, 1988, ISBN 0-931837-80-4

Volume 113—Fly Ash and Coal Conversion By-Products: Characterization, Utilization, and Disposal IV, G. J. McCarthy, D. M. Roy, F. P. Glasser, R. T. Hemmings, 1988, ISBN 0-931837-81-2

Volume 114—Bonding in Cementitious Composites, S. Mindess, S. P. Shah, 1988, ISBN 0-931837-82-0

Volume 115—Specimen Preparation for Transmission Electron Microscopy of Materials, J. C. Bravman, R. Anderson, M. L. McDonald, 1988, ISBN 0-931837-83-9

Volume 116—Heteroepitaxy on Silicon: Fundamentals, Structures,and Devices, H.K. Choi, H. Ishiwara, R. Hull, R.J. Nemanich, 1988, ISBN: 0-931837-86-3

Volume 117—Process Diagnostics: Materials, Combustion, Fusion, K. Hays, A.C. Eckbreth, G.A. Campbell, 1988, ISBN: 0-931837-87-1

Volume 118—Amorphous Silicon Technology, A. Madan, M.J. Thompson, P.C. Taylor, P.G. LeComber, Y. Hamakawa, 1988, ISBN: 0-931837-88-X

Volume 119—Adhesion in Solids, D.M. Mattox, C. Batich, J.E.E. Baglin, R.J. Gottschall, 1988, ISBN: 0-931837-89-8

Volume 120—High-Temperature/High-Performance Composites, F.D. Lemkey, A.G. Evans, S.G. Fishman, J.R. Strife, 1988, ISBN: 0-931837-90-1

Volume 121—Better Ceramics Through Chemistry III, C.J. Brinker, D.E. Clark, D.R. Ulrich, 1988, ISBN: 0-931837-91-X

Volume 122—Interfacial Structure, Properties, and Design, M.H. Yoo, W.A.T. Clark, C.L. Briant, 1988, ISBN: 0-931837-92-8

Volume 123—Materials Issues in Art and Archaeology, E.V. Sayre, P. Vandiver, J. Druzik, C. Stevenson, 1988, ISBN: 0-931837-93-6

Volume 124—Microwave-Processing of Materials, M.H. Brooks, I.J. Chabinsky, W.H. Sutton, 1988, ISBN: 0-931837-94-4

Volume 125—Materials Stability and Environmental Degradation, A. Barkatt, L.R. Smith, E. Verink, 1988, ISBN: 0-931837-95-2

Volume 126—Advanced Surface Processes for Optoelectronics, S. Bernasek, T. Venkatesan, H. Temkin, 1988, ISBN: 0-931837-96-0

Volume 127—Scientific Basis for Nuclear Waste Management XII, W. Lutze, R.C. Ewing, 1989, ISBN: 0-931837-97-9

Volume 128—Processing and Characterization of Materials Using Ion Beams, L.E. Rehn, J. Greene, F.A. Smidt, 1989, ISBN: 1-55899-001-1

Volume 129—Laser and Particle-Beam Modification of Chemical Processes on Surfaces, A.W. Johnson, G.L. Loper, T.W. Sigmon, 1989, ISBN: 1-55899-002-X

Volume 130—Thin Films: Stresses and Mechanical Properties, J.C. Bravman, W.D. Nix, D.M. Barnett, D.A. Smith, 1989, ISBN: 1-55899-003-8

Volume 131—Chemical Perspectives of Microelectronic Materials, M.E. Gross, J. Jasinski, J.T. Yates, Jr., 1989, ISBN: 1-55899-004-6

Volume 132—Multicomponent Ultrafine Microstructures, L.E. McCandlish, B.H. Kear, D.E. Polk, and R.W. Siegel, 1989, ISBN: 1-55899-005-4

Volume 133—High Temperature Ordered Intermetallic Alloys III, C.T. Liu, A.I. Taub, N.S. Stoloff, C.C. Koch, 1989, ISBN: 1-55899-006-2

Volume 134—The Materials Science and Engineering of Rigid-Rod Polymers, W.W. Adams, R.K. Eby, D.E. McLemore, 1989, ISBN: 1-55899-007-0

Volume 135—Solid State Ionics, G. Nazri, R.A. Huggins, D.F. Shriver, 1989, ISBN: 1-55899-008-9

Volume 136—Fly Ash and Coal Conversion By-Products: Characterization, Utilization and Disposal V, R.T. Hemmings, E.E. Berry, G.J. McCarthy, F.P. Glasser, 1989, ISBN: 1-55899-009-7

Volume 137—Pore Structure and Permeability of Cementitious Materials, L.R. Roberts, J.P. Skalny, 1989, ISBN: 1-55899-010-0

Volume 138—Characterization of the Structure and Chemistry of Defects in Materials, B.C. Larson, M. Ruhle, D.N. Seidman, 1989, ISBN: 1-55899-011-9

Volume 139—High Resolution Microscopy of Materials, W. Krakow, F.A. Ponce, D.J. Smith, 1989, ISBN: 1-55899-012-7

Volume 140—New Materials Approaches to Tribology: Theory and Applications, L.E. Pope, L. Fehrenbacher, W.O. Winer, 1989, ISBN: 1-55899-013-5

Volume 141—Atomic Scale Calculations in Materials Science, J. Tersoff, D. Vanderbilt, V. Vitek, 1989, ISBN: 1-55899-014-3

Volume 142—Nondestructive Monitoring of Materials Properties, J. Holbrook, J. Bussiere, 1989, ISBN: 1-55899-015-1

Volume 143—Synchrotron Radiation in Materials Research, R. Clarke, J. Gland, J.H. Weaver, 1989, ISBN: 1-55899-016-X

Volume 144—Advances in Materials, Processing and Devices in III-V Compound Semiconductors, D.K. Sadana, L. Eastman, R. Dupuis, 1989, ISBN: 1-55899-017-8

Volume 145—III-V Heterostructures for Electronic/Photonic Devices, C.W. Tu, V.D. Mattera, A.C. Gossard, 1989, ISBN: 1-55899-018-6

Volume 146—Rapid Thermal Annealing/Chemical Vapor Deposition and Integrated Processing, D. Hodul, J. Gelpey, M.L. Green, T.E. Seidel, 1989, ISBN: 1-55899-019-4

Volume 147—Ion Beam Processing of Advanced Electronic Materials, N.W. Cheung, A.D. Marwick, J.B. Roberto, 1989, ISBN: 1-55899-020-8

Volume 148—Chemistry and Defects in Semiconductor Heterostructures, M. Kawabe, T.D. Sands, E.R. Weber, R.S. Williams, 1989, ISBN: 1-55899-021-6

Volume 149—Amorphous Silicon Technology-1989, A. Madan, M.J. Thompson, P.C. Taylor, Y. Hamakawa, P.G. LeComber, 1989, ISBN: 1-55899-022-4

Volume 150—Materials for Magneto-Optic Data Storage, C.J. Robinson, T. Suzuki, C.M. Falco, 1989, ISBN: 1-55899-023-2

Volume 151—Growth, Characterization and Properties of Ultrathin Magnetic Films and Multilayers, B.T. Jonker, J.P. Heremans, E.E. Marinero, 1989, ISBN: 1-55899-024-0

Volume 152—Optical Materials: Processing and Science, D.B. Poker, C. Ortiz, 1989, ISBN: 1-55899-025-9

Volume 153—Interfaces Between Polymers, Metals, and Ceramics, B.M. DeKoven, A.J. Gellman, R. Rosenberg, 1989, ISBN: 1-55899-026-7

Volume 154—Electronic Packaging Materials Science IV, R. Jaccodine, K.A. Jackson, E.D. Lillie, R.C. Sundahl, 1989, ISBN: 1-55899-027-5

Volume 155—Processing Science of Advanced Ceramics, I.A. Aksay, G.L. McVay, D.R. Ulrich, 1989, ISBN: 1-55899-028-3

Volume 156—High Temperature Superconductors: Relationships Between Properties, Structure, and Solid-State Chemistry, J.R. Jorgensen, K. Kitazawa, J.M. Tarascon, M.S. Thompson, J.B. Torrance, 1989, ISBN: 1-55899-029

Volume 157—Beam-Solid Interactions: Physical Phenomena, J.A. Knapp, P. Borgesen, R.A. Zuhr, 1989, ISBN 1-55899-045-3

Volume 158—In-Situ Patterning: Selective Area Deposition and Etching, R. Rosenberg, A.F. Bernhardt, J.G. Black, 1989, ISBN 1-55899-046-1

Volume 159—Atomic Scale Structure of Interfaces, R.D. Bringans, R.M. Feenstra, J.M. Gibson, 1989, ISBN 1-55899-047-X

Volume 160—Layered Structures: Heteroepitaxy, Superlattices, Strain, and Metastability, B.W. Dodson, L.J. Schowalter, J.E. Cunningham, F.H. Pollak, 1989, ISBN 1-55899-048-8

Volume 161—Properties of II-VI Semiconductors: Bulk Crystals, Epitaxial Films, Quantum Well Structures and Dilute Magnetic Systems, J.F. Schetzina, F.J. Bartoli, Jr., H.F. Schaake, 1989, ISBN 1-55899-049-6

Volume 162—Diamond, Boron Nitride, Silicon Carbide and Related Wide Bandgap Semiconductors, J.T. Glass, R.F. Messier, N. Fujimori, 1989, ISBN 1-55899-050-X

Volume 163—Impurities, Defects and Diffusion in Semiconductors: Bulk and Layered Structures, J. Bernholc, E.E. Haller, D.J. Wolford, 1989, ISBN 1-55899-051-8

Volume 164—Materials Issues in Microcrystalline Semiconductors, P.M. Fauchet, C.C. Tsai, K. Tanaka, 1989, ISBN 1-55899-052-6

Volume 165—Characterization of Plasma-Enhanced CVD Processes, G. Lucovsky, D.E. Ibbotson, D.W. Hess, 1989, ISBN 1-55899-053-4

Volume 166—Neutron Scattering for Materials Science, S.M. Shapiro, S.C. Moss, J.D. Jorgensen, 1989, ISBN 1-55899-054-2

Volume 167—Advanced Electronic Packaging Materials, A. Barfknecht, J. Partridge, C-Y. Li, C.J. Chen, 1989, ISBN 1-55899-055-0

Volume 168—Chemical Vapor Deposition of Refractory Metals and Ceramics, T.M. Besmann, B.M. Gallois, 1989, ISBN 1-55899-056-9

Volume 169—High Temperature Superconductors: Fundamental Properties and Novel Materials Processing, J. Narayan, C.W. Chu, L.F. Schneemeyer, D.K. Christen, 1989, ISBN 1-55899-057-7

Volume 170—Tailored Interfaces in Composite Materials, C.G. Pantano, E.J.H. Chen, 1989, ISBN 1-55899-058-5

Volume 171—Polymer Based Molecular Composites, D.W. Schaefer, J.E. Mark, 1989, ISBN 1-55899-059-3

Volume 172—Optical Fiber Materials and Processing, J.W. Fleming, G.H. Sigel, S. Takahashi, P.W. France, 1989, ISBN 1-55899-060-7

Volume 173—Electrical, Optical and Magnetic Properties of Organic Solid-State Materials, L.Y. Chiang, D.O. Cowan, P. Chaikin, 1989, ISBN 1-55899-061-5

Volume 174—Materials Synthesis Utilizing Biological Processes, M. Alper, P.D. Calvert, P.C. Rieke, 1989, ISBN 1-55899-062-3

Volume 175—Multi-Functional Materials, D.R. Ulrich, F.E. Karasz, A.J. Buckley, G. Gallagher-Daggitt, 1989, ISBN 1-55899-063-1

Volume 176—Scientific Basis for Nuclear Waste Management XIII, V.M. Oversby, P.W. Brown, 1989, ISBN 1-55899-064-X

Volume 177—Macromolecular Liquids, C.R. Safinya, S.A. Safran, P.A. Pincus, 1989, ISBN 1-55899-065-8

Volume 178—Fly Ash and Coal Conversion By-Products: Characterization, Utilization and Disposal VI, F.P. Glasser, R.L. Day, 1989, ISBN 1-55899-066-6

Volume 179—Specialty Cements with Advanced Properties, H. Jennings, A.G. Landers, B.E. Scheetz, I. Odler, 1989, ISBN 1-55899-067-4

Recent Materials Research Society Proceedings listed in the front.